Children's
Mathematical
Development

Children's Mathematical Development

RESEARCH AND PRACTICAL APPLICATIONS

American Psychological Association, Washington, DC

David C. Geary

First printing November 1994
Second printing January 1996

Published by the
American Psychological Association
750 First Street, NE
Washington, DC 20002

Copies may be ordered from
APA Order Department
P.O. Box 2710
Hyattsville, MD 20784
Order Number: 431-6340

In the UK and Europe, copies may be ordered from
American Psychological Association
3 Henrietta Street
Covent Garden, London
WC2E 8LU England

This book was typeset in Futura and New Baskerville by Easton Publishing Services, Inc., Easton, MD

Printer: Data Reproductions Corporation, Rochester Hills, MI
Cover designer: Grafik Communications, Ltd., Alexandria, VA
Technical/production editor: Kathryn Lynch

Library of Congress Cataloging-in-Publication Data
Geary, David C.
    Children's mathematical development : research and practical applications / David C. Geary.
      p. cm.
    Includes bibliographical references and index.
    ISBN 1-55798-258-9 (acid-free paper)
    1. Mathematics—Study and teaching (Elementary)   2. Mathematical ability.
  I. Title.
QA135.5.G398 1994
370.15′651—dc20
                                            94-21047
                                             CIP

British Library Cataloguing-in-Publication Data
A CIP record is available from the British Library.

Printed in the United States of America

Contents

Preface

In recent years, issues surrounding the mathematical development of children and adolescents have attracted the attention of the scientific community as well as the general public. In the United States, for instance, the news media frequently publicizes research findings that show that the mathematical skills of American children lag behind the skills of their peers from many other nations. Poor mathematical skills, in turn, will most likely influence the individual's later employability, wages, and standard of living. Thus, it is not surprising that children's numerical and mathematical development is a growing area of scientific interest and an area of national concern (U.S. Department of Education, 1991). Because of this growing interest, it seemed to be a good time to pull together the research findings on the diverse areas of children's numerical and mathematical development.

Reflecting the diversity of research in children's mathematics, the material presented in this book covers a wide range of material, from the numerical competencies of animals to instructional approaches to mathematics. The basic aim was to provide a comprehensive treatment of the extant literature on the mathematical growth of children and adolescents and to draw some conclusions about important and oftentimes controversial issues in this area. Some of these issues include gender differences in mathematical abilities and the factors that underlie the consistent advantage of East Asian children over American children in mathematical performance (e.g., are these differences related to differences in intelligence or to cultural factors, such as schooling?). With its comprehensive scope, the book should be of interest to cognitive, developmental, and educational psychologists, as well as educators and other professionals interested in mathematical development and related issues (e.g., gender differences, mathematical disabilities, or mathematics instruction).

I would like to thank Peter Frensch for providing insightful comments on each of the eight chapters in this book; Craig Anderson, Mark Ashcraft, Christy Bow-Thomas, Harris Cooper, Linda Coutts, Leslie Geary, Jeff Gilger, Webster Kher, and Fred vom Saal for thoughtful comments on one or more chapters; Judy Nemes, APA development editor, for helpful suggestions on some organizational features of the book; and Kathy Lynch, APA technical/production editor, for her work on various aspects of the final manuscript. Finally, I would like to acknowledge support from the Research Board of the University of Missouri during the preparation of portions of this book.

Introduction

The area of children's mathematical development is diverse and exciting. The associated research programs address very basic issues about human cognition and development, as well as more applied concerns about children's mathematical learning. For instance, one current debate focuses on whether number skills, like basic language skills, are an integral part of a person's biological makeup. At the same time, other researchers in this area are comparing the classroom practices of teachers in East Asia and the United States as part of an effort to identify effective teaching methods and to better understand the source of international achievement differences in mathematics. In all, the various research endeavors in children's mathematics have much to offer American education and American society and, at the same time, add to our understanding of the developing mind.

The purpose of this book is threefold. The first general goal is to provide a comprehensive overview of what is known about children's mathematical skills and the development of these skills. This goal is addressed in the first three chapters, which cover early numerical abilities, arithmetical development, and mathematical problem solving. The second general goal of the book is to provide a thorough discussion of specific issues within the area of children's mathematical development—including individual differences, gender differences, and cross-national differences in mathematical achievement. The goal is not simply to provide a summary of one theory or another, but rather to integrate what is known about the area and to provide a substantive conclusion about the issue. The final goal is to present some ideas on how research in children's mathematics might be used to improve the mathematical instruction of American children and to consider how this research might be used in the remediation of children's learning problems in mathematics.

For each chapter, a final conclusion section highlights the basic area discussed and major issues within the area. Moreover, rather than writing chapters that only provided a breezy overview of the associated material, I decided to present the details and provide a summary section for the longer or more technical sections within each chapter. (Sections that contain summaries are indicated in the note at the beginning of each chapter.) Therefore, if one is interested in the details, the entire section should be read, whereas the summary section only needs to be read if one is interested simply in the gist of the associated research findings. This style has led to some redundancies in parts of the book, but this seemed preferable to providing too little information for some readers and not enough information for others.

The remainder of this introduction contains a brief overview of some of the more salient and important issues addressed in each of the chapters of this book.

Chapter 1 starts with a consideration of the basic numerical skills of human infants. The results of the human infancy studies are at times quite surprising in that they suggest that infants are born with a sensitivity to the numerosity, or quantity, of small sets of items. In the 1st week of life, infants appear to know that, for instance, three sets of items differ in quantity from two sets of items (Antell & Keating, 1983). By 5 months of age, perhaps sooner, infants appear to understand the effect that the addition or subtraction of one item has on the quantity of a small set of items (Wynn, 1992a). By 18 months of age, infants respond to ordinal relationships; that is, they appear to understand that a set of three items is more than a set of two items (R. G. Cooper, 1984).

During the preschool years, children learn basic number names, their culture's counting system, and all of the different contexts (e.g., measurement) within which numbers are used (Fuson, 1988). Many features of children's unfolding number skills appear to be universal, although there are important and intriguing cultural variations in the ways in which this knowledge is expressed (Saxe, 1982a).

The first chapter closes with a consideration of the numerical competencies of animals. The finding that many species of animals show numerical competencies that are similar to those found in human infants supports the argument that many basic numerical skills are fundamental, or biologically primary. For instance, the basic addition skills found in human infants and toddlers are highly similar to those found in the common chimpanzee (*Pan troglodytes*; Boysen & Berntson, 1989).

Many arithmetical skills, like basic number skills, are found throughout the world. For instance, children throughout the world apply their counting skills and counting knowledge to the solving of basic arithmetic problems. At the same time, there are other features of arithmetic, such as fractions, that are very difficult for children to master (Gelman, 1992). The first two sections of chapter 2 provide an overview of children's *informal* (learned before school) and *formal* (learned in school) knowledge of arithmetic. This overview is followed by a thorough consideration of children's developing skills in addition, subtraction, multiplication, and division. In each section, cross-cultural similarities and differences in children's developing competencies are noted. Chapter 2 closes with a consideration of the cognitive and social mechanisms that underlie children's arithmetical development.

An essential component of children's mathematical development is learning how to effectively problem solve. Chapter 3 contains an introduction to research on children's ability to solve arithmetical and algebraic word problems. Some of the topics addressed in this chapter include the relationship between problem structure and children's problem-solving strategies, the development of problem-solving schema, and metacognitive processes in mathematical problem solving. For instance, one fascinating line of research in this area has shown that the strategies young children use to solve arithmetic word problems mirror the semantic structure of the problem (Carpenter & Moser, 1983). Research in this area shows that key words such as *more* can strongly influence whether a child will use one type of strategy or another to solve the associated problem.

Processes that are internal to the child rather than reflected in the word problem itself are also extensively considered for both arithmetical and algebraic problem solving. These processes include the ability to mentally represent the meaning of the quantitative relationships presented in the problem and the ability to translate the representations into appropriate equations (A. B. Lewis, 1989). The sections on arithmetical and algebraic problem solving end with a consideration of developmental patterns, that is, the mechanisms that might govern improvements from one grade to the next. The associated materials reflect the influence of reading skills and working memory, as well as a consideration of expert-novice differences, on developmental changes in children's mathematical problem-solving ability.

Chapter 4 is concerned with why both children and adults differ in mathematical abilities. This question is approached from three perspectives: psychometric, cognitive, and behavioral genetic. The psychometric studies suggest that there are two general domains of mathematical ability:

(a) number and arithmetic and (b) mathematical problem solving. An equally important finding is that by high school, these two classes of ability are essentially independent of one another (Dye & Very, 1968), suggesting that the processes that contribute to good number and arithmetic skills differ from those that contribute to mathematical problem solving.

The cognitive studies are concerned with identifying the processes underlying individual performance differences on psychometric tests. These studies suggest that children who score high on achievement and ability tests in arithmetic are very skilled at choosing the most effective strategy available to them to solve each individual problem (Geary & Burlingham-Dubree, 1989; Siegler, 1988a). For example, to solve simple arithmetic problems (e.g., 3 + 5), these children will, at times, simply retrieve the answer from memory; for other problems they might count on their fingers; and for still other problems they might count in their head. The ability to solve arithmetical word problems, in contrast, appears to be more strongly related to *working-memory resources*—that is, the ability to keep important information in mind while performing basic calculations—than to strategy choices.

There is not an extensive literature on the genetic contributions to individual differences in arithmetical and mathematical problem-solving abilities. However, those studies that have been conducted are intriguing. For example, it appears that roughly one half of the variability in basic arithmetical skills is due to differences in the underlying constellation of genes that support these skills (Vandenberg, 1966). In other words, basic arithmetical skills appear to be partly heritable.

The issue of whether children have specific learning disabilities in mathematics has been largely ignored or misunderstood by the scientific and educational communities. For instance, one recent textbook in developmental psychology discusses reading disabilities but does not even mention mathematical disabilities (Steuer, 1994). Worse yet, another recent textbook argues that mathematical disabilities do not exist at all (Krantz, 1994). In fact, about 6% of school-age children appear to have some form of learning disorder in mathematics (Badian, 1983). In chapter 5, I provide a comprehensive review of the cognitive, neuropsychological, and potential genetic components of mathematical disabilities (a more technical review is presented in Geary, 1993). Here I argue that there are three subtypes of arithmetic-related disabilities, as well as specific cognitive deficits associated with difficulties in mathematical problem solving. For each of the arithmetic-related subtypes, cognitive and neuropsychological correlates are discussed—along with developmental patterns, their potential relationship to reading disabilities, and potential genetic contributions.

The issue of gender differences in mathematical abilities is complex and controversial. By chapter 6, it should be clear that there are many different types of mathematical skills and many different influences on mathematical development. Broad statements, such as "males are better at mathematics," about such a complex topic are not warranted. In fact, the research reviewed in chapter 6 suggests the existence of a consistent male advantage in some areas of mathematics, beginning as early as the first grade, but no gender differences at all in other areas. One of the more fascinating, and heretofore ignored, issues in this area is whether there are gender differences in those numerical skills that appear to be inherent, or biologically primary. The consistent finding, across cultures, of no gender differences in these basic skills suggests that there is no "male math gene."

Recent meta-analytic reviews of gender differences in more complex mathematical abilities (those learned in school) have suggested that any gender differences that are found do not emerge until adolescence, are small, and are disappearing (i.e., have gotten smaller over the last several decades; Hyde, Fennema, & Lamon, 1990). These results are widely cited and have been highly influential but are probably wrong. In chapter 6, the issue of gender differences in a variety of mathematical skills is addressed, with a focus on cross-cultural patterns. Across cultures, boys have an advantage over girls in geometry, measurement, and the solving of word problems (arithmetical and algebraic), beginning as early as first grade and continuing into adulthood. The conclusions drawn by Hyde, Fennema, and Lamon were based primarily on studies conducted in the United States. Cross-cultural research shows that gender differences in mathematics tend to be the smallest in those countries with the lowest achieving children, which include the United States (Husén, 1967). The finding of small and disappearing gender differences in mathematical abilities in the United States appears to be largely due to a "floor effect": The performance of American children is too low to detect any differences.

After an overview of these cross-cultural data, a consideration of cognitive, psychosocial, and biological influences on the magnitude of the gender difference in mathematical abilities is presented. Topics such as the relationship between spatial abilities and mathematical performance, gender differences in perceived mathematical competence and career aspirations, the classroom experiences of boys and girls, and related issues are thoroughly discussed. This section of chapter 6 closes with a consideration of gender differences in mathematical abilities within the broader context of human evolution, which in turn is integrated within a formal biosocial model of mathematical gender differences.

The final section of chapter 6 focuses on the cognitive correlates of mathematical giftedness. The research on gender differences and the mathematically gifted is presented in tandem, because the magnitude of the gender difference in mathematical problem solving tends to be largest in gifted samples (Benbow, 1988). Thus, some of the skills that contribute to mathematical precocity might also be an important source of gender differences in mathematical abilities. For instance, apparently, mathematically gifted junior high school students have better memories for spatial information than university students, but they do not have exceptional memories for verbal information (Dark & Benbow, 1991).

Chapter 7 addresses an issue that is of continuing political, social, and scientific concern: the nature and source of international differences in mathematical achievement. The political nature of this issue was reflected in *America 2000: An Education Strategy* (U.S. Department of Education, 1991). Here, one of the primary goals stated that "U.S. students will be first in the world in science and mathematics achievement" by the year 2000 (p. 3). In chapter 7, I provide an extensive review of international studies of mathematics achievement. These studies consistently show that American children are among the most poorly educated children in mathematics in the industrialized world.

In the second general section of chapter 7, potential causes of these international differences in mathematics achievement are examined. For instance, the achievement gap is most dramatic when the performance of American children is compared with that of their peers from East Asia (i.e., Japan, Taiwan, Korea, and mainland China). Some have argued that the associated achievement differences reflect racial differences in intelligence (Lynn, 1983). This issue and the associated evidence are critically reviewed in chapter 7, as are parallel achievement differences across ethnic groups within the United States. The influence of general cultural factors—such as values, schooling, and family—on international achievement differences is also considered in turn. The chapter ends with my conclusion as to why East Asian children perform so much better in mathematics than their American peers and with a judgment regarding whether the *America 2000* goal will be achieved.

Chapter 8 deals directly with children's mathematics education—specifically, issues in instruction and remediation. The chapter opens with a critique of current philosophical approaches to educational reform within the United States (for a fuller discussion of this topic, see Geary, in press). From there, social policy and direct instructional issues are addressed. For instance, the question of whether the United States should adopt a national curriculum in mathematics is considered. More practical information can also be found in this section of the chapter. Suggestions, which are based on psychological research, for how to better teach mathematical procedures and concepts are provided. The chapter closes with a discussion of mathematics anxiety and suggestions for how to remediate the different forms of mathematical disability discussed in chapter 5.

Children's Early Numerical Abilities

Research conducted during the past 25 years has provided a wealth of information about the basic numerical competencies of human infants and preschool children. The basic findings in these areas are highlighted in this chapter, beginning with a consideration of the numerical skills of infants. After this discussion is a presentation of the developing number skills of preschool children, focusing primarily on the development of counting skills and number knowledge. Many of the experiments discussed in the first two major sections of this chapter strongly suggest that number is a natural domain in humans—that is, humans are born with a fundamental sense of quantity. To bolster this argument, the chapter concludes with a consideration of the numerical skills of many species of animals, especially as these skills are related to the early competencies of human infants and preschool children.

Our journey into the mind of the child will begin with a brief introduction to Piaget's (1965) seminal studies of the development of number skills in young children. Piaget developed a variety of tasks to assess the child's conception of number. For instance, in one task, two rows of four marbles were presented to the child, and the child was asked which row had more. If the marbles in the two rows were aligned in a one-to-one fashion, then 4- to 5-year-old children almost always stated that there was the same number of marbles in each row. Next, Piaget would spread out one of the rows, so that the one-to-one correspondence between the rows was not obvious, and ask again which row had more. This time the child almost always stated that the longer row had more marbles. This type of

---

Summaries are provided at the end of each of the main sections: Numerical Competencies in Human Infants (p. 11), Number and Counting (p. 28), and Numerical Competencies in Animals (p. 33).

justification led Piaget to argue that the child's understanding of equivalence, or that there were four in each row, was based on how the rows looked rather than on a conceptual understanding of *number*. An understanding of *number* would require the child to state that after the transformation, the number in each row was the same, even though the rows now looked different from one another. With Piaget's tasks, children do not typically provide this type of justification until they are 7 or 8 years old, leading Piaget to argue that younger children do not possess a conceptual understanding of *number* and that any number-related activities, such as counting, are largely done by rote.

Mehler and Bever (1967) challenged Piaget's argument that preschool children do not have a conceptual understanding of number by demonstrating that under some circumstances even 2½-year-old children use quantitative, rather than perceptual, information to make judgments about *more than* or *less than*. In one procedure, they first presented 2½- to 4½-year-old children with two rows of four M&Ms and asked if the two rows were the same. Next, one or both of the rows were transformed. For instance, in one scenario the experimenter added two M&Ms to the second row and then made the row shorter. The child was then told, "Take the row you want to eat, and eat all the M&M's in that row" (Mehler & Bever, 1967, p. 141). The 2½-year-olds chose the row with more M&Ms more than 80% of the time, suggesting that they understood *more than* and *less than*. In a replication, Beilin (1968) showed that the Mehler and Bever result was probably due to the act of adding M&Ms and transforming the length of the row at the same time, because the performance of 3-year-olds deteriorated when adding the M&Ms and transforming the row were done in separate steps. Nevertheless, both studies did show that even 2½- to 3-year-olds were sensitive to the fact that adding M&Ms to the row increased the quantity in that row, suggesting that the standard number-concept tasks developed by Piaget (1965) underestimated the numerical competencies of young children.

Numerical Competencies in Human Infants

The research of Mehler and Bever (1967), as well as Beilin (1968), provided the motivation for contemporary psychologists to begin to examine the cognitive skills of preschool children, and even infants, more closely. The results of the infancy studies are intriguing and oftentimes quite surprising. The world of the infant, rather than being a "great, blooming, buzzing, confusion" (James, 1890/1950; I always thought that this better described the behavior of new parents), is in fact rather orderly. The infant is able to extract meaningful information from his or her environment probably from the moment of birth. An infant's sensitivity to important aspects of the environment, such as the facial features or smell of her or his mother is not too surprising, but a sensitivity to more abstract features of the environment, such as quantity, is surprising. Unlike Mom's eyes or nose, quantitative features of the environment are not an attribute of what is perceived. As both Mom and Dad hover around the infant, the infant not only perceives their physical features, such as the shape of the face, but also seems to be aware that there are two, not one or three, people milling about. "Twoness" is abstract because it is not an attribute of either parent, but rather represents an important feature of the set of parents, that is, how many.

When considering the numerical competencies of infants, there are three general issues that have been addressed (R. G. Cooper, 1984; Strauss & Curtis, 1984). The first concerns the infant's understanding of *numerosity*, that is, the ability to discriminate arrays of objects on the basis of the quantity of items presented—for example, an awareness that a group of three items differs in quantity from a group of two items. The second issue concerns an awareness of *ordinality*, for example, that three items are more than two items. The final and most recent area of study in infancy research concerns the infant's ability to add and subtract. The basic findings in each of these areas are reviewed below and are followed by a description of theoretical positions on what develops.

Numerosity

One of the first contemporary investigations of the basic numerical competencies of infants was conducted by Starkey and Cooper (1980). Starkey and Cooper sought to determine if infants could tell the difference between sets of items that differed in quantity by using a habituation procedure. Generally, the *habituation procedure* involves presenting a stimulus, an array of two to six dots in this study, to the infant. When the infant is first presented with an array of dots, the information is novel. In this circumstance, the infant will typically be attracted to the array. With repeated presentations, the amount of time spent looking at the array declines, as the infant habituates to it. If the infant again is attracted to the array when the number of presented dots changes (dishabituation), then it is assumed that the infant discriminates between the two quantities.

If, however, the infant does not change his or her viewing time, then it is assumed that the infant cannot discriminate between the two quantities. Starkey and Cooper found that infants between the ages of 4 months and 7½ months were able to discriminate two items from three items, but not four items from six items. Working independently, Strauss and Curtis (1981) found that 10- to 12-month-old infants discriminated two items from three items, but not four items from five items. Under some conditions, some of the infants were able to discriminate three items from four items.

The infant's sensitivity to the numerosity of an array of one to three, and sometimes four, items has since been replicated many times and under various conditions, such as homogeneous versus heterogeneous collections of objects (Antell & Keating, 1983; Starkey, 1992; Starkey, Spelke, & Gelman, 1983, 1990; van Loosbroek & Smitsman, 1990). Among the most notable and exciting of these findings is that infants show a sensitivity to differences in the numerosity of small sets during the 1st week of life (Antell & Keating, 1983), with displays (i.e., sets of rectangular figures) in motion (van Loosbroek & Smitsman, 1990) and intermodally (Starkey et al., 1983). The intermodal studies deserve further elaboration, because the results of these studies suggest that the infant's sensitivity to numerosity is based on an abstract representation rather than on modality-specific representations. In other words, the infant's knowledge that two items differ somehow from three items is not dependent on whether the items are seen or heard (e.g., as in a series of two or three drumbeats).

In the first of these studies, Starkey et al. (1983) presented 7-month-old infants with two photographs, one of two items and the other of three items, and simultaneously presented either two or three drumbeats. Infants looked longer at the photograph with the number of items that matched the number of drumbeats, suggesting that the infants somehow extracted quantity from the visual information (i.e., the photographs) and matched this with the quantity extracted from the auditory information (i.e., the drumbeats). This pattern of results suggests that infants have a cognitive system that abstractly codes for numerosities up to three or four items and that this system is neither visually nor auditorily based (but see D. Moore, Benenson, Reznick, Peterson, & Kagan, 1987).

In all, the results of these studies add up; infants are able to represent and remember quantities of up to three and sometimes four. The ability to represent the numerosity of small sets appears to exist in the 1st week of life (Antell & Keating, 1983); is not dependent on a specific modality (Starkey, 1992; Starkey et al., 1983); and is not influenced by factors such as whether dots or household items are presented, whether the presented items are static or moving, or by the density of the displays (e.g., bunched together or spread out). These results should not be taken to mean that the infant's understanding of numerosity is equivalent to that of an older child (Gelman & Gallistel, 1978), nor do the results inform us about how the infant is able to make these discriminations. Rather, these findings indicate that a sensitivity to numerosity, at least for arrays of up to three or four items, is likely to be innate (Gelman, 1990).

Ordinality

Even though infants are able to detect and represent small quantities, the results of the just-described studies should not be taken to mean that infants are necessarily sensitive to which set has more or less items. That is, infants appear to represent numerosity but do not rank order these representations. A sensitivity to ordinality (i.e., somehow knowing that 3 follows 2 and 2 follows 1 in terms of numerosity) appears to develop during the first 1½ years of life (R. G. Cooper, 1984; Strauss & Curtis, 1984).

For instance, to assess when infants begin to understand that there is a relationship between numerosities, Strauss and Curtis (1984) first taught infants, by means of operant conditioning, to touch the side of a panel that contained the smaller or larger number of two arrays of dots. For the smaller condition, the smaller array might contain three dots and the larger array four dots; the infant would then be rewarded for touching the panel associated with the smaller array.

Next, the infant might be presented with arrays of two and three dots. If the infant were simply responding to the value that was rewarded, then she or he would touch the panel associated with three. On the other hand, if the infant were responding on the basis of the ordinal relationship—responding to the smaller array in this example—then he or she would touch the panel associated with two. In this study, 16-month-old infants responded to the two, suggesting a sensitivity to *less than* in this example (Strauss & Curtis, 1984). Ten- to 12-month-olds seemed to notice that the numerosities in the arrays had changed but did not discriminate between *less than* and *more than*; it was simply *different from* (R. G. Cooper, 1984). These studies suggest that infants begin to develop an awareness of ordinal relationships across small values (up to three or four) in the first 18 months of life.

Arithmetic

The results of a recent and intriguing study suggest that infants as young as 5 months of age can even add and subtract small quantities (Wynn, 1992a). In one procedure, the infants were shown a single item, a Mickey Mouse doll, in a display area. A screen was then raised and blocked the infant's view of the doll. Next, the infant watched the experimenter place a second doll behind the screen. The screen was then lowered and showed either one or two dolls. Infants tend to look longer at unexpected events. Thus, if they were aware that adding a doll to the original doll would result in two dolls, then when the screen was lowered, they would look longer at one-doll displays than at two-doll displays. This is exactly the pattern that was found! Further experimental manipulations suggested that these infants were also aware that $2 - 1 = 1$, and that these results were not simply due to the infants' noticing a change in numerosity but rather that they appeared to have been aware of the precise results (for at least $1 + 1 = 2$ and $2 - 1 = 1$). Of course, these results need to be replicated in other laboratories before it can be said with certainty that infants have an innate sense of the effects of addition and subtraction on quantity, although Starkey (1992) did find similar abilities in 18-month-olds (see Informal Arithmetic section of chapter 2).

Developmental Mechanisms for Early Numerical Abilities

Once a basic understanding of infant numerical skills is achieved, then the next set of questions to be answered is, How are infants able to represent numerosity, understand ordinal relationships, or add and subtract? Initial speculation about the mechanisms underlying their early numerical abilities was based on the performance of adults (e.g., Mandler & Shebo, 1982). The goal of early research in this general area was to better understand how adults could determine the numerosity of briefly presented sets of items. On the basis of this research, two general processes have been suggested as underlying the representation of numerosity: perception-based subitizing and preverbal counting (Gallistel & Gelman, 1992; Klahr & Wallace, 1973; Mandler & Shebo, 1982; Starkey & Cooper, 1980). *Subitizing* refers to the ability to rapidly report the numerosity of briefly presented arrays of typically up to three or four items. Of course,

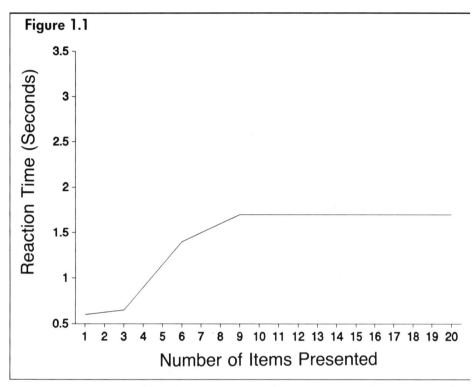

Figure 1.1

Reaction time patterns for making numerosity judgments. (From "Subitizing: An Analysis of Its Component Processes," by G. Mandler and B. J. Shebo, 1982, *Journal of Experimental Psychology: General, 111,* p. 8. Copyright 1982 by the American Psychological Association. Adapted with permission of the author.)

adults and children can determine the numerosity of an array of items by verbally counting the items (Silverman & Rose, 1980). In these experiments, however, the items are often presented very briefly (often for only 200 ms), which makes counting difficult.

The studies of Mandler and Shebo (1982) suggest that adults can use at least three different types of processes for determining the numerosity of arrays: subitizing, counting, and estimating (Klahr & Wallace, 1973). This conclusion was based, in part, on the amount of time (termed *reaction time*) needed to judge the numerosity of arrays ranging from 1 to 15 or 20 items. With this procedure, an array of, say, three dots is presented for 200 ms. The subject is required to state the number of dots in the array; reaction time is measured from the point of presentation to the point of response. When reaction times are plotted against the number of items in the array, as shown in Figure 1.1, then the resulting pattern is consistent with three distinct ways to determine numerosity.

Viewing Figure 1.1 from left to right, we first see a relatively flat slope for quantities that range from 1 to 3 items; that is, it requires very little additional time to determine the numerosity of a display of 2 items in relation to a display of 3 items. The slope increases dramatically from 3 to 9 items, but for 10 or more items the slope is again rather shallow. Mandler and Shebo (1982) argued that the shallow slope for the array sizes of 1 to 3 was the result of subitizing. The increased slope from 3 to 9 items was thought to reflect the use of implicit counting to determine numerosity. The flat slope for larger values was based on estimating the number of items in the display. Even though the slopes are similarly flat for the 1 to 3 range and the greater-than-9 range, the error rates differed significantly. Errors were rare for arrays with 4 or fewer items but were greater than 50% for arrays with more than 7 items. The divergence in error rates for small and large arrays is consistent with the argument that different processes are being used to determine the numerosity of these arrays.

Of particular interest are the processes underlying subitizing, or the rapid determination of numerosities up to three items, because these same processes might underlie the infant's early sensitivity to numerosity. Mandler and Shebo (1982) argued that subitizing was a basic perceptual process that allowed adults to extract numerosities of two or three on the basis of frequently occurring patterns in the natural environment. For instance, an array of three objects is perceived as a triangle. The quantity of three is quickly and automatically, in adults at least, extracted from the perception of a triangle. Similarly, an array of two objects can be perceived as a straight line, with the quantity of two automatically extracted from the perception of straight lines. Mandler and Shebo (1982) argued further that these seemingly automatic processes in adults develop gradually in children. In other words, the automatic association of the basic straight line and triangle patterns with the respective numerosities of two and three only occurs after much experience. Thus, it is not likely that such perception-based processes underlie the infant's sensitivity to numerosity.

Indeed, Gallistel and Gelman (1992; see Wynn, 1992b, 1992c) argued that subitizing in infants and adults, among other things, was based on an innate preverbal counting and timing mechanism rather than on basic perceptual cues (Meck & Church, 1983). Meck and Church (1983) developed a model to explain numerosity- and time-estimation skills in animals, which consists of three important features: a pacemaker, a gate, and an accumulator. It is assumed that the pacemaker generates a con-

Figure 1.2

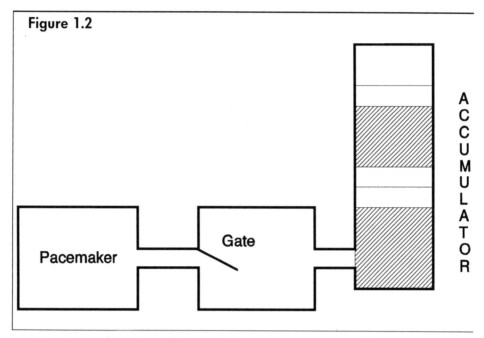

Schematic representation of the preverbal counting and timing mechanism.

tinuous flow of pulses. With the processing of the first item in a numerosity task, the gate is opened, allowing for a flow of pulses toward the accumulator. The pulses are then gathered into the accumulator. When the first discrete item is perceived, the gate closes. The pulses deposited in the accumulator at that point represent the numerosity of one. When the second item is perceived (e.g., visually scanned), the gate is opened for a fixed duration and then closed, again allowing for the depositing of pulses into the accumulator, as shown in Figure 1.2.

After all of the items have been processed, the information stored in the accumulator is read into memory and compared with innate, preverbal representations of numerosity. In other words, the child is born with an understanding that two pulses in the accumulator represent two items, three pulses represent three items, and so on. It is further assumed that the accuracy of the preverbal counting and timing mechanism for making numerosity judgments declines as the number of items to be judged increases. The use of this preverbal mechanism would, therefore, enable fast and accurate judgments for small numerosities, as is found in infants, but rather inaccurate judgments for sets of many items. Although Gallistel and Gelman (1992) argued that a sensitivity to quantity is innate, it does not follow that the infant is born with a mature understanding of number. Rather, the preverbal counting and timing mechanism, among other things, orients the infant's attention to quantitative features of the environment and automatically processes these features. A more mature understanding of *number* will emerge from this innate system and the child's number-related activities.

Moreover, the Gallistel and Gelman (1992) model predicts an early sensitivity to ordinal relationships. It is assumed that the innate preverbal counting and timing system also provides information on the relative quantities of sets of items. So the infant knows that a set of two items is less than a set of three items, for instance. R. G. Cooper (1984), on the other hand, argued that infants' understanding of ordinal value develops gradually as they experience changes in the quantity of items in their natural environment. These changes typically involve some form of addition or subtraction. If both parents are playing with the infant and then one of the parents leaves, the "numerosity detectors are activated in sequence" (R. G. Cooper, 1984, p. 164), the detectors for two and one, in this example. This pattern is repeated many times in the infant's everyday activities and results in the frequent sequential activation of the underlying numerosity detectors and in the gradual discovery of the ordinal relationships among detectors for the numerosities of one, two, three, and so on. It is likely that features of both Gallistel and Gelman's (1992) model and R. G. Cooper's (1984) model are correct. Basically, as noted above, it is likely that a child's understanding of ordinal values develops from an interaction between innate sensitivities to numerosity and the child's experiences (Piaget, 1965; Steffe, von Glasersfeld, Richards, & Cobb, 1983).

At first glance, the results of the previously described study of infant arithmetic skills (Wynn, 1992a) would appear to contradict the findings of R. G. Cooper (1984; Strauss & Curtis, 1984). This is because adding and subtracting typically require an understanding of ordinality (Brainerd, 1979; Gelman & Gallistel, 1978; Siegel, 1974). An understanding of ordinality appears to emerge by 1½ years of age, but the infants in Wynn's (1992a) study appeared to understand the effects of addition and subtraction by 5 months of age. However, it is not at all clear what processes might have been used by the infants in Wynn's study to add and subtract. It is possible that infants are able to change their representation of the numerosity of a display (Mickey Mouse dolls in this study) without an understanding that two is more than one. In other words, it does not necessarily follow from Wynn's study that infants have an understanding of ordinal relationships. At this point, it is not clear how infants understand that adding results in more whereas subtracting results in less.

Summary

Infants show an amazing variety of basic numerical skills, including a basic understanding of quantity, or numerosity; ordinal value; and the effects of addition and subtraction on quantity. The finding that infants are sensitive to quantity in the 1st week of life (Antell & Keating, 1983) and under many different conditions makes it very likely that infants are born with a sensitivity to numerosity, at least for arrays of up to three or four items. The infant's understanding of ordinal relationships, for small values at least, appears to emerge by about 18 months of age. That is, at this age, infants appear to understand that three is more than two and that two is more than one. More recent evidence suggests that by 5 months of age, infants are sensitive to the effects that adding or subtracting one or two items has on quantity; they seem to know that adding $1 + 1 = 2$ and that $2 - 1 = 1$.

Gallistel and Gelman's (1992) preverbal counting and timing model is an intriguing attempt to integrate animal research on number skills with research on the early skills of human infants. They argued that the same processes that allow for timing and counting in animals are involved in the human infant's basic numerical skills. In fact, it is very reasonable to assume a continuity between the numerical competencies of some animals and the numerical competencies of human infants (see Numerical Competencies in Animals section), although the mechanisms underlying these skills are not entirely clear at this point. In any event, the research in this area is consistent with the view that many basic numerical skills are innate (Starkey, Spelke, & Gelman, 1991) and provide the skeletal principles on which later numerical skills emerge (Dehaene & Mehler, 1992; Gelman, 1990).

Number and Counting

Counting and number-related activities are a natural human enterprise (Crump, 1990). Saxe, Guberman, and Gearhart (1987), for instance, showed that children as young as 2 years of age regularly engage in number-related activities (see also Ginsburg & Russell, 1981). These activities include solitary episodes, such as counting toys or number of snacks, as

Figure 1.3

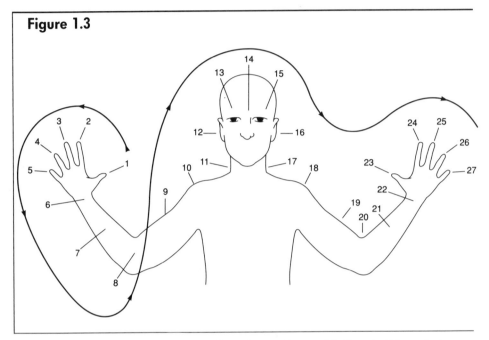

The Oksapmin counting system. (From "Developing Forms of Arithmetical Thought Among the Oksapmin of Papua New Guinea," by G. B. Saxe, 1982, *Developmental Psychology*, *18*, p. 585. Copyright 1982 by the American Psychological Association. Reprinted with permission of the author.)

well as social play. Parents and children frequently engage in number-related play, such as singing songs with number references (e.g., "One, two, buckle my shoe") and counting toes and fingers, as well as activities that require an understanding of the uniqueness of individual digits (e.g., "Please turn the TV to Channel 5"). The natural occurrence of number-related activities and the development of methods for representing number and for counting are not confined to industrialized cultures. Saxe, for instance, has studied the rather interesting representational and counting systems of the Oksapmin, a horticultural society in Papua New Guinea (Saxe, 1981, 1982a, 1982b). Here, counting and numerical representations are mapped onto 27 body parts, as shown in Figure 1.3. "To count as Oksapmins do, one begins with the thumb on one hand and enumerates 27 places around the upper periphery of the body, ending on the little finger of the opposite hand" (Saxe, 1982a, pp. 159–160). This system is used not only for counting but also for representing ordinal position and for making basic measurements. Similarly, Zaslavsky (1973) described many common number-related activities and games engaged in by children in Africa.

Given the naturalness of number-related activities and the importance of basic number skills in arguably all societies, it is not surprising that a primary task of preschool and early elementary school children is to gain an understanding of number concepts and to master counting (L. B. Resnick, 1983). To be sure, these are effortless and seemingly inconsequential tasks for most adults. The child's acquisition of these skills, however, is a slow and often difficult process. For most children, this process spans a 6-year period, from the ages of 2 to 8 years (Fuson, 1988; Piaget, 1965). To better gain an appreciation of these skills, consider the knowledge of typical 8-year-old children. First, they understand a variety of important features of numbers: that each number word is unique and represents a unique quantity and that numbers are *serially ordered*, that is, successive numbers represent successively larger quantities (Brainerd, 1979; Fuson, 1988; Gelman & Gallistel, 1978). The hallmark of their understanding is reflected in the knowledge that each number reflects groups of smaller numbers. For instance, they now understand that 8 can be decomposed into 6 and 2, 3 and 5, or several other combinations. The understanding that numbers can be represented by groups of other numbers is an essential step in conceptually understanding the addition and subtraction of relatively large numbers (i.e., beyond the 1 or 2 items that infants can "add or subtract"; Fuson, 1988). They are also able to use numbers for measurements, such as for age: "I'll be 8 next summer."

Children not only must develop a conceptual understanding of numbers themselves but also must learn to appropriately use numbers in a variety of contexts. As aptly noted by Fuson (1988), numbers are used not only in the traditional sense—to note the cardinal value of a set of items or ordinal rankings (e.g., first, second, . . .)—but also in many less traditional ways. For example, the phone number 547-9258 does not represent the 5,479,258th phone, nor does it represent a numerical value. Similarly, the child who rides Bus 39 to and from school understands that this is not the 39th bus to stop by each day and that Bus 39 is not somehow less than Bus 83 but greater than Bus 12. How about "Wait a second"? *Second* in this context pertains to a continuous feature of the environment (time), whereas *second* in the lunch line refers to the ordering of different people. Clearly, the child's understanding of number reflects not only concepts in the traditional sense but also an appreciation of the many different ways, numerical and nonnumerical, and the many different contexts within which numbers are used.

To further complicate the situation, the child must master a procedure for *enumerating*, or assigning cardinal or ordinal meanings to items.

This procedure is, of course, counting (Gelman, 1978; Greeno, Riley, & Gelman, 1984; Miller & Gelman, 1983; Silverman & Rose, 1980; Steffe et al., 1983). As with the acquisition of number concepts, verbal counting skills develop over the course of several years, beginning some time between 2 and 3 years of age (Gelman & Gallistel, 1978). An essential feature of enumeration involves assigning word tags (such as *one*) to the items as they are being counted. To do this in an understandable manner, the child needs to learn (i.e., memorize) the number words; in some cultures tagging involves finger patterns or gestures (Zaslavsky, 1973). A more difficult process involves mapping the number words onto the growing concept of number (Fuson, Richards, & Briars, 1982; Wynn, 1992b). For instance, children must gain an appreciation that enumeration and number words serve many purposes—such as determining the numerosity of a set of items, ordering the items within a set (e.g., first, second, third), comparing the quantity of two or more sets of items, or abstractly representing quantity (e.g., Saxe, 1977).

The remainder of this section begins with a descriptive overview of how the child's counting skills change during the preschool years. These changes include the acquisition of number words, number concepts, counting errors, and an understanding of the use of numbers and counting for representing cardinality and ordinality and for making measurements. The section closes with a consideration of developmental models, that is, theoretical views on the processes underlying the acquisition of number concepts and counting.

Number Words

To use counting effectively, the child must master several basic skills (Fuson, 1988; Gelman & Gallistel, 1978; Schaeffer, Eggleston, & Scott, 1974; Wynn, 1990). First, she or he must learn to create a one-to-one correspondence between number names and counted items, such that each item is assigned only a single number name. The child must also learn to order the number names in the correct sequence (i.e., one, two, three, . . .) and come to understand that the last number named in the count, that is, the cardinal number, holds special meaning—it represents the total number of counted items.

To master these skills, children must learn, by rote, their culture's number words (Fuson, 1991). Most 3- to 4-year-olds know the number words, in the correct sequence, from *one* to *ten* (Fuson, 1988; Siegler & Robinson, 1982). For English-speaking children, and for children speaking most European-derived languages, learning the words for numbers greater than 10 is particularly difficult (Fuson & Kwon, 1991; K. F. Miller, 1992; K. F. Miller & Stigler, 1987). This is so because the number words for values up to the hundreds are irregular in these languages; that is, the names for these words do not map onto the underlying base-10 structure of the number system. In contrast, in most Asian languages there is a direct one-to-one relationship between the number words greater than *ten* and the underlying base-10 values represented by those words (Fuson & Kwon, 1991; Hatano, 1982; Miura, 1987; Miura, Kim, Chang, & Okamoto, 1988; Miura, Okamoto, Kim, Steere, & Fayol, 1993). For instance, the Chinese, Japanese, and Korean words for 11 are translated as *ten one*. The use of *ten one*, rather than *eleven*, to represent 11 has two advantages. First, children do not need to memorize additional word tags, such as *eleven* and *twelve*. Second, the fact that 11 is composed of a single tens value and a single units value is obvious in Chinese but is not at all obvious in English, or in most other European-derived languages. For English-speaking children, *eleven* is simply another number word in a continuous string of number words. Its special status, reflecting the repetition of the basic-unit-value numbers (1, 2, and so on) within the base-10 system is not evident from the number word itself.

To make matters worse, for English-speaking children the units value is spoken before the tens value for the teen words, as with *fifteen*, but after the tens value thereafter, as with *thirty-five* (Fuson & Kwon, 1991). Number words in most Asian languages do not have these confusing features. For instance, in Chinese, 15 is simply *ten five*, and 35 is *three ten five*. These irregularities in English, and many other European-derived languages, not only slow the learning of number words but also make tagging errors, such as writing 51 when hearing 15, common well into the elementary school years (Ginsburg, 1989). Nevertheless, by second grade, most children know the basic number words up to *one hundred* reasonably well. This is not to say that children in different cultures have an equally good understanding of what the different numbers represent but rather that children in many cultures are able to count correctly up to 100 by this age.

Number Concepts

In addition to learning the number words of their language, children must also develop an understanding of how these number words relate to number concepts and how number words are used to count. Children must first understand that each number word refers to a different quantity and then map specific number words onto specific representations of quantity—or, in the case of the Oksapmin, for instance, map body part names onto specific quantities (Saxe, 1981). I now discuss children's understanding of how number words map onto number concepts. In the next section, errors in the application of number words, that is, counting errors, are briefly reviewed.

Some time between 2 and 3 years of age, children begin to use number names when counting (Gelman & Gallistel, 1978). As any parent knows, the child's counting at this age is less than perfect, even for relatively small set sizes. In particular, young children often do not use the standard order of word tags (i.e., *one, two, three*) and sometimes do not use word tags at all. For instance, the child might state "three, five" to count two items and "three, five, six" to count three items. Sometimes children of this age use letters of the alphabet to count, stating "A, B" to count two items and "A, B, C" to count three items. There are two important aspects of this pattern: First, each number word, or letter, is used only once during each count, and second, the sequence is stable across counted sets. The overall pattern suggests that the child implicitly understands that different number words (or different letters) represent different quantities and that the sequence with which they are stated is important (Gelman & Gallistel, 1978).

Many children as young as 2½ years of age also seem to understand that number words are different from other descriptive words, such as *red* or *happy*. For instance, if a child is asked to count a row of three red toy soldiers, then he or she will typically, though not always, use number words to count the set. Moreover, the child understands that *red* describes an attribute of the counted items but that the number assigned to each soldier does not describe an attribute of the soldier itself but somehow refers to the collection of soldiers (Gelman & Gallistel, 1978; Markman, 1979). Even though many 2-year-olds seem to understand that different number words refer to different quantities, they do not know which specific quantity each number word refers to (Wynn, 1992b). For instance, many 2½-year-olds can discriminate four-item sets from three-item sets and know that 4 is more than 3 (Bullock & Gelman, 1977) but might not be able to correctly label the sets as containing *four* and *three* items, respectively. Wynn (1992b) argued that it might require as long as a year of counting experience, from 2 to 3 years of age, for children to begin to associate specific word tags, or number words, with their mental representations of quantity and then to use this knowledge in counting tasks.

Thus, it is likely that many 3-year-olds are beginning to associate number words with specific quantities, but this knowledge probably only extends to small values. Thus, by 3 years of age many children can correctly label a set of 3 items as *three* but might label a set of 6 items as *ten*.

In fact, mapping number words onto the specific representations for quantities is a difficult task for children and is quite likely not complete for many 4-year-olds, even for numbers less than 10 (Gallistel & Gelman, 1992). More important, when we consider the earlier-described irregularities in the number names in many European-derived languages, including English, then the relationship between learning number names and the conceptual understanding of *number* is even more complex. The fact that number words for teens do not map directly onto the base-10 structure of the Arabic number system (i.e., 1, 2, 3, . . .) in most European-derived languages not only makes learning specific number words for teens difficult but also appears to impede the child's conceptual understanding of the base-10 system (Fuson & Kwon, 1991, 1992b).

In contrast, because the Asian number-word system maps directly onto the base-10 structure, the associated number words appear to facilitate the understanding of the base-10 structure. Asian children do not need to memorize arguably unnecessary number words, such as *twelve* or *thirty*, which leads to fewer counting errors and an acquisition of basic counting skills at a younger age for Asian children than for American children (Fuson & Kwon, 1991; K. F. Miller, 1992). Anyway, the point is that children first learn number words and then associate each word with a specific quantity. This process begins with smaller valued numbers and is gradually extended to larger valued numbers (Fuson, 1988). For numbers greater than 10, the speed with which number names and the underlying concepts become associated appears to vary across languages.

Counting Errors

Even after children memorize the standard sequence of number names, counting errors are still common. Gelman and Gallistel (1978) argued that counting, specifically mapping number names onto counted objects, involves the processes of tagging and partitioning, as well as the coordination of these acts. *Tagging* involves assigning a number word to the counted item, whereas *partitioning* involves breaking the items to be counted into two parts: those that have already been counted and those that still need to be counted. During the process of tagging and partitioning, children typically either point to the objects as they are being counted (adults use discrete eye fixations) or move the item from one location to the next as it is counted. The use of finger pointing or moving the objects helps the child to keep track of her or his counting. Tagging errors, which are relatively uncommon, would involve assigning the same word tag to two or more items, such as stating "one, one, one" during the counting of the three toy soldiers.

A partitioning error might involve, for example, counting an item but not moving it from the to-be-counted pile to the already-counted pile. Gelman and Gallistel (1978) found that the most common error involved the coordination of tagging and partitioning—that is, failing to stop the tagging and partitioning simultaneously. For instance, during the counting of the three toy soldiers, the child might continue to state word tags even after he or she has pointed at the last item. Fuson (1988) also extensively studied young children's counting errors and documented many different types of errors. Some of the more common types of counting errors are displayed in Figure 1.4. In Example A, the child points to each item once but speaks two or more word tags with each point. In Example B, the child correctly assigns one word tag to each of the circles but speaks a third tag, without pointing, in between. As shown in Example C, sometimes children point and tag items with each syllable of a number word, and at other times they might tag and point to each item two or more times. Despite all of the different types of errors that can be made during counting, by the time the child reaches kindergarten, he or she is typically a proficient counter, especially for smaller set sizes.

Cardinality and Ordinality

There is more to counting than simply assigning appropriate word tags in the right order (e.g., Becker, 1993; Fuson, Pergament, Lyons, & Hall, 1985; Wynn, 1990, 1992b). The child must come to understand that the word tags provide important information about the counted items. In particular, the child needs to develop an understanding of cardinality and ordinality (Brainerd, 1979). The child must learn that the number word assigned to the last counted object can be used to represent the total number of counted objects (*cardinality*) and that successive number words represent successively larger quantities (*ordinality*). With regard to the former, a variety of methods can be used to gain some insight into the child's understanding of cardinality. One method involves simply asking a child to count her or his fingers, or some other set of objects, and then asking, "How many fingers do you have?" Children who do not

Figure 1.4

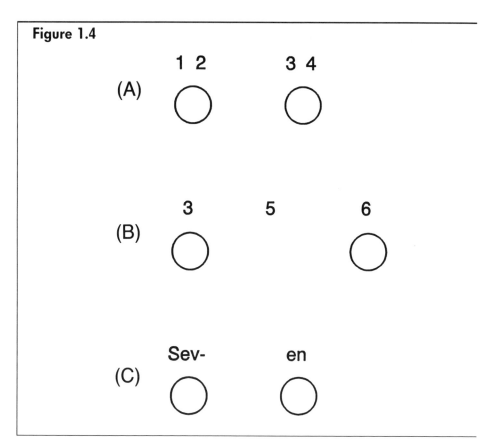

Young children's common counting errors. (From *Children's Counting and Concepts of Number* [p. 182] by K. C. Fuson, 1988, New York: Springer-Verlag. Copyright 1988 by Springer-Verlag New York Inc. Adapted with permission.)

understand the significance of the last word tag, *five* in this example, will recount the fingers rather than simply restating "five" (Fuson, 1991).

Although most 3- and 4-year-olds recount when asked "how many," some 3½-year-old children do respond to the "how many" question by restating the last word tag (Wynn, 1990). Though suggestive, this in itself does not necessarily indicate that the child has an understanding of cardinality. Some children appear to use a last-word rule without an understanding of cardinality, that is, "The last counted word is the answer to a how-many question" (Fuson, 1991, p. 33). One way that can be used to distinguish between children who are simply using the last-word rule and those who are developing an understanding of cardinality is to ask them to hand you two, three, or seven items. Children who have a basic understanding of cardinality will count out the requested number of items and then hand them over. Children who do not understand cardinality usually grab a handful of items, without counting (Wynn, 1990).

Although some 3- and many 4- and 5-year-olds perform well on these types of tasks, suggesting a developing sense of cardinality, the child's understanding of cardinality is typically not mature. This is because many children of this age are easily confused by how the items are arranged (Piaget, 1965). For instance, even if the child is presented with two rows of five toys, counts both of the rows, and then states that there is the same amount in each row, he or she can often be talked out of that conclusion. If one of the rows is spread out and it is pointed out that "the top row is longer," preschool children will often conclude that the top row now has more toys. Thus, even though the child seems to understand that the last number in a count stands for the number of toys, he or she will often disregard this information and base all judgments on perceptual cues. Becker (1993) recently demonstrated that under some conditions, 4- to 5-year-old children demonstrate an understanding of cardinality in the absence of perceptual cues. Thus, children's understanding of cardinality is not necessarily dependent on perceptual cues (i.e., having actual items to count), but perceptual cues (i.e., appearance) can sometimes lead children to disregard this cardinal knowledge. It is not until the child is 7 or 8 years old that she or he will consistently favor cardinality information over perceptual information when making judgments about the quantity of counted items (Piaget, 1965).

The concept of *ordinality*, or order relationships, refers most generally to the child's knowledge of *equivalence* (that two quantities are the same) and *greater than* and *less than* (that two quantities are not the same). As is the case with most number skills, the demonstration of a young child's knowledge of ordinal relationships requires the use of special techniques. Bullock and Gelman (1977), for instance, assessed the ordinal knowledge of 2½- to 5-year-old children by using a "magic" game. In the first of two phases, the children were shown two plates of toys. One plate contained a single toy animal, and the other plate contained two toy animals. The child was told either that the two-toy plate was the winner (*more* condition) or that the one-toy plate was the winner (*less* condition). The child was then shown a series of one- and two-toy plates and was asked to pick the winner. In Phase 2, "the experimenter surreptitiously added one animal to the two-toy plate and three animals to the one-toy plate" (Bullock & Gelman, 1977, p. 429). The critical question was whether the children would choose the winner on the basis of the relation, that is, more or less, that was reinforced in the first phase. The majority of the 3- and 4-year-olds based their responses on the relational information, but less than one half of the 2-year-olds did.

However, in Phase 2 of a second experiment, when the memory demands of the task were reduced, more than 90% of the 2-year-olds based their responses on the relational information. The results of this study suggest that children as young as 2½ years of age have an understanding of ordinal relationships, at least for small set sizes (see also Siegel, 1971a, 1971b, 1974). That is, they understand that 4 is greater than 3 and that 2 is greater than 1. These results are consistent with the ordinality studies with infants described by R. G. Cooper (1984) and Strauss and Curtis (1984). Nevertheless, there are many unresolved issues in this area. Although many young children understand ordinal relationships across small values, it is not clear how far this extends (e.g., to 9 or to 12?).

Moreover, it is not yet clear whether a child's understanding that 10 is greater than 9 reflects the same skill that underlies an infant's understanding that 3 is greater than 2 (Fuson, 1988, 1991). In other words, it is not clear whether the development of ordinal knowledge for larger numbers involves simply linking number words to innate preverbal magnitudes (Gallistel & Gelman, 1992) or whether the child's knowledge for larger numbers is induced from learning and use of the standard sequence of number words (Fuson, 1988). Either way, we can count on many more studies on this topic in the years to come.

Finally, consider the tendency of young children to use nonnumerical strategies to determine if two sets contain the same or a different number of items (Fuson, 1988). For example, to determine which of two sets has more items, 3-year-olds will typically match the items one-for-one. Many 4- and 5-year-olds will count the number of items in each set and then compare the cardinal values of the sets, a strategy that works only if the child understands that the relative quantities are represented by the cardinal values. As noted earlier, it is not until the child is about 7 years of age that he or she can consistently use this type of counting strategy to determine ordinal relationships without being confused by perceptual cues, such as length and density. When appropriate testing conditions are used, young children and perhaps even infants (R. G. Cooper, 1984) show a sensitivity to *greater than* and *less than* and show some understanding of the ordering of the quantities represented by number words. The child's understanding of ordinal relationships between larger numbers probably develops more gradually, as does the use of counting strategies to make ordinal judgments (Fuson, 1988).

Measurement

Another important use of numbers and ordinal scales, such as rulers, is for measurement. Measurements tend to be very sensitive issues for children: "HE GOT MORE COOKIES THAN ME!" So it is not surprising that children would develop strategies for assessing relative quantity, volume, length, and so on (K. Miller, 1984). K. Miller found that the measurement strategies of preschool children tend to be prequantitative but nonetheless do reflect an understanding of equality and of *more than* and *less than*. For instance, when dividing up a cupful of candy, many 2- and 3-year-olds use a simple "dumping" strategy. They dump a few pieces of candy into their friends' cups; everybody gets some, but there is no concern for equating the pieces of candy across children (Gelman & Gallistel, 1978; K. Miller, 1984). A more sophisticated approach involves dumping and then checking by counting. Most older preschool and early elementary school children divide the candies by using distributive counting: "One for me and one for you." To divide the candy among three people, a more sophisticated approach would be to count the pieces of candy, divide this total by 3, and then distribute the candy equally. K. Miller found that not even 9-year-old children used this type of strategy.

A clearer developmental trend is evident in children's strategies for dividing length, as in dividing a hot dog among three people (K. Miller, 1984; Piaget, Inhelder, & Szeminska, 1960). A common strategy among 3-year-olds is to cut the hot dog into, say, six pieces, without regard to the length of each piece, and then distribute two pieces to each person (K. Miller, 1984). The total amount of hot dog given to each person will almost always differ when this strategy is used, with the distributor usually getting the biggest pieces. Nevertheless, the finding that each person gets the same number of pieces suggests a rudimentary understanding of the quantification requirements of the task. K. Miller (1984) described an interesting reaction of some of the young children who used this strategy: "To fix the inequality, they took a piece from the [person] who had not gotten enough, cut it in two, and returned it to the same [person]" (p. 202), as if increasing the number of pieces somehow compensated for differences in the length of the pieces. By the time children are 9 years old, they use numerical procedures to solve the same task: They use a ruler to measure equal portions before cutting and dividing those proportions. This skill, of course, requires an understanding of cardinal value as well as ordinal relationships.

In all, it is clear that a concern for measurement is common in children and represents an important and practical use of numerical knowledge and numerical systems (Saxe, 1977). Many 2- and 3-year-olds seem to have a basic understanding of the requirements of measurement tasks, though they lack the numerical skills to perform these tasks in the same manner as adults (K. Miller, 1984; Piaget et al., 1960; Siegel, 1971a, 1971b). Young children appear to rely on prequantitative strategies to determine equivalence, for example, to apportion shares. However, as illustrated above in the quote from K. Miller, they still seem to understand the inherently quantitative nature of many measurement tasks. Gradually, children increasingly rely on explicitly quantitative procedures, such as using a ruler, to make measurements (Petitto, 1990). K. Miller's study suggests that most 9-year-olds apply quantitative procedures to make most simple types of measurements (i.e., quantity, length, and area), although mature strategies for more complex measurements, such as volume, do not emerge until adolescence (Piaget et al., 1960; Saxe, 1981). More work in this area is clearly needed to better understand how and when children come to appreciate the utility of number knowledge in measurement situations.

Developmental Mechanisms for Counting and Number Knowledge

The fact that children become better counters with age and practice is important in itself, but the real issue is the development of the child's conceptual understanding of number and counting and the factors that govern this development. Although most researchers in this area would probably agree that a sensitivity to numerical information is, at least in part, inborn, there is considerable disagreement over the relative importance of this innate sensitivity, as contrasted with the child's experiences with numbers and counting (Briars & Siegler, 1984; Fuson, 1988; Gallistel & Gelman, 1992; Gelman & Gallistel, 1978; Siegler, 1991; Wynn, 1992b). Much of the theoretical debate in this area concerns children's counting. There are two general positions. The first is that innate principles guide the development of counting skills and counting knowledge (Gelman & Gallistel, 1978; Gelman & Meck, 1983). This position is often termed the *principles-first model.* The second general position is that children first count by rote and gradually induce counting concepts (Briars & Siegler, 1984), which might be termed the *procedures-first model.*

Principles First

In a highly influential book, Gelman and Gallistel (1978) argued that the counting behavior of young children was guided by five implicit principles: *one-one correspondence, stable order, cardinality, abstraction*, and *order irrelevance*. The one-one correspondence principle emphasizes that only one number word can be assigned to each counted object. Implicit knowledge of this rule would be reflected not in the use of the standard counting sequence, but by the child's use of different number words to tag each separate item. Thus, a child's counting is said to be guided by this principle if she or he never uses the same number word to refer to different items during a count (e.g., never states, "one, three, three").

The stable-order principle would be reflected in the child's counting, if the child used the same sequence of number words for counting different sets of objects. So the child might count a set of three objects by stating, "two, four, five," and count, "two, four, five, eight," for a set of four objects. The cardinality principle is reflected in the child's understanding that the number word associated with the last counted item has a special meaning. The last number, whether or not it is the correct number in terms of the standard sequence, represents the numerosity of the set. Knowledge of this principle might be reflected in the child's counting if the child emphasizes the last number word or if the child restates the last number word when asked "how many" in a just-counted set. According to Gelman and Gallistel (1978), knowledge of the cardinality principle is not manifested by the child until he or she has had some experience in the use of the one-one correspondence and stable-order principles (but see Fuson, 1988).

The abstraction principle refers to the child's awareness of *what* is countable. That is, it does not guide the act of counting in itself, but rather defines the domain onto which counting can be applied. A child who understands that counting can be applied to any set of items should be able not only to count sets of homogeneous items but also to show skill at counting mixed sets (e.g., toys that differ in color, shape, size, and arrangement) and even actions or sounds. Indeed, many 2- and 3-year-old children are able to use counting to enumerate mixed sets of objects as well as actions and sounds (Gelman & Gallistel, 1978; Schaeffer et al., 1974; Wynn, 1990). The final principle, order irrelevance, reflects the child's understanding that no matter what order the items are counted in, from left to right or right to left, the result is the same. Consistent use of this principle does not seem to emerge until 4 or 5 years of age (Gelman & Gallistel, 1978).

The first three principles, one-one correspondence, stable order, and cardinality, define the *how-to-count* principles. The interrelationship of these rules forms the child's initial counting scheme, in the Piagetian sense (Piaget, 1965). "The principles guide and structure the child's counting behavior, serve as a reference against which the child can evaluate this actual counting behavior, and motivate that behavior" (Gelman & Gallistel, 1978, p. 208). In other words, this implicit knowledge focuses the child on numerical features of the environment and places constraints on the types of behaviors that the child uses to gain information about these numerical features. This is not to say that the overt behavior of the child will always be in accord with these principles. A child might fail to use a stable counting order for larger set sizes, for example, because he or she does not understand that the stable-order principle applies to all sets, small and large, or because she or he has not yet memorized enough number words (Fuson, 1988). The child might "fail" the task because the required counting performance is too difficult, not because the child has no understanding of counting principles. In other words, if a child's behavior does not appear to reflect underlying knowledge or an implicit principle, then the child might not know the principle or the test might be too difficult (i.e., the child knows what to do but cannot express it; Kreutzer, Leonard, & Flavell, 1975).

Gelman and Meck (1983) sought to test this possibility for the assessment of the how-to-count principles with 3- to 5-year-old children by removing the performance demands of counting. They did this by having the child watch a puppet who was learning how to count. Sometimes the puppet counted correctly, and at other times the puppet made an error that violated one of the three how-to-count principles. For instance, one error that violated the one-one principle involved the puppet's double counting one of the items. It was assumed that if the child detected the violation, then he or she understood, at least implicitly, the underlying principle. Implicit knowledge, in this situation, means that the child might not be able to explicitly state the counting principle but nonetheless knows that "you can't count that way." In a series of studies, Gelman and Meck demonstrated that even 3-year-old children readily detect violations of the how-to-count principles. In this research, 75% of the 3-year-olds detected violations of the one-one principle, and some provided insightful explanations as to why, for example, the double count was wrong—such as the puppet "did it again."

If children as young as 3 years of age seem to understand the basic principles that underlie counting, then why doesn't the average child develop mature number knowledge and counting skills until age 8? Even if the how-to-count scheme is innate, this does not mean that the associated knowledge and the ability to apply that knowledge are mature.

Indeed, the how-to-count scheme not only motivates the child to apply this knowledge (i.e., count in natural settings; Piaget, 1965) but also provides the knowledge against which she or he gauges whether the counting is done correctly (cf. Ohlsson & Rees, 1991). Initially, counting errors are expected. With experience and practice, the behaviors that express counting knowledge become refined, less demanding, and increasingly accurate (Case, 1985; Gelman & Gallistel, 1978). Experience and practice also lead to the elaboration and modification of the initial how-to-count scheme. So, with experience the child understands that counting can be applied in many different contexts and, for instance, that the order in which items are counted is not important. Gelman and Gallistel's how-to-count scheme provides the skeletal principles for counting (Gelman, 1990), and experience and practice flesh out these principles.

Procedures First

The alternative view is that children first learn to count largely by rote, through the imitation of parents or siblings, for example. In this circumstance, the child's counting is not guided by conceptual knowledge, but rather the child induces basic principles by noticing regularities in the outcome of counting (Briars & Siegler, 1984; Fuson, 1988; Schaeffer et al., 1974; Siegler, 1991; Wynn, 1992b). This position should not be taken to mean that infants are not born with an innate sensitivity to quantity (Fuson, 1988; Siegler, 1991). The infant's innate sensitivity to quantity provides the foundation for later number development rather than the skeletal structure, as in the principles-first model. For instance, the ability to subitize numerosities of two and three might provide the foundation for inferring that number words refer to unique quantities (Schaeffer et al., 1974). So 2-year-olds are able to detect, without verbal counting, that two cookies differ in quantity from three cookies, whether or not they can label the sets *two* and *three*. These sets of cookies, as well as other things, are labeled by people in the child's environment: "You can have *two* cookies, but not *three*," as Mom takes one of the cookies back. As the child hears people label the different numerosities with different words, he or she induces that different words need to be used to represent different numerosities (see also Wynn, 1992b).

Briars and Siegler (1984) experimentally tested the hypothesis that children induce counting principles by noticing regularities in the outcome of counting. The authors argued that if the child had induced knowledge of counting principles, then the child should be sensitive to counting that violated the standard (i.e., 1, 2, 3, . . .) counting. Working independently of Gelman and Meck (1983), Briars and Siegler argued that the child's sensitivity to nonstandard correct counts and counting errors would indicate their underlying conceptual knowledge. A sensitivity to the correctness of nonstandard counting and to counting errors suggests at least partial counting knowledge, whereas a lack of sensitivity to these types of counts suggests that all counting is done completely by rote with no underlying counting knowledge. It was further assumed that if young children had any partial counting knowledge and if this knowledge were largely induced rather than innate, then that knowledge would be reflected in a particular sensitivity to counting features that were commonly observed in the environment, such as counting from left to right.

Thus, the child's knowledge of counting should mirror the types of features that the child commonly observes in counting acts rather than mirroring Gelman and Gallistel's (1978) how-to-count principles. The essence of this position is that the child first counts by rote and gradually induces essential and unessential features of counting, with the child's conceptual knowledge mirroring these essential and unessential features (Briars & Siegler, 1984). Briars and Siegler (1984) argued that the one essential feature of counting is the word-object correspondence rule: "Given a correctly ordered list of number words, assigning one and only one number word to each object during counting is both necessary and sufficient to determine a set's cardinality" (p. 608). The word-object correspondence rule encompasses the one-one correspondence, order-irrelevance, and stable-order principles proposed by Gelman and Gallistel. Briars and Siegler also described four common but unessential features of counting: standard direction (counting starts at one of the end points of a set of objects), adjacency (contiguous objects are counted in succession), pointing (counted objects are typically pointed at only once), and start at an end (counting proceeds from left to right).

Briars and Siegler (1984) assessed 3- to 5-year-old children to determine whether the children's counting knowledge reflected the pattern of essential and unessential features of counting. As with the Gelman and Meck (1983) study, the children watched a puppet who was learning how to count. The puppet systematically violated the essential and unessential counting features, and the children's sensitivity to these violations was noted. Most of the 4-year-olds and all of the 5-year-olds believed that the word-object correspondence rule was an essential feature of counting. However, as many as half of the 5-year-olds also believed that counting had to start at one of the end points and that adjacent items had to be counted in consecutive order (skipping an item and then coming back and counting it later was not acceptable). This pattern is what would be expected if children observed counting and then induced the rules of counting by noting common features. The common counting features, some essential and some not, then become the basis for the child's conceptual understanding of counting.

Summary

Although there is general agreement that the acquisition of counting skills in children involves both inborn and experiential factors, there is much debate about the relative contributions of these factors (Fuson, 1991; Gallistel & Gelman, 1992; Siegler, 1991; Wynn, 1992b). The principles-first model holds that inborn sensitivities to quantitative features of the environment, manifested in the how-to-count scheme, provide the skeletal framework for later numerical competencies (Gelman, 1990); experiences refine and flesh out latent numerical knowledge. The alternative view is that the infant's basic numerical competencies provide the springboard for later development, but once the child has left this springboard, so to speak, then conceptual development is primarily governed by an inductive process: The child largely constructs his or her conceptual understanding of counting (Siegler, 1991). There is certainly much to be learned in this area, and it will probably be many more years before the developmental process is fully understood. Regardless of the outcome of this debate, research from both positions is making it abundantly clear that preschool children, and even infants, have considerable, if not immature, numerical competencies.

Numerical Competencies in Animals

In this final section, the basic numerical competencies of animals are reviewed. The basic question is whether animals manifest numerical skills in some of the same ways that human infants and children do; in particular, whether animals are sensitive to numerosity, respond to the ordinal nature of numerosities, can count, or can even add. A consideration of this research is important for two reasons. First, if numerical competencies are evident in animals, then such findings will provide further evidence that many basic numerical skills might be inborn in human infants (Gallistel & Gelman, 1992). Second, limitations of the numerical competencies of animals, especially primates, might inform us about the likely constraints on the early numerical skills of human infants (Boysen & Capaldi, 1993; Wasserman, 1993). In this section, studies that bear on each of these issues are briefly reviewed, as they relate to the numerical skills of young children.

Numerosity

Evidence for the ability to represent numerosity has been found in many species, ranging from the laboratory rat to the chimpanzee (*Pan troglodytes*; Boysen & Berntson, 1989; Davis & Memmott, 1982; Matsuzawa, 1985; Pepperberg, 1987). In an influential review, Davis and Memmott concluded that laboratory rats were able to discriminate small numerosities, such as three from two or four. Matsuzawa was able to train a chimpanzee to use Arabic numerals to label collections of one to six items. Boysen (1993) reported similar results with several other chimpanzees. Here the chimpanzees were trained to associate Arabic numerals with the corresponding number of items in a food dish, as shown in Figure 1.5. Pepperberg (1987, 1993) showed that numbers are for the birds, in particular an African grey parrot (*Psittacus erithacus*) named Alex. Alex was taught to verbally label, in English, collections of two to six items.

These and other studies converge on the conclusion that many species are able to represent specific small numerosities. However, the use of these numerical representations typically does not emerge except with extensive training (but see Pepperberg, 1987), leading Davis and Memmott (1982) to conclude that the estimation of quantity might be more easily achieved in natural settings with the use of other cues, such as density. Regardless, a basic sensitivity to number is evident in many species and in some cases appears to be similar to the human infants' sensitivity to numerosity. In fact, as noted earlier, Gallistel and Gelman (1992) argued that the basic number skills of human infants and the number skills of animals reflect the operation of analogous preverbal counting and timing mechanisms.

Ordinality

Washburn and Rumbaugh (1991) were among the first to demonstrate an understanding of ordinality in nonhuman primates. In a series of experiments, two rhesus monkeys (*Macaca mulatta*) were initially taught to select the larger of two Arabic numerals, ranging inclusively from 0 to 5. The macaques were rewarded with a number of food pellets equal to

Figure 1.5

Procedure for testing chimpanzees' ability to associate numbers with quantities. (From "Numerical Competence in a Chimpanzee [*Pan troglodytes*]," by S. T. Boysen and G. G. Berntson, 1989, *Journal of Comparative Psychology, 103*, p. 25. Copyright 1989 by the American Psychological Association. Reprinted with permission of the author.)

the value chosen. After 400 trials, one of the macaques, Abel, selected the larger number on more than 88% of the subsequent trials. The performance of the second macaque, Baker, was not this good but was still better than chance. In a second experiment, the numerals 6 to 9 were introduced. An important feature of this experiment was the presentation of never-before-seen pairs of numerals. For instance, in the initial stages of the study, the pairs 8, 7 and 9, 8 were presented, but the pair 9, 7 was not presented. If Abel and Baker had learned to discriminate the numbers on the basis of ordinal value, then they would be able to choose the larger number in a novel pairing (e.g., 9, 7). Indeed, Abel chose the larger number for seven of the seven novel pairings (i.e., on the first trial that the pairs were presented); Baker correctly responded to five of the seven novel pairings.

In a final study, Abel and Baker were presented with a series of five unique numbers. The task was to select, in successive order, the largest of the five numbers (e.g., 5, 6, 3, 8, or 1), then the largest of the four remaining numbers, and so on. Both Abel and Baker tended to choose the largest number available in sequence; that is, first they chose the 8, then the 6, and then 5, 3, and 1, respectively. The results of these studies suggest that some primates apparently can learn not only the cardinal value of Arabic numbers but also the ordinal value of these numbers. In other words, Abel and Baker appeared to understand that the quantity represented by 9 was greater than the quantity represented by 8, the quantity represented by 8 was greater than the quantity represented by 7, and so on. This skill does not appear to be restricted to rhesus monkeys, as Boysen reported similar results for several chimpanzees (Boysen, 1993; Boysen & Berntson, 1989).

Counting

There is also some evidence that chimpanzees can count in a manner similar to that of young children (Boysen, 1993; Rumbaugh & Washburn, 1993). For instance, consider another study conducted by Rumbaugh and Washburn. In this experiment, an Arabic number 1, 2, 3, or 4 was presented on a computer screen to an adult chimpanzee, Lana. Lana's task was to tag, by using a cursor on the computer screen, a number of boxes equal to the value of the presented number and then return the cursor to another location on the screen to indicate that she was finished. The authors argued that this task was analogous to counting, because Lana had to tag, in successive order, a number of boxes equal to the presented number, which is similar to a young child's pointing at objects as they are being counted (Gelman & Gallistel, 1978). Lana's performance on this task was above chance levels for all of the presented numbers; though, as with children (Fuson, 1988), accuracy did decline for larger numbers. Boysen described a similar study, except that her chimpanzee, Sheba, was required to point to the Arabic number that corresponded to the number of food pellets on a food tray. During this task, Sheba often pointed to the food pellets in succession and then pointed to the corresponding Arabic numeral. As noted earlier, the pointing to, or tagging of, objects as they are counted is commonly seen in preschool children (Gelman & Gallistel, 1978). Nevertheless, though suggestive, it is not yet clear whether Sheba and Lana were actually (nonverbally) counting.

Figure 1.6

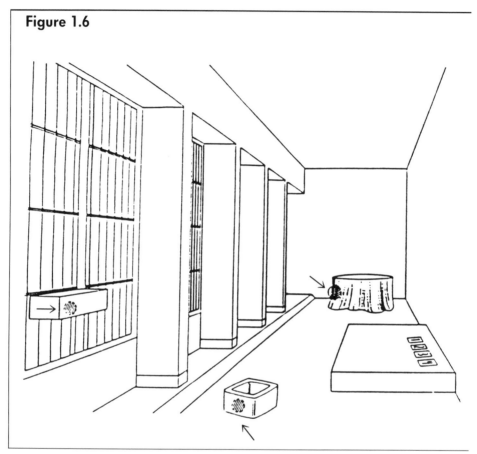

Experimental setting for testing Sheba's addition skills. Small circles represent locations of food sources. (From "Numerical Competence in a Chimpanzee [*Pan troglodytes*]," by S. T. Boysen and G. G. Berntson, 1989, *Journal of Comparative Psychology, 103*, p. 27. Copyright 1989 by the American Psychological Association. Reprinted with permission of the author.)

Addition

Another set of experiments suggested that Sheba could add small numbers (Boysen & Berntson, 1989). In one of these experiments, a total of one to four oranges were placed in two of three separate locations. The setting for this experiment is depicted in Figure 1.6. Sheba's task was to visit each of the three locations and then report the total number of oranges in the three locations to the experimenter. Sheba reported the sum by pointing to or picking up a card with the corresponding Arabic numeral. Sheba was not allowed to take the oranges from any of the locations, so she had to remember the quantity in each location and then add them to provide the answer. For sums up to and including 4, Sheba provided correct answers on 84% of the trials. In another experiment, cards with Arabic numbers on them were placed in two of the food sites. The task was to visit each site and then add the two numbers. Sheba provided correct answers on 76% of the trials, for sums up to and including 4. Boysen and Berntson's study clearly suggests that some chimpanzees are able to add very small quantities (see also Boysen, 1993). Boysen and Berntson argued that Sheba might have been using her counting skills to solve the addition tasks. They suggested that Sheba might have solved these problems by using counting strategies similar to those seen in young children (Groen & Resnick, 1977)—for example, counting 1, 2, 3, 4, in a nonverbal manner, of course, to solve 3 + 1. Note, however, that the conclusion that Sheba counted to solve the addition problems is rather speculative at this point.

Summary

The just-described studies indicate that many species of animals possess a variety of numerical competencies (Boysen & Capaldi, 1993). These competencies include the ability to mentally represent the numerosity of small numbers of items or events (Davis & Memmott, 1982), as well as the ability to label the numerical representation with English words (Pepperberg, 1987) or Arabic numbers (e.g., Matsuzawa, 1985). Many nonhuman primate species also appear to be able to determine the cardinal value of small set sizes, to understand ordinal relationships, and perhaps even to nonverbally count and add (Boysen, 1993; Boysen & Berntson, 1989; Washburn & Rumbaugh, 1991). More important, the pattern of skills exhibited in many of these studies is qualitatively similar to the skills exhibited by human infants and young children (Gallistel & Gelman, 1992), although these basic numerical skills are easier to demonstrate in human infants than in animals. The demonstration of basic number skills in primates, for example, often requires extensive training and experience, whereas the ability to represent numerosities appears to be evident in the first few days of life in humans (e.g., Antell & Keating, 1983). Nevertheless, the animal studies suggest a fundamental continuity between the numerical competencies of animals and humans and certainly support the position that number skills represent an innate system that is at least partially independent of other biological systems, such as those underlying language acquisition (Starkey et al., 1991).

Conclusion

Research conducted since the early 1980s indicates that human infants have a basic and fundamental sense of quantity (e.g., Starkey & Cooper, 1980). The infant's sense of quantity has been demonstrated in the first week of life, under many different experimental conditions, and is reflected in a sensitivity to numerical features in the environment and the ability to mentally represent numerosities of up to four (Antell & Keating, 1983; Starkey et al., 1990). More recent research suggests that 5-month-old infants are even aware of the effects that adding and subtracting have on the quantity of items in the environment, at least for quantities of one and two (Wynn, 1992a). Finally, some time around 18 months of age, infants begin to respond to ordinal relationships, that is, they seem to know, for example, that three items are more than two items. Note that these *more than* or *less than* decisions, at this age, appear to be based on an understanding of ordinal relationships rather than on other types of information, such as density (R. G. Cooper, 1984; Strauss & Curtis, 1984).

During the preschool years, children's use of language to represent numerical information and relationships becomes increasingly important. During this time, they must learn their culture's number words; map these number words onto their existing number knowledge; and then learn the many ways in which number words can be used in quantitative situations, such as for counting or measurement (Fuson, 1988; Gelman & Gallistel, 1978). The development of basic number skills spans about a 6-year period for most children, ranging from the ages of 2 to 8 years (Fuson, 1991; Piaget, 1965). During this period, children memorize the number words, come to understand that each word represents a different quantity, and develop their counting skills. During the refinement of these counting skills, children learn ordinal and cardinal meanings, first for small sets and then later for larger sets (Brainerd, 1979). These skills, in turn, are used in many different contexts, such as for measurements or for arithmetic. The hallmark of children's basic-number-skill development, which is achieved by 7 or 8 years of age, is the knowledge that numbers reflect the groupings of smaller sets of numbers and the use of number knowledge rather than perceptual cues, such as length or density, to make quantitative decisions (Piaget, 1965).

The numerical skills of human infants and preschool children, combined with the apparent ubiquity of number-related games and activities across cultures (Ginsburg, Posner, & Russell, 1981b; Saxe, 1981, 1982a; Saxe et al., 1987; Zaslavsky, 1973), provide strong evidence that number, like language, is a natural domain of human cognition and activity. Further support for this position is found with studies of the numerical competencies of animals across many species (Boysen & Capaldi, 1993). In all, these studies indicate that animals of many species are sensitive to and can represent small numerosities; can be taught to label these numerosities; often demonstrate an awareness of ordinal and cardinal values; and, finally, at times can use this numerical knowledge in rather sophisticated ways, such as to add (Boysen & Berntson, 1989; Davis & Memmott, 1982; Pepperberg, 1987; Rumbaugh & Washburn, 1993). Although the specifics of how number development occurs, and the relative contributions of innate sensitivities and knowledge as opposed to inductive processes, will most likely occupy researchers in this domain for many years to come (Briars & Siegler, 1984; Gelman & Meck, 1983), it has nevertheless become clear that human infants and preschool children are much more quantitatively sophisticated than was thought even 15 years ago.

2 Developing Arithmetical Skills

Arithmetic, like counting, is an activity that appears in one form or another in many cultures (Ginsburg, 1982, 1989; Ginsburg, Posner, & Russell, 1981a, 1981b; Saxe, 1985, 1988, 1991; Zaslavsky, 1973). As one early convert observed long ago, "in using that tool (arithmetic) there is no difficulty, hardship nor heavy cost for him who is prepared for it, noting its little toil and great benefit" (Al-Uqlidisi, 952/1978, p. 35). Throughout this chapter, cross-cultural similarities, as well as differences, in children's arithmetical development are highlighted. The chapter begins with a discussion of children's informal knowledge of arithmetic (Baroody & Ginsburg, 1986) and how this flows naturally from the child's counting and number knowledge. In the second section, the child's understanding of more formal arithmetic concepts, such as commutativity, is described. The next section discusses children's acquisition of addition, subtraction, multiplication, and division skills. The chapter ends with an overview of general theoretical models of arithmetical development. A discussion of children's skill at solving arithmetic word problems is presented in chapter 3.

Informal Arithmetic

Piaget (1965) argued that children did not have a conceptual understanding of basic arithmetic until the age of 7 or 8 years. He came to this conclusion partly on the basis of results of the following experiments.

Summaries are provided at the end of the following subsections on arithmetic operations: Addition (p. 59), Subtraction (p. 69), and Multiplication (p. 78). In the Developmental Models section, they are provided for both subsections: Strategy-Choice Model (p. 86) and Schema-Based Model (p. 90).

In one study, children of various ages were presented with two rows of eight pieces of candy each. It was noted that there were the same number of candies in each row. One of the rows was then divided into two sets of four candies, and the other was divided into sets of one and seven candies. The candies were to be eaten in the morning and the afternoon on successive days. So on the 1st day, four candies were to be eaten in the morning and four in the afternoon; the morning and afternoon split was one and seven for the 2nd day. Piaget then asked the children on which day would they get to eat the most candy.

Five- and 6-year-old children almost invariably stated that they would get to eat more candy on the 2nd day, because the group of seven candies was more than the group of four candies. Somewhat more sophisticated children stated that the number of candies would be the same for both days but came to this conclusion only after counting each array or aligning them into a one-to-one correspondence. A mature conceptual understanding of addition, in this instance, was reflected in judgments stating that there were the same number of candies for both days, without counting or using one-to-one correspondence. These children noted that even though seven was more than four, one was less than the other four, and therefore it "comes out the same." This type of judgment reflects the child's understanding that, for instance, $1 + 7 = 2 + 6, = 3 + 5$, and so on. On the basis of this pattern of results, Piaget (1965) argued that a mature conceptual knowledge of arithmetic required an understanding that numbers are composed of groups of smaller numbers and that a variety of different combinations can result in the same quantity.

Piaget's (1965) definition of a mature conceptual understanding of arithmetic is illustrated in another experiment. Here, Piaget presented two rows of differing numbers of beads to children and asked them to make the rows the same. The first row consisted of 8 beads, the second of 14 beads. In solving this task, one 5-year-old, Jac, first removed 5 beads from the second row and then added them to the first, producing rows of 13 and 9 beads. Noticing that there were now more beads in the first row, he removed 7 beads from this row and added them to the second, producing rows of 6 and 16 beads. After several more transformations of this sort, he gave up. Again, it is not until children are 7 or 8 years of age that they solve this task with little difficulty. An 8-year-old child might notice that $14 - 8 = 6$, divide the 6 into two equal parts, and then move one of those parts (i.e., 3) to the smaller set of beads. Again we see that mature performance, as defined by Piaget, on this task requires an understanding that numbers are composed of groups of other numbers. Any indication of arithmetical skills before the child understands number groups, such as knowing that 1 + 2 = 3, was assumed to reflect, for the most part, rote learning; it was assumed that there was little or only rudimentary conceptual knowledge underlying these skills.

However, more recent studies, using different methods, show that children have considerable informal knowledge of arithmetic (Starkey, 1992; Starkey & Gelman, 1982). Children younger than 7 or 8 years of age performed poorly on Piaget's tasks not because they lacked an understanding of addition and subtraction, but rather because they too often relied on spatial cues (i.e., how the array looked) rather than arithmetic knowledge to make the judgments. This is not to say that the results of Piaget's experiments were wrong but rather that the nature of the tasks resulted in children using spatial information rather than their informal knowledge of arithmetic to make judgments about numerical equivalence. Indeed, more recent studies suggest that children in many different cultural groups develop roughly comparable informal number and arithmetical skills before they are 7 years old (e.g., Ginsburg, 1982; Ginsburg et al., 1981b). These groups include children from Western and agricultural societies as well as mercantile societies. Recall that Wynn's (1992a) study suggested that infants as young as 5 months of age understood the consequences of very simple additions (1 + 1) and subtractions (2 − 1). In all, these studies suggest that young children possess a very basic understanding of arithmetic, an understanding that is not dependent on formal instruction or social need, and that the tasks used by Piaget greatly underestimated the arithmetic knowledge of preschool children.

To better illustrate this point and to gain an understanding of the limits of children's fundamental knowledge of arithmetic, consider a recent study that was conducted by Starkey (1992). Starkey tested 18-month-old to 4-year-old children with a task designed to uncover the children's understanding of the effects of addition and subtraction on numerosity. The goal was not simply to demonstrate that young children could add and subtract, but rather to determine if they understood how addition and subtraction affected numerosity *without* the use of verbal counting. To achieve this end, Starkey developed the searchbox:

> [The searchbox] consisted of a lidded box with an opening in its top, a false floor, and a hidden trap-door in its back. This opening in the top of the box was covered by pieces of elastic fabric such that a person's hand could be inserted through pieces of the fabric and into the chamber of the searchbox without visually revealing the chamber's contents. (Starkey, 1992, p. 102)

In the first of two experiments, children placed between 1 and 5 table tennis balls, one at a time, into the searchbox. Immediately (1 or 2 s later) after the child had placed the last ball into the searchbox, he or she was told to take out all of the balls. An assistant removed the balls that the child had placed in the box and then replaced them one at a time as the child searched for the balls. So after the child reached into the box and retrieved one ball, the assistant placed another ball in the box in anticipation of the child's next retrieval attempt.

The question was whether the children would attempt to retrieve the same number of balls as were originally placed in the searchbox. For instance, if the child placed three balls in the searchbox and then stopped searching once three balls were removed, it could be argued that the child was able to represent *and* remember the number of balls deposited and use this representation to guide her search. The results showed that 24-month-olds were able to mentally represent and remember numerosities of one, two, and sometimes three. Thirty-month-olds were able to represent and remember numerosities up to and including three, whereas 36- and 42-month-olds could sometimes represent numerosities as high as four. The most important finding was that for most of the children, these representations did not appear to be achieved through verbal counting. Thus, in keeping with the infancy studies described in chapter 1, Starkey (1992) showed that young children could mentally represent quantities of up to three to four items without the use of language. But could they add or subtract from these quantities without verbally counting? The issue of whether children solve such problems by means of verbal counting is crucial, because if the problems are solved nonverbally, then the results would suggest that children have a basic understanding of arithmetic that is independent of language skills. The second experiment sought to address just this issue.

The same general procedure was followed in this second experiment, except that once all of the original balls were placed in the searchbox, the experimenter placed an additional one to three balls in the box or removed from one to three balls. The question was whether the children would search for the same number of balls that they had placed in the searchbox or whether the search would take into consideration the balls added or removed by the experimenter. If the children understood the effect of addition, then they would search for more balls than they had originally placed in the searchbox. Conversely, for subtraction they would stop searching before this point. Nearly all 24- to 48-month-olds responded in this manner. Moreover, many but not all of the 18-month-old children also showed this pattern! The results of this experiment indicate that many 18-month-old children, and nearly all 2-year-olds, understand that addition increases the numerosity of a set, whereas subtraction decreases the numerosity of the set (see also Sophian & Adams, 1987; Starkey & Gelman, 1982).

Examination of the accuracy data indicated that similar to Wynn's (1992a) findings with 5-month-olds, many of the 18-month-olds were accurate for addition and subtraction with sums or *minuends* (the first number in a subtraction problem) less than or equal to 2 (e.g., 1 + 1 or 2 − 1) but were not accurate for problems with larger numbers. Most 24-month-olds were accurate with values up to and including 3 (e.g., 2 + 1 or 3 − 1). None of the children in the study were accurate with sums or minuends of 4 or 5. As with the first experiment, most of the children did not appear to rely on verbal counting to solve these addition and subtraction problems. Starkey (1992) argued that the children might have been using some form of image-based strategy to represent the number and to keep track of the additions and subtractions performed by the experimenter. Regardless of the specific cognitive skills used by these children, the results of this intriguing set of experiments provide further evidence that humans have a fundamental sense of number and quantity. More important, this number sense includes an intuitive understanding of simple addition and subtraction and appears to be largely independent of the language system (see also Jordan, Huttenlocher, & Levine, 1992; S. C. Levine, Jordan, & Huttenlocher, 1992).

Nevertheless, by 4 or 5 years of age, children of many different cultures rely on verbal counting to solve simple arithmetic problems (Baroody & Ginsburg, 1986; Ginsburg, 1982; Groen & Resnick, 1977; Hatano, 1982; Saxe, 1985; Siegler & Jenkins, 1989). Baroody and Ginsburg argued that the use of counting by young children to solve arithmetic problems represents informal knowledge, because children use these skills without formal instruction. The most important finding across all of these studies is that children adapt their already-existing counting skills and knowledge to situations that require addition or subtraction. The specific strategies used by children for solving addition or subtraction problems depend on the culture's counting system. Addition by young children in the United States usually involves the counting of sets of objects (Steffe, Thompson, & Richards, 1982); Korean and Japanese children apparently use a similar strategy, although counting tends to be discouraged in Japan (Hatano, 1982; Song & Ginsburg, 1987). Saxe (1982b) showed that Oksapmin adults, of Papua New Guinea, with no experience with trade or commerce solve simple addition problems through the use of their counting system and also use objects to represent and solve problems.

To illustrate, a 4-year-old who gets three cookies from his or her mother and then on the sly gets three more from his or her dad, "Mom said I could have some," will count to enumerate the total cache. The strategy will typically involve laying out all of the cookies and then counting them in succession starting from 1 (Fuson, 1982). Similarly, when presented with verbal problems, such as "how much is 3 + 4," children of this age will typically represent each addend with a corresponding number of objects and then count all of the objects starting with 1; for instance, counting out three blocks and then four blocks and then counting all of the blocks. If objects are not available, then fingers are often used as a substitute (Siegler & Shrager, 1984). Somewhat older Oksapmin use an analogous strategy that is based on their body part counting system (see chapter 1). Here, 3 + 3 might be solved by counting from the thumb (1) to the wrist (6), with the word for wrist with an appropriate suffix representing the sum (Saxe, 1982a).

In all, these studies suggest that young children, and perhaps even infants, have a fundamental understanding that addition and subtraction influence quantity—that addition produces more whereas subtraction produces less (Starkey, 1992; Wynn, 1992a). The early implicit addition and subtraction knowledge, however, appears to be limited to quantities up to and including three items and is accomplished nonverbally. Starkey argued that his subjects might have been using images to represent numerosities and then added or subtracted items from this representation as appropriate. Gallistel and Gelman (1992), on the other hand, would argue that these skills reflected the use of the preverbal counting and timing mechanism described in chapter 1. Either way, the fact that very young children implicitly understand basic arithmetic is quite amazing. By the time they are 4 to 5 years of age, most children use verbal counting in situations that require addition or subtraction. Note that this age range is for children in the United States. Although children in other cultures show a similar pattern of development, the age at which skilled counting emerges varies, apparently in response to differences in the relative frequency of number and counting activities across cultures (Ginsburg et al., 1981b; Saxe, 1985; Song & Ginsburg, 1987). Nevertheless, in all cultures, the use of verbal counting strategies flows naturally from the child's developing knowledge of counting. Verbal counting seems to be a very reasonable arithmetic strategy, given that the basic nonverbal skills of children appear to be applicable to only small values.

Arithmetic Knowledge

Unfortunately, there has not been a great deal of research on the development of a child's knowledge of specific arithmetical concepts, in comparison with the research efforts that have been devoted to the child's understanding of number and counting (Fuson, 1988; Gelman & Gallistel, 1978). Nevertheless, there has been research on children's understanding of commutativity (Baroody, Ginsburg, & Waxman, 1983), the base-10 system (Fuson & Briars, 1990), and fractions (Clements & Del Campo, 1990). The basic findings in each of these areas are presented below.

Commutativity

The child's understanding of *commutativity*—that the order with which the numbers are added together does not affect the sum (e.g., 3 + 4 = 4 + 3)—has been studied by Baroody and his colleagues (Baroody, 1987b; Baroody & Gannon, 1984; Baroody et al., 1983). Baroody et al. assessed the understanding of commutativity in first-, second-, and third-grade children. The children were presented with a series of addition problems in the context of a game. Across some of the trials, the same addends were presented but with their positions reversed. For instance, the children were asked to solve 13 + 6 on one trial and 6 + 13 on the next trial. If the children understood that addition was commutative, then the strategy used to solve 13 + 6, in this example, would differ from that used to solve 6 + 13. In particular, they might count "13, 14, . . . 19" to solve 13 + 6 but then state, without counting, "19" for 6 + 13 and argue, "It has the same numbers, it's always the same answer" (Baroody et al., 1983, p. 160). In this study, 72% of the first-grade children and 83% of the second- and third-grade children consistently (on at least three of four trials) showed this type of pattern. This finding is especially interesting, because the first-grade children had not been explicitly taught commutativity.

In a later study, Baroody and Gannon (1984) assessed kindergartners' understanding of commutativity by using a task that was similar to that used in the just-described study and a second commutativity task. In the second commutativity task, each child was first presented with a problem to solve, such as 6 + 4. Next, the experimenter wrote down the same addends in the reverse order, 4 + 6, and asked if this would add up to 10 (or whatever answer the child provided for the first problem). If the child understood that addition was commutative, then she or he would answer "yes" without having to recalculate the answer. A child who did not understand that addition was commutative would add 4 + 6, usually by counting, before answering the question. In this study, 40% of the children showed an understanding of commutativity on both tasks, 29% did not show an understanding of commutativity on either task, and the remaining children fell in between these two extremes.

On the basis of these results and others, Baroody and Gannon (1984; see also Baroody, 1987b) argued that most children develop an understanding of commutativity some time between the ages of 4 and 6 years and that children discover this principle by noting the outcome of simple addition—noting, for example, that 3 + 2 = 5 and so does 2 + 3. This position does not rule out the possibility that some children might simply make the assumption that addend order does not matter and act accordingly, without a conceptual understanding of commutativity. Such an assumption might arise out of the child's counting knowledge; items can be counted, and therefore added, in any order (Gelman & Gallistel, 1978). Regardless, it seems that many kindergarten children and most first-grade children understand that addition is commutative and use this knowledge when solving addition problems. Moreover, it appears that many children do not need to be taught the principle of commutativity but rather will induce this knowledge by noticing the outcome of basic addition.

Base-10 Knowledge

Children's understanding of the base-10 system and the importance of this knowledge for arithmetical development have been studied by Fuson and her colleagues (e.g., Fuson, 1990; Fuson & Briars, 1990). Knowledge of the base-10 system is important for arithmetical development for several reasons. First, the child's understanding of the conceptual meaning of spoken and written multidigit numbers is dependent on knowledge of the base-10 system. The word *thirteen* does not simply refer to a collection of 13 objects; it also represents groups of tens and units values, 1 ten and 3 ones in this example. Second, the understanding that multidigit numbers represent groups of hundreds, tens, and ones, for example, influences the sophistication of the problem-solving strategies that the child can use to solve complex arithmetic problems. One strategy that is based on this knowledge involves a type of regrouping, so that 43 + 24 is solved by adding the tens values together (40 + 20), then the ones values (3 + 4), and finally 60 + 7; this strategy is described later in the chapter, under Complex Addition. A child without this knowledge might need to rely on more time-consuming and error-prone counting to solve this problem. Finally, an understanding of the base-10 system is important for the child's understanding of trading (i.e., carrying and borrowing) and place value (Fuson & Kwon, 1992b).

Many elementary school children in the United States do not fully understand the base-10 structure of multidigit number words (Fuson, 1990). For many American elementary school children, *thirteen* is not conceptually different from *eight*. Both numbers simply reference collections of items. For many American children, there is little understanding of the decade structure of multidigit numbers—that is, they do not understand that *thirteen* represents 1 ten and 3 ones. In contrast, most second-grade Korean children, for example, have a clear understanding of the base-10 structure of numbers and use this knowledge in arithmetical problem solving (Fuson & Kwon, 1992b). In comparison with their American peers, Korean children are able to use more sophisticated strategies to solve arithmetic problems, such as the regrouping strategy described above, and have a better understanding of trading and place value. As noted in the Number Words section of chapter 1, one reason for these cultural differences appears to be differences in multidigit number words between the Korean language (and many other Asian languages) and the English language. The decade structure of complex numbers is transparent in most Asian languages but is not at all obvious in English, at least up until the hundreds values. Miura et al. (1993) recently showed that a poor conceptual understanding of place value is also evident for children in France and Sweden. As with English, place value is not obvious for French and Swedish number words.

Despite this difficulty with English number words, Fuson and Briars (1990) showed that American children can be easily taught the meaning of multidigit numbers if this meaning is made obvious. One of the methods used in their approach is shown in Figure 2.1. Here the values of thousands, hundreds, tens, and ones are represented concretely for the child. In a set of experiments, Fuson and Briars demonstrated that making the decade structure of multidigit numbers obvious to American children facilitated their conceptual understanding and their problem-solving skills. For example, the frequency of trading errors was greatly reduced after the children were instructed on the meaning of multidigit numbers. In all, when the decade structure of multidigit numbers is made obvious, either with the language's words for numbers or with concrete representations, most second-grade children and some advanced first-grade children will readily acquire knowledge of the base-10 structure. For Asian

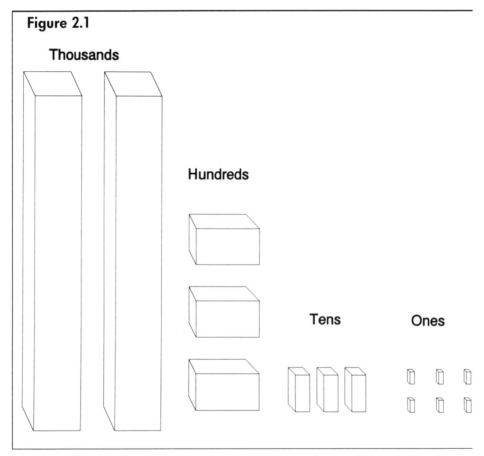

Figure 2.1

Concrete representation of the base-10 system.

children, it is likely that their language's number words make induction of the base-10 system relatively easy and certainly make direct teaching of this structure straightforward. For other children, those whose language does not facilitate the induction of the base-10 structure of multidigit numbers, direct teaching of this structure appears to be necessary and can be successful in the second grade.

Fractions

Preschool children have many opportunities to engage in activities in which knowledge of fractions might be induced. For instance, with a little parental cajoling, young children can share; they can divide eight pieces of candy between two people, so that each person gets half of the total. Yet acquiring knowledge about fractions is a very difficult task for most children (Clements & Del Campo, 1990). Children do not appear to have an intuitive understanding of fractions, as they do of counting and simple arithmetic. In fact, most children treat fractions in terms of their informal knowledge of counting and arithmetic (Gallistel & Gelman, 1992). For example, to solve $\frac{1}{5} + \frac{1}{6}$, most children simply add the numerators and denominators, yielding an answer of $\frac{2}{11}$. Even though children are able to share and can name the pieces of an item that has been cut in half as *one half*, they still do not understand that a fraction is a numeral that represents some part of a whole (Gelman, 1992). Knowledge of fractions is not something that most children will easily acquire through their everyday numerical activities. In fact, children's informal knowledge of counting and simple addition probably makes the conceptual understanding of fractions difficult to grasp and difficult to teach (Gelman, 1992). For instance, because elementary school children understand that 3 is greater than 2, they appear to assume, therefore, that $\frac{1}{3}$ is greater than $\frac{1}{2}$.

Arithmetic Operations

In the next sections, research on the development of problem-solving strategies for simple and complex arithmetic is presented. The focus is on the solution of nonword problems, such as $3 + 5$, and very simple word problems, such as "How many are three apples and five apples?" Skill development for solving more complex word problems is presented in chapter 3. There is one section for each of the four arithmetic operations. Within each section, basic strategies and their development are described, accompanied by a description of common errors and the associated conceptual knowledge needed to use the various strategies.

Early on in the child's development, the solving of arithmetic problems is based heavily on counting and number knowledge and on application of the culture's counting system. With instruction, children learn more formal and presumably more powerful procedures for solving arithmetic problems (Ginsburg, 1982; Saxe, 1982b). Cultural variations in arithmetic strategies—in particular, counting strategies—are most apparent in the early elementary school years. However, these strategies become more similar across cultures with increasing years of formal education (e.g., Ginsburg et al., 1981b). This is not to say that achievement levels converge across cultures (see chapter 7) but rather that the use of culturally idiosyncratic procedures, such as the use of the Oksapmin counting system to add, begins to be replaced with formally taught procedures during the elementary school years.

Even before the influence of formal instruction, there are important cross-cultural similarities in children's arithmetical development. Within all cultural groups that have been studied and across all four arithmetic operations, arithmetical development is not simply a matter of switching from the use of a less mature problem-solving strategy to the use of a more adultlike strategy. Rather, at any given time, most children use a variety of strategies to solve arithmetic problems. They might count on their fingers to solve one problem, retrieve the answer to the next problem, and count verbally to solve still other problems. Arithmetical development involves a change in the mix of strategies, as well as changes in the accuracy and speed with which each strategy can be executed. A more intriguing finding is that children do not randomly choose one strategy, such as counting, to solve one problem and then a different strategy, such as retrieval, to solve the next problem. Rather, children "often choose among strategies in ways that result in each strategy's being used most often on problems where the strategy's speed and accuracy are advantageous, relative to those of other available procedures" (Siegler, 1989a, p. 497).

Within this mix, there are typically one or two strategies that are used more frequently than are other strategies. For example, when first learning to solve addition problems, children rely heavily on the use of manipulatives. As the child gains experience and maturity, the favored strategy shifts to finger counting (Ashcraft & Fierman, 1982; Carpenter & Moser, 1984; Siegler & Jenkins, 1989; Siegler & Shrager, 1984). The following discussion of the different arithmetic strategies and the associated ages reflects when the strategies are most commonly used by children and should not be taken to mean that these strategies are exclusively used at these ages. For the most part, the same general progression appears to be the norm for children in different cultures, although the rate with which children progress from one favored strategy to the next varies across cultures (e.g., Geary, Fan, & Bow-Thomas, 1992; Saxe, 1985; Svenson, Hedenborg, & Lingman, 1976; Svenson & Sjöberg, 1983). Chinese children, for instance, shift from one favored strategy to the next at a younger age than do children in the United States, whereas the Oksapmin appear to shift at later ages than do U.S. children (Geary, Fan, & Bow-Thomas, 1992; Saxe, 1982a).

For the most part, these cultural differences appear to reflect differences in the amount of experience with solving arithmetic problems (e.g., Stevenson, Lee, Chen, Stigler, et al., 1990) and, to a lesser extent, language differences in number words. Support for this argument is presented in chapter 7; nevertheless, a brief preview is in order. Ilg and Ames (1951) provided a normative study of arithmetic development for American children between the ages of 5 and 9 years. These children received their elementary school education at a time when basic skills were emphasized much more in the mathematics curriculum than today. The types of skills and problem-solving strategies that they described for American children at that time are very similar to arithmetic skills we see in present-day, same-age Asian children (e.g., Fuson & Kwon, 1992a; Geary, Fan, & Bow-Thomas, 1992). Today, however, the arithmetic skills of elementary school Asian children are much more developed than those of their same-age American peers.

Addition

The discussion of addition skills is presented in three general sections. In the first, children's problem-solving strategies and common errors, as well as associated conceptual knowledge for solving simple addition problems, are presented. The second section presents children's skills for solving more complex addition problems, and the third section presents a summary of the first two sections.

Simple Addition

For solving simple addition problems, such as 3 + 2, children use five general classes of strategies: using manipulatives; finger counting; verbal counting without the use of manipulatives (i.e., mentally); derived facts; and fact retrieval (Carpenter & Moser, 1983; Ilg & Ames, 1951; Siegler, 1987). Each of these general classes of strategies, and the associated conceptual knowledge, is presented in turn. Roughly the same pattern of development is evident across cultures, although cultural differences do emerge in the relative frequency with which the strategies are used and in the age at which the strategies first appear (e.g., Ginsburg, 1982; Hatano, 1982; Saxe, 1985). The strategies described in this section focus on research conducted in the United States. Cross-national studies are included, where appropriate, to highlight cultural variations within these basic categories. Before starting, I should note that the strategies described below are the most common, but not the only, strategies used by children to solve simple addition problems.

Some children as young as 3 years of age can use their counting skills to add in everyday contexts (Fuson, 1982; Saxe et al., 1987). At this age, addition is typically aided by the use of objects or manipulatives. So if the child is asked, "How many are three cookies and two cookies," then, as stated earlier, he or she will typically count out three objects, then count two objects, and finally count all of the objects, starting from 1. The child states, while pointing at each object in succession, "1, 2, 3, 4, 5; the answer is 5." The use of manipulatives serves at least two purposes. First, the sets of objects represent the numbers to be counted. The meaning of the abstract number, 3 in this example, is literally represented by the objects. Second, pointing to the objects during counting helps the child to keep track of the counting, that is, she or he knows to stop once the last object is tagged with a number word (Carpenter & Moser, 1983). The use of manipulatives is even seen in some 4- and 5-year-olds, depending on the complexity of the problem and whether manipulatives are readily available (Fuson, 1982).

Nevertheless, most 4- and 5-year-old children, if they cannot directly retrieve an answer from memory, use a combination of finger counting and verbal counting to solve addition problems (Geary & Burlingham-Dubree, 1989; Siegler & Shrager, 1984). Fingers, like objects, can be used to represent the numbers to be counted but seem to be used primarily to help the child to keep track of his or her counting (Fuson, 1982); the Oksapmin use the body parts associated with their counting system, rather than fingers, to keep track of their counting (Saxe, 1982b, 1985). For simple problems, those with sums less than 10, children represent the numbers to be counted by uplifting a corresponding number of fingers. The value of the *augend* (the first number) is represented on one hand, the value of the *addend* (the second number) is represented on the other hand, and then children move their fingers in succession as they count, "1, 2, 3, . . ." For problems with sums greater than 10, such as 7 + 8, a common strategy involves lifting seven fingers, then lifting eight fingers, and then moving the fingers in succession while counting "8, 9, 10, 11, 12, 13, 14, 15." Finger counting is also common for kindergarten children and is used occasionally by older elementary school children (Geary & Brown, 1991; Geary, Brown, & Samaranayake, 1991; Ilg & Ames, 1951).

Young Asian children also count on their fingers to solve addition problems. However, in comparison with American children, they use a more sophisticated finger-counting procedure (Fuson & Kwon, 1992a; Geary, Bow-Thomas, Fan, Mueller, et al., 1992). Fuson and Kwon (1992a), for instance, studied the strategies used by first-grade children in Korea for solving addition problems with sums less than 19 and subtraction problems with minuends less than 19. The different types of finger-counting strategies used by these children are presented in Figure 2.2. With one such strategy, the "hands were held up facing the counter and the thumbs faced out. Counting began with the thumb ... it moved linearly across all of the fingers to the little finger, continued to the other thumb, and moved across to the little finger on that hand" (Fuson & Kwon, 1992a, p. 152). This procedure is illustrated in the first panel of Figure 2.2; variants of this procedure are illustrated in the next two sequences. Note that this pattern differs from the typical American pattern in that the Korean children represent addends on their fingers continuously rather than on one hand and then on the other hand. Chinese kindergarten children use a finger-counting procedure that is similar to the procedure used by Korean children. Fuson and Perry (1993) found that some Hispanic-American elementary school children also use a variant of this Asian counting method.

The advantage of this continuous representation becomes more obvious when the solution of problems with sums greater than 10 is considered. The Korean strategy for solving such problems is depicted in the fourth panel of Figure 2.2. Again both hands are held up, facing the counter. Counting involves folding down a number of fingers, representing the augend—8 in this example—and then continuing with the very next finger until all of the fingers are folded down. The remaining values of the addend are represented by unfolding the corresponding number of fingers, 4 for this problem. Recall that number words for teens in most Asian languages, including Korean, are *ten* and some value. Thus, all that the child has to do to answer, when using this approach, is to state *ten* in front of the word for the number of unfolded fingers, *ten four* in this problem. More important, this strategy clearly capitalizes on the base-10 system. The strategy is taught by Korean and Chinese teachers and flows easily from Asian language number words for teens. As is noted later, the use of this type of finger-counting strategy quite likely facilitates the understanding and use of the relatively sophisticated derived-facts strategy.

The transition from finger to verbal counting is primarily dependent on the child's ability to mentally keep track of the numbers that have already been counted and those that still need to be counted (Fuson, 1982). For most children, the shift from finger to verbal counting is gradual. Kindergarten children in the United States use finger counting about as often as they use verbal counting (Baroody, 1987a; Geary & Burlingham-Dubree, 1989; Siegler & Shrager, 1984). Finger counting is

Figure 2.2

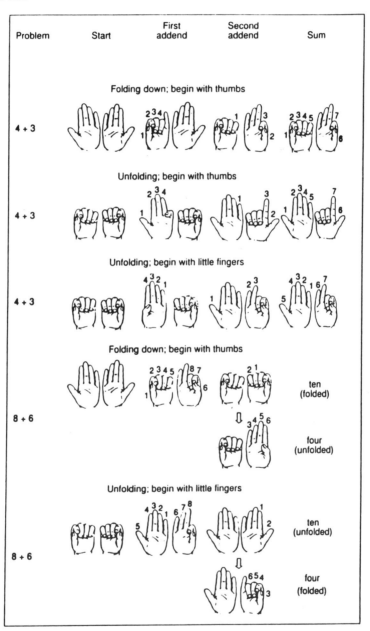

Korean children's finger-counting methods for solving simple addition problems. (From "Korean Children's Single-Digit Addition and Subtraction: Numbers Structured by Ten," by K. C. Fuson and Y. Kwon, 1992, *Journal for Research in Mathematics Education, 23,* p. 153. Copyright 1992 by the National Council of Teachers of Mathematics. Reprinted with permission.)

primarily used to solve problems with larger addends, for example, 6 + 4, where keeping track of the counting process is more difficult than for problems with smaller addends (Geary, 1990). Chinese kindergarten children, on the other hand, abandon finger counting early in the kindergarten year, in favor of verbal counting. This is apparently because number words (e.g., *yi, er*, for *one, two*) are shorter (i.e., quicker to pronounce) in Chinese than in English, which makes keeping track of the numbers, while verbally counting, easier for Chinese children than for American children (Geary, Bow-Thomas, Fan, & Siegler, 1993). Regardless, by the end of first grade, verbal counting is used much more frequently than finger counting by most American children (Geary et al., 1991).

There are three basic procedures that children use when counting verbally: *counting all, counting on from the first number*, and *counting on from the larger number* (Baroody, 1987a; Carpenter & Moser, 1984; Fuson, 1982). Cognitive psychologists identified the same procedures but call them the *sum, first*, and *min* procedures, respectively (Ashcraft, 1982; Groen & Parkman, 1972). The counting all, or sum, procedure is analogous to counting manipulatives except that the child does not rely on objects to keep track of the count. The procedure involves counting both the augend and the addend starting from 1 (Baroody, 1984a). To solve the problem 4 + 3, for example, the child will count "1, 2, 3, 4, 5, 6, 7, the answer is 7." The counting on from the first number, or first, procedure involves stating the value of the first number and then counting a number of times equal to the value of the second number, for example, counting "4, 5, 6, 7," to solve 4 + 3. A child using this procedure will count, "3, 4, 5, 6, 7," to solve 3 + 4.

The adoption of this more sophisticated procedure requires that the child understand that stating the cardinal value of the first number is in a sense a shortcut to counting that number and that counting does not have to start from 1 (Fuson, 1982; Geary, Bow-Thomas, & Yao, 1992). The most sophisticated, counting on from the larger—or min—procedure, requires not only an understanding of how the cardinal value of the addends can be used to make verbal counting more efficient but also an understanding that the order with which numbers are added together does not affect the result. This does not necessarily mean, however, that the child has a formal understanding of commutativity—she or he may simply assume that order does not matter (Baroody & Gannon, 1984). Either way, the child needs to identify and count on from the larger number. As noted earlier, verbal counting is a common strategy for most kindergarten and first-grade children. Kindergarten children typically use the counting all procedure, whereas first-grade children typically use the counting on from the larger procedure. More important, children do not need to be taught these different ways of solving addition problems but rather discover for themselves the most efficient procedure (Baroody, 1984a; Groen & Resnick, 1977; Siegler & Jenkins, 1989).

Counting errors, whether the child counts on his or her fingers or verbally, tend to be systematic (Ginsburg, 1989). The majority of counting errors involve either undercounting or overcounting by one, more often undercounting by one (Fuson, 1982; Geary & Burlingham-Dubree, 1989; Siegler & Shrager, 1984). Counting errors are often due to the child either losing track of which values have and have not been counted or making a procedural error. If the child is counting on from 4 to solve 4 + 3, then she or he must state "4" and then count up three times. Losing track might result in the child counting up four or five times, for example. The procedural error occurs with the counting-on methods (Fuson, 1982). Here the child counts the correct number of times but starts with the wrong number. To solve 4 + 3, the child counts up three times but includes 4 as representing the value of the augend and as the first number in the addend count: counting 4, 5, 6 rather than 4, 5, 6, 7.

The derived-facts strategy involves using memorized addition facts as the basis for solving more difficult problems (Carpenter & Moser, 1983). Cognitive psychologists sometimes call this strategy *decomposition* (Siegler, 1987). Children tend to memorize *doubles*, or *tie problems*, sooner than other combinations; doubles include 1 + 1, 2 + 2, 3 + 3, and so on (Ashcraft, 1992; Groen & Parkman, 1972). These memorized facts can then serve as the basis for solving other addition problems. To solve the problem 6 + 7, the child might retrieve the answer to 6 + 6 (i.e., 12) from long-term memory and then add 1 to the provisional sum. Another common derived-facts strategy is structured around an understanding of the base-10 system. So 6 + 7, for example, might be solved by decomposing the 6 into two 3s, adding one of these 3s to 7 to get 10, and then adding 10 + 3. This last step might involve retrieving the answer to 10 + 3, or counting "10, 11, 12, 13." Fuson and Kwon (1992a) called this derived-facts strategy the *up-over-tens* method.

The use of derived-facts strategies appears to be much more common in Asian children than in American children, at least in more recent studies (Fuson & Kwon, 1992a; Geary, Fan, & Bow-Thomas, 1992). Recall that Asian children tend to use finger-counting strategies that reflect an understanding of the base-10 structure of the Arabic number system. The use of the up-over-tens method seems to flow easily from the finger-counting strategies used by Asian children. Indeed, this derived-facts strategy involves the same operations, albeit mentally, as in Asian children's finger counting. Even though these strategies tend to be taught in Asian schools, it is likely that many children construct this derived-facts strategy on their own (Fuson & Kwon, 1992a).

Evidence that many children discover the derived-facts strategies on their own can be found even within the United States. Although the derived-facts strategies are not often used by American children these days, this was not always the case. Ilg and Ames (1951), in their longitudinal study of arithmetic development, noted that the derived-facts strategies were commonly used by young elementary school children in the United States at that time (1940s). Because the English number words have not changed in the ensuing 50 or so years between the Ilg and Ames research and contemporary studies, a decreased emphasis on basic mechanics, through practice, might have contributed to the disappearance of this strategy in American children. With practice, children do not simply memorize facts, but rather execute many different types of procedures, such as counting on, to solve the presented problems. Through the repeated execution of these procedures, children quite likely come to understand addition concepts and then use this knowledge to construct new, more sophisticated strategies (Siegler & Crowley, 1991; Siegler & Jenkins, 1989). These days, American children might not be getting enough practice with basic addition to construct the derived-facts strategies for themselves. Even with equal amounts of practice, however, Asian children would quite likely still construct this strategy earlier than their American counterparts, simply because of the difference in Asian- and English-language number words for teens. Finally, derived-facts errors seem to involve either retrieving the wrong provisional answer or miscounting (e.g., undercounting by one).

The final type of process that children use to solve simple addition problems is fact retrieval. Here, children quickly produce the answer, without overt signs of counting, and state that they just "knew it," "know it by heart," or "remembered." Children come to memorize answers through the execution of the counting and derived-facts strategies (Siegler, 1986; see the Strategy-Choice Model section). Even preschool children can retrieve a few answers from memory; $1 + 2 = 3$ seems to be especially easy, due to the fact that 3 follows the 1, 2 sequence in the counting string (Siegler & Shrager, 1984). For the most part, doubles, or tie problems, are among the first to be memorized, followed by problems with smaller valued addends, such as $2 + 1$, $3 + 2$, and so on. Problems with larger valued addends, such as $7 + 8$, are the last to be memorized.

Educators and psychologists have been studying why some problems are easier to learn or memorize than others for decades (e.g., Washburne & Vogel, 1928; Wheeler, 1939) and continue to do so to this day. One contributing factor is the frequency of problem presentation. Smaller valued problems are presented much more frequently in textbooks, and tie problems, or doubles, are presented more frequently by parents for children to solve than are larger valued problems (Ashcraft & Christy, in press; Hamann & Ashcraft, 1986; Siegler & Shrager, 1984). Another contributing factor appears to be the ease with which counting strategies can be used to solve the problem. The more easily counting can be used to solve the problem, the quicker the correct answer is memorized (Geary et al., 1991; Siegler, 1986). Others have argued that people have an innate understanding of different quantities and that as the quantities get larger, the differences among these quantities become more fuzzy (Dehaene & Mehler, 1992; Gallistel & Gelman, 1992). It is easier to discriminate one from two than it is to discriminate eight from nine. Thus, applying mental procedures, such as fact retrieval, to larger numbers is necessarily more error-prone and time-consuming.

Even after all of the basic facts have been memorized, some facts are easier to retrieve than others. This phenomenon, as related to memory retrieval, is called the *problem-size effect*. It was discovered by Ashcraft and his colleagues (Ashcraft, 1992; Ashcraft & Battaglia, 1978). The problem-size effect has been found for subtraction and for multiplication as well as for addition (nobody has tested it for division; Geary, Frensch, & Wiley, 1993; Geary, Widaman, & Little, 1986; K. Miller, Perlmutter, & Keating, 1984; Stazyk, Ashcraft, & Hamann, 1982). Basically, the larger the numbers in the problem, the slower and more error-prone the retrieval. So even after children are at the point where they can retrieve all of the basic facts from memory, larger problems will remain more difficult to solve. Nevertheless, committing basic facts to memory seems to be a good idea, because at this point the solution of simple problems occurs more or less automatically (LeFevre, Bisanz, & Mrkonjic, 1988). The automatic processing of basic operations makes the solution of more complex problems much faster and less error-prone (Geary & Burlingham-Dubree, 1989; Geary & Widaman, 1992; Kaye, 1986; L. B. Resnick & Ford, 1981). Finally, note that some addition facts seem to be retrieved from long-term memory with the use of rules (Baroody, 1984c). The use of rules, however, appears to be limited to problems in the form of $n + 0 = n$.

Retrieval errors fall into four general categories: *wild guesses, near misses, operation confusions*, and *table errors* (Ashcraft, 1992; Baroody, 1989; Siegler & Shrager, 1984). Wild guesses are fairly common for kindergarten children, for example, stating that $4 + 1 = 41$ or simply restating one of the addends as the answer (Baroody, 1989). Near misses involve retrieving an answer that is 1 or 2 higher than the correct sum and mirror the child's earlier counting errors (Siegler & Shrager, 1984). If the child frequently counts "4, 5, 6" to solve $4 + 3$, eventually that child will begin to retrieve 6 as the answer. Operation confusions are easy to understand; the child simply retrieves the correct answer to an analogous problem with a different arithmetic operation. For example, the child might retrieve 12 when asked to solve $4 + 3$.

These types of retrieval errors are most common when children are first learning a new arithmetic operation. K. F. Miller and Paredes (1990) found that children were most likely to retrieve multiplication products when solving addition problems when first learning multiplication. Operation-confusion errors occur not only because children misread, or fail to read, operation signs but also because addition and multiplication facts are represented together in the same memory system (e.g., Campbell & Graham, 1985). Table errors involve retrieving the answer to a related problem, for instance, retrieving 12 for the problem $6 + 7$ (12 is the correct answer to $5 + 7$). Table errors also seem to be related to the way in which arithmetic facts are represented in long-term memory (Ashcraft, 1992).

Complex Addition

Children typically learn to solve multicolumn problems, such as $23 + 4$ or $47 + 64$, in elementary school. Learning to solve such problems involves developing not only efficient problem-solving strategies but also an understanding of place value and how to carry or trade. Children in Japan, mainland China, Taiwan, the former Soviet Union, and the United States are introduced to two-column problems without trades, for instance, $23 + 45$, in the second semester of first grade (Fuson, Stigler, & Bartsch, 1988). Problems with trading are also introduced at this time in China and the former Soviet Union. Problems with trades are presented early in the second grade in Japan and Taiwan, whereas these same problems are presented in the second semester of Grade 2 in the United States. Similarly, more complex addition problems are presented between 1 to 3 years later in the United States than they are in most other countries. For instance, problems such as $234{,}547 + 487{,}523$ are presented in third

---

Figure 2.3

(A)   46
      +58
       94

(B)   46
      +58
     914

(C)   22
      +64
      96

Common carrying or trading errors in complex addition. Problem A illustrates a failure to complete trading. Problem B shows a trading error that resulted from a lack of conceptual understanding of place value. Problem C illustrates an unnecessary trade.

grade in China but not until the fifth or sixth grade in the United States (Fuson et al., 1988).

Thus, the acquisition of more complex addition skills spans much of the elementary school years in the United States, whereas the same topics are covered in Grades 1 to 3 in many other countries. As might be expected, when solving more complex problems, children initially rely on the knowledge and skills acquired for solving simple addition problems. Strategies for solving more complex problems include counting and decomposition, or regrouping, as well as the formally taught columnwise procedure (Fuson et al., 1988; Ginsburg, 1989). Counting strategies typically involve counting on from the larger number (Siegler & Jenkins, 1989). For instance, 23 + 4 would be solved by counting "23, 24, 25, 26, 27." As described earlier, the regrouping strategy is dependent on the child's knowledge of the base-10 system and involves adding the tens values and the units values separately. So the problem 23 + 45 would involve 20 + 40, 3 + 5, and then 60 + 8 (Fuson & Kwon, 1992b).

The most difficult process in complex addition involves carrying, or trading, as in the problem 46 + 58. Two factors appear to make trading difficult for children, and for some adults. First, trading often involves manipulating numbers mentally. For the problem 46 + 58, the child must first add 6 + 8. Then she or he must mentally note that a trade has taken place and retain this notation in working memory while writing the units-column answer (Hamann & Ashcraft, 1985; Hitch, 1978; Widaman, Geary, Cormier, & Little, 1989; Widaman, Little, Geary, & Cormier, 1992). While writing the units-column answer, it is easy to forget the trade. Second, skilled trading involves an understanding of place value. In particular, the child needs to understand that the 1 traded from the units to the tens column actually represents 10 and not 1; failure to understand this contributes to children's trading errors. Common trading errors for complex addition are shown in Figure 2.3 (Fuson & Briars, 1990). Problem A illustrates a failure to complete trading. The units-column sum is correct, but the traded tens value was not added to the tens-column information, most likely because the child simply forgot to complete the trade. Problem B shows that the child did not really understand trading. The 14 is treated as if it were a single number rather then a combination of 1 ten and 4 ones. Sometimes children trade when they do not need to trade, as shown with Problem C. Although these are common errors for American children, they are rather rare in Korea and Japan (Fuson & Kwon, 1992b; Miura et al., 1993; Song & Ginsburg, 1987). Again, it seems that Asian-language words for complex numbers, such as *two ten one* for 21, make the place value of the written numbers obvious. This in turn seems to facilitate the understanding of place value and the associated procedures, such as carrying and borrowing.

Summary

In apparently all cultures in which addition is common, the first problem-solving strategy to emerge is based on counting (Geary, Fan, & Bow-Thomas, 1992; Ginsburg, 1982; Saxe, 1982a; Zaslavsky, 1973). Of course, the surface structure—so to speak—of the counting sometimes differs across cultures, but the same deep structure (the knowledge on which these strategies are developed) appears throughout the world. This deep structure, the child's fundamental understanding of number and quantity, provides the foundation on which the child learns his or her culture's counting system. This counting system, combined with a fundamental understanding of simple arithmetic (Starkey, 1992), provides the seed for the development of many of the different types of problem-solving strategies that are used by young children throughout the world to solve addition problems: counting manipulatives, finger counting, and verbal counting. Culture makes its own impositions on these emerging addition skills. Children in Asian countries, aided by their number words, develop their early addition strategies around the base-10 structure of the Arabic system (Fuson & Kwon, 1992a, 1992b), whereas the Oksapmin base their early addition strategies on their body part number system (Saxe, 1982a, 1985).

Despite differences in surface structure, many cross-cultural similarities exist in the types of early strategies used by young children to solve addition problems. Across cultures, the counting of manipulatives and body parts, such as fingers, are the first types of strategies to emerge with any frequency in preschool children (i.e., at this age, they are the favored strategy of most children). The use of manipulatives and fingers, or body parts in the Oksapmin, helps the child to concretely represent the numbers to be added and helps the child to keep track of the counting process. As children become better at mentally keeping track of counting, verbal counting becomes more common than finger counting as a problem-solving strategy (Fuson, 1982). Counting errors, whether associated with finger or verbal counting, are systematic and often involve undercounting by one. Here, the counting-on procedures are executed correctly but are started at the wrong number (Fuson, 1982). With the repeated use of these reconstructive strategies (L. B. Resnick, 1989), children gain an understanding of how addition works; are able to execute basic processes, such as counting on, more quickly and with fewer demands on their working-memory resources (Case, 1985); and finally begin to learn or memorize basic facts (Siegler, 1986).

The memorization of basic facts makes the derived-facts strategies and direct-retrieval processes possible. The use of the derived-facts strategies is also dependent on the child's understanding that numbers can be decomposed into groups of smaller numbers and appears to be used more readily in children who understand the base-10 structure of the number system (Fuson & Kwon, 1992a). Finally, note that even though direct retrieval is not a strategy in itself (Bisanz & LeFevre, 1990) and retrieval does not require any special conceptual knowledge, the memorization of basic facts is important. This is because once basic facts have been memorized, the solution of simple problems occurs more or less automatically. The automatic processing of basic features makes the solution of more complex arithmetic problems, such as word problems, quicker and less error-prone (Geary & Burlingham-Dubree, 1989; Geary & Widaman, 1992; Kaye, 1986; L. B. Resnick & Ford, 1981).

Even though the solution of simple addition problems is possible and common before formal instruction (Groen & Resnick, 1977), children are not typically asked to solve more complex, multicolumn problems until elementary school. Again, the solution of such problems is initially based on existing counting and addition skills. So to solve the problem 45 + 3, first-grade children will typically count on from the larger number, just as they would to solve 5 + 3. Children who understand that numbers can be decomposed into groups of smaller numbers, especially those children who can decompose numbers with respect to the base-10 number system, often use a regrouping strategy to solve these problems (Fuson & Kwon, 1992a). Trading, or carrying, is probably the most difficult feature of complex addition that children need to learn. Trading requires an understanding of the base-10 structure and places demands on the child's working memory resources. Thus, it is not surprising that the most common complex addition errors involve trading and that the trading errors more often than not reflect a lack of understanding of the base-10 structure or forgetting to complete the trade (Fuson & Briars, 1990; Hamann & Ashcraft, 1985). To make all of this easier to follow, the most common problem-solving strategies in addition are presented in Table 2.1.

Subtraction

As with the discussion of addition, the discussion of subtraction skills is presented in three sections. Children's problem-solving strategies; common errors; and associated conceptual knowledge for solving simple problems, such as $9 - 3$, are presented in the first section. The second section discusses children's skills for solving more complex subtraction problems, such as $14 - 6$ or $56 - 29$. The third section contains a summary of the first two sections.

Simple Subtraction

The same general findings that were described for simple addition also apply to children's strategies for solving simple subtraction problems. Early on, subtraction is aided by the use of manipulatives, followed by the development of verbal counting strategies and then by the decomposition (for subtraction, this term is better than derived facts) and retrieval strategies. Sometimes subtraction problems are solved by referring to complementary addition problems (i.e., *addition reference*; Baroody, 1984b; Carpenter & Moser, 1984; Siegler, 1989a; Steinberg, 1985; Svenson & Hedenborg, 1979; Woods, Resnick, & Groen, 1975). Again, as with addition, common features of each of these different classes of strategy are presented, although not in the depth presented in the section on addition. Not to take anything away from the importance of subtraction but rather for the sake of brevity, complementary concepts or developmental trends are not treated in as much detail for subtraction as they were for addition. Overall, the general trend is for children to initially rely heavily on manipulatives and finger counting, then on verbal counting strategies to solve subtraction problems. Next, children tend to rely on the decomposition strategy (for some children) and finally on direct retrieval (Carpenter & Moser, 1984).

Many 4- and 5-year-old children can solve formally presented subtraction problems (Siegler & Shrager, 1984); for instance, "If you had three cookies and gave one to your brother, how many would you have

Table 2.1

Commonly Used Addition Strategies

Strategy	Description	Example
Simple Addition		
Counting manipulatives	The problem's augend and addend are represented by objects. The objects are then counted, starting from 1.	To solve 2 + 3, two blocks are counted out, then three blocks are counted out, and finally all five blocks are counted.
Counting fingers	The problem's augend and addend are represented by fingers. The fingers are then counted, usually starting from 1.	To solve 2 + 3, two fingers are lifted on the left hand, and three fingers are then lifted on the right hand. The child then moves each finger in succession as he or she counts them.
Verbal counting Counting all (sum)	The child counts the augend and addend in succession starting from 1.	To solve 2 + 3, the child counts "1, 2, 3, 4, 5; the answer is 5."
Counting on first	The child states the value of the augend and then counts a number of times equal to the value of the addend.	To solve 2 + 3, the child counts "2, 3, 4, 5; the answer is 5."
Counting on larger (min)	The child states the value of the larger addend and then counts a number of times equal to the value of the smaller addend.	To solve 2 + 3, the child counts "3, 4, 5; the answer is 5."
Derived facts (decomposition)	One of the addends is decomposed into two smaller numbers, so that one of these numbers can be added to the other addend to produce a sum of 10. The remaining smaller number is then added to 10.	To solve 8 + 7: Step 1. 7 = 5 + 2 Step 2. 8 + 2 = 10 Step 3. 10 + 5 = 15

continued

Table 2.1, continued		
Strategy	**Description**	**Example**
Simple Addition, continued		
Fact retrieval	Direct retrieval of basic facts from long-term memory	Retrieving 5 to solve 2 + 3.
Complex Addition		
Verbal counting Counting on larger	Same as above.	To solve 23 + 2, the child counts "23, 24, 25; the answer is 25."
Regrouping	The addends are decomposed into tens and units values. The tens and units values are summed separately. The two provisional sums are then added together.	To solve 25 + 42: Step 1. 25 = 20 + 5 Step 2. 42 = 40 + 2 Step 3. 20 + 40 = 60 Step 4. 5 + 2 = 7 Step 5. 60 + 7 = 67
Columnar retrieval	The problem is solved by retrieving columnwise sums.	To solve 27 + 38: Step 1. 7 + 8 = 15 Step 2. Note trade (carry) Step 3. 2 + 3 = 5 Step 4. 5 + 1 (from trade) = 6 Step 5. Combined 6 from tens column to 5 from ones column to produce 65

left?" For solving this type of problem, Carpenter and Moser (1984) described three common procedures that involve the use of manipulatives. The first of these procedures is called *separating from.* For this problem, the child first gets three blocks to represent the three cookies, removes one block, and then counts or simply states the number of remaining blocks. The second procedure, *adding on,* involves starting with a number of blocks stated by the *subtrahend* (the smaller number) and then adding a number of blocks until the value of the *minuend* (the larger number) is reached. The answer is represented by the number of blocks added to the subtrahend. So, for this example, the child would place one block in front of him- or herself and add two more blocks while counting "2, 3."

Because two blocks were added to the first block, the answer would be 2. The final procedure involves matching in a one-to-one fashion the number of blocks represented by the minuend and subtrahend. The answer is represented by the number of unmatched blocks. For this example, the child would have one row consisting of a single block aligned with a second row consisting of three blocks. The two unmatched blocks would represent the answer. The use of these different procedures varies with how the problem is presented to the child (see the Semantic Structure section of chapter 3).

Most 5- and 6-year-olds typically use counting to solve simple subtraction problems (e.g., Siegler & Shrager, 1984). As with addition, counting is sometimes done with the aid of fingers and sometimes done without fingers. Again, finger counting allows the child to represent the numbers to be manipulated and to keep track of the subtraction process (Baroody, 1984b) and is more likely to be used to solve problems with larger numbers, such as $7 - 3$, than for problems with smaller numbers, such as $3 - 1$. For solving subtraction problems, counting—whether on fingers or verbally—can involve one of two procedures, *counting up* and *counting down*. Counting down involves counting backward from the minuend a number of times represented by the value of the subtrahend. To solve the problem $7 - 3$, the child would count, "6, 5, 4; the answer is 4." If the child cannot keep track of how many values have been counted while counting backward, then she or he will first lift seven fingers and then fold down three fingers in succession. He or she might count backward, "6, 5, 4," while folding down the fingers or first fold them down and then count the remaining fingers, "1, 2, 3, 4" (Baroody, 1984b).

The counting-up procedure involves stating the value of the subtrahend and then counting until the value of the minuend is reached. The number of times counted is the answer. So $9 - 7$ is solved by counting "8, 9." Because two numbers were counted, the answer is 2. This procedure can also be executed using fingers, if the child has enough fingers to represent the difference. In other words, counting up using fingers works well when the difference is less than 11 but not for larger problems, such as $25 - 12$. Some studies have suggested that when children count to solve subtraction problems, they choose the most efficient procedure, that is, the procedure involving the fewest counts (Svenson & Hedenborg, 1979; Woods et al., 1975). Counting down would be used to solve $9 - 2$, whereas counting up would be used to solve $9 - 7$. This makes sense, because such choices reduce the number of incrementations that the child needs to make, thereby reducing the chances of committing an error.

Nevertheless, not all studies have found this choice pattern (Siegler, 1989a). In fact, unless manipulatives are used, children rarely use the counting-down procedure to solve simple subtraction problems (Carpenter & Moser, 1982). This is because counting backward and simultaneously keeping track of the counting are difficult for children. Nevertheless, counting down is often used when children solve more complex subtraction problems, such as 23 − 4, because counting up requires many more incrementations for these problems than does counting down (Siegler, 1989a). Regardless, when children count verbally to solve simple subtraction problems, they almost always use the counting-up procedure.

Finally, counting errors for simple subtraction appear to parallel the types of errors commonly seen in children's addition (Ilg & Ames, 1951; Siegler & Shrager, 1984). As with addition, these errors appear to result from the child's losing track of the counting or incorrectly executing the procedure. For instance, when counting up to solve 9 − 5, the child might count, "5, 6, 7, 8, 9; the answer is 5," rather than "6, 7, 8, 9; the answer is 4." Siegler and Shrager found that for simple subtraction, 5- and 6-year-old children were much more likely to make an error when counting verbally, as compared with counting using their fingers. This pattern suggests that verbal-counting errors also occur because children lose track of the counting process.

Another strategy that is often used to solve simple subtraction problems is addition reference (Ilg & Ames, 1951; Siegler, 1989a). With this strategy, a subtraction problem is solved by reference to the complementary addition problem. For instance, 8 − 2 would be solved by retrieving 6 + 2 = 8. Siegler found that second-grade children used addition reference to solve about 2% of the subtraction problems he presented but that this increased to 21% by the fourth grade. The addition-reference strategy tends to be quicker and less error-prone than counting. Of course, to use this strategy, children need to have many basic addition facts committed to memory. The jump in usage from second to fourth grade probably reflects differences in the number of addition facts committed to memory rather than differences in the understanding that addition and subtraction are inverse operations. Indeed, Ilg and Ames (1951) stated that addition reference was a commonly used strategy by American first-grade children at that time.

The final process used to solve simple subtraction problems is direct retrieval. Compared with research on direct retrieval for addition (Ashcraft, 1992; Ashcraft, Fierman, & Bartolotta, 1984; Ashcraft & Stazyk, 1981), little is known about the retrieval process in subtraction. Nevertheless, it seems likely that the same types of processes govern retrieval for addition, subtraction, and the two other arithmetic operations as well. For instance, Ilg and Ames (1951) found that a common retrieval error for subtraction was retrieving the answer to the complementary addition problem, for instance, retrieving 12 for 8 − 4. Geary, Frensch, and Wiley (1993) found the same problem-size effect in subtraction retrieval as is found for addition and multiplication retrieval (Ashcraft & Battaglia, 1978; Campbell, 1987). As was argued for addition, it seems to be a good idea to commit basic subtraction facts to long-term memory. Again, if basic operations can be executed quickly and with little effort, the solution of more complex subtraction problems will almost certainly be easier (Kaye, 1986). The American children in the Ilg and Ames (1951) study knew most, if not all, of the basic subtraction facts "by heart" by the third grade. In contrast, Geary, Frensch, and Wiley found that only one out of three contemporary college undergraduates could retrieve all of the basic simple subtraction facts (minuend and subtrahends less than 10) from long-term memory; 83% of a comparison group of older adults (with a mean age of 72 years) could retrieve all of these basic facts. Either way, more research on the factors that influence the representation and retrieval of basic subtraction facts is clearly needed.

Complex Subtraction

When first learning to solve multicolumn subtraction problems, such as 17 − 3 or 48 − 27, children rely on the knowledge and strategies developed for solving simple subtraction problems (Ilg & Ames, 1951; Siegler, 1989b). In particular, children count and refer to related addition problems. For example, to solve 17 − 3, the child might count down, "16, 15, 14; the answer is 14." Children also use the addition-reference strategy, if they have the complementary addition fact memorized. Moreover, with the introduction of these more complex problems, children also begin to use a problem-solving rule: *the delete-10s rule*.

> Deleting 10s involved a type of decomposition in which children treated the 10s value separately from the 1s value. For example, on 15 − 3, they might explain their answer by saying, "5 − 3 = 2, and you put back the 1, so 12." (Siegler, 1989a, p. 500)

Children also use a columnar-processing strategy, in which the units-column information is processed first, followed by the tens-column information (Fuson & Kwon, 1992b). To solve 48 − 27, the child might first retrieve, or count, to get the answer to 8 − 7 and then process in a similar manner the 4 − 2 in the tens column.

In keeping with the research on children's addition, the decomposition strategy is also used to solve subtraction problems. Again, this strategy is used much more frequently by Asian children than by American children (Fuson & Kwon, 1992a; Hatano, 1982). Fuson and Kwon (1992a) described two common decomposition strategies used by first-grade Korean children. Both procedures are based on the base-10 structure of the number system. The first is called the *down-over-the-ten method* and is used to solve problems with minuends greater than 10, such as 14 − 6. Here, 10 is first subtracted from the minuend, 14 − 10; the difference, 4, is then subtracted from the subtrahend, 6 − 4; and this difference is subtracted from 10 to yield the answer, 10 − 2 = 8. Remember, the Asian language term for 14 is *ten four*, so the first step for these children does not really require the subtraction operation, 14 − 10. This is because the children can simply subtract the word after the ten from the minuend, which would not be possible if the procedure were to be used in English.

The second decomposition procedure described by Fuson and Kwon (1992a) is called the *take-from-the-ten method*. Here, the first operation involves subtracting the subtrahend from 10. So for the problem 14 − 6, the first operation would involve 10 − 6. The child then notes the difference, 4 in this example, and then subtracts 10 from the minuend, 14 − 10. Finally, the two provisional differences are added together to give the answer, 4 + 4 = 8. Both of these methods are taught in Asian schools in first grade, and their use suggests that Asian children have a good conceptual understanding of subtraction—which is reflected in the ability to decompose and recombine groups of numbers (Piaget, 1965). Although differences in Asian- and English-language number words might contribute to this cultural difference in the use of these strategies, this is not the whole story. From the research of Ilg and Ames (1951), it appears that 40 or 50 years ago a form of decomposition was commonly used by American 6-year-olds. For instance, to solve 10 − 4, Ilg and Ames (1951) described the following sequence, "10 − 4 = 6 because 5 + 5 = 10, 5 − 1 = 4, so 5 + 1 = 6" (p. 6). Moreover, Steinberg (1985) showed that contemporary second-grade American children could be taught to use this type of strategy.

In keeping with findings for complex addition, Fuson and Kwon (1992b) found that Asian children use more sophisticated problem-solving strategies and show a better conceptual understanding of complex subtraction than their American peers. One area in which these differences are especially pronounced is for the borrow operation, as with the problem 63 − 48. Borrow errors, which have been studied extensively by cognitive

Figure 2.4

(A)	$\begin{array}{r} 73 \\ -35 \\ \hline 42 \end{array}$	(B)	$\begin{array}{r} 5\,2 \\ 73 \\ -35 \\ \hline 20 \end{array}$
(C)	$\begin{array}{r} 8\ \ 10 \\ 903 \\ -405 \\ \hline 408 \end{array}$	(D)	$\begin{array}{r} 10 \\ 73 \\ -54 \\ \hline 29 \end{array}$
(E)	$\begin{array}{r} 60 \\ -27 \\ \hline 47 \end{array}$	(F)	$\begin{array}{r} 1010 \\ 505 \\ -287 \\ \hline 328 \end{array}$

Common borrowing or trading errors in complex subtraction. Problem A shows the result of always subtracting the smaller from the larger number regardless of position. Problem B shows a trade of 2 from the tens column. Problem C shows the result of ignoring zeros. Problem D shows a failure to decrement the tens-column minuend after a correct trade. Problem E again shows the result of ignoring zeros. Problem F shows several failures to decrement after correct borrowing.

scientists (VanLehn, 1990; Young & O'Shea, 1981), are very common in American children but not common at all in Korean children (Fuson & Kwon, 1992b). Again, this cultural difference seems to be related, in part, to differences in Asian- and English-language number words. Recall that the number words in Asian languages make the base-10 structure of multidigit numbers obvious. When borrowing from the tens column to the units column, American children often treat the borrowed value as a one rather than a ten, whereas Korean children rarely make this type of mistake. Korean children describe the borrow operation as a lend or give, implying an understanding of an even trade, that is, one ten for ten ones. The English term, *borrow*, implies no such trade, which is why mathematics educators now use the term *trade* instead of *carry* or *borrow*.

In keeping with studies of addition, complex subtraction errors often reflect the misapplication of a procedure rather than inattention (Fuson & Kwon, 1992b; Ginsburg, 1989; VanLehn, 1990). The misapplication of a procedure means that the procedure might be correct for some problems but has been inappropriately applied to other problems, resulting in errors. These types of errors are sometimes referred to as *bugs* (Young & O'Shea, 1981). Some of the more common subtraction bugs for American children, as described by VanLehn, are shown in Figure 2.4. The most common bug is depicted in Problem A. Here, the child subtracts the smaller number from the larger number regardless of position. In the next example, B, the child borrows a 2 from the tens column to subtract 5 from 5 in the units column. Another common borrowing error is shown in Problem C, where the child ignores the 0 in the tens column and borrows directly from the hundreds to the units column. To make matters worse, most children treat the borrowed hundreds value as 10 rather than 100.

Next, in Problem D, we see correct borrowing, in terms of the units column, but a failure to decrement the tens column value. This type of error suggests that the child simply forgot to complete the borrow operation. Sometimes children ignore zeros, as shown in Problem E. Finally, in Problem F, multiple borrows have been executed, but none of the associated decrements have been completed. Most of these borrow errors reflect a lack of understanding of place value and the associated correspondence between tens and ones, that is, that 1 ten = 10 ones. Given that Korean children, and presumably children in other Asian countries, have a good understanding of place value and the base-10 system, it is not surprising that they do not often make these types of errors.

Summary

Many of the same strategic and developmental trends described for children's addition also apply to children's subtraction (Carpenter & Moser, 1982). Early on in development, children count to solve simple subtraction problems and rely on the use of manipulatives and fingers to help them represent the problem and to help them keep track of the counting (see also Saxe, 1982a, 1985). With experience and maturation, children are better able to mentally keep track of the counting process and thus abandon the use of manipulatives and fingers for verbal counting. Children also rely on their knowledge of addition facts to solve subtraction problems (addition reference). Here, a child might solve $8 - 3$ by retrieving $5 + 3 = 8$ (Ilg & Ames, 1951; Siegler, 1989a). The most sophisticated problem-solving strategy that can be used to solve subtraction problems involves decomposing the problems into a series of simpler problems (Fuson & Kwon, 1992b). This type of decomposition strategy is used much more frequently by Asian children than by American children, at least in more recent studies. The final process that can be used to solve subtraction problems involves simply retrieving the difference or columnar difference (e.g., for $56 - 23$) from long-term memory. At this point, the solving of simple subtraction problems, and columnar subtraction problems without borrows, occurs more or less automatically. To make all of this easier to follow, the most commonly used subtraction strategies are described in Table 2.2.

The same types of errors that were described for children's addition are also evident in children's subtraction. Counting errors are often due to losing track of the counting process, which often results in under- or overcounting. Another common counting error involves starting the counting procedure at the wrong value. For instance, to solve $9 - 7$, the child counts, "7, 8, 9; the answer is 3" rather than "8, 9; the answer is 2." The child includes the number representing the value of the subtrahend as part of the counting-up process. As with addition, retrieval errors in subtraction are often the result of operation confusions, for instance, retrieving 16 to solve $8 - 8$ (Ilg & Ames, 1951). Errors in complex subtraction typically result from the inappropriate use of a subtraction procedure (VanLehn, 1990). The procedure is often appropriate for some subtraction problems, but is indiscriminately applied to problems where it is not appropriate. These types of errors are much more common in American children than in Asian children and reflect a general lack of understanding of place value and the base-10 structure of the number system (Fuson & Kwon, 1992b).

Multiplication

Developmental trends in children's simple multiplication mirror the trends described for children's addition and subtraction, although formal skill acquisition begins in the second or third grade for multiplication, at least in the United States, rather than during the preschool years. Unlike the sections on addition and subtraction, the discussion of multiplication skills focuses primarily on simple problems, such as 4×7, because little psychological research has been conducted on how children solve more complex problems (e.g., 34×47). Nevertheless, for the sake of completeness, a brief presentation of the strategies that adults use to solve complex multiplication problems is included.

Simple Multiplication

The early acquisition of multiplication skills is heavily dependent on the child's knowledge of addition and counting (Cooney & Ladd, 1992; Cooney, Swanson, & Ladd, 1988; Ilg & Ames, 1951; K. F. Miller & Paredes, 1990; Siegler, 1988b). Two strategies that are used by many children when first learning to multiply include *repeated addition* and *counting by n*. The repeated-addition strategy involves representing the *multiplicand* (the

Table 2.2

Commonly Used Subtraction Strategies

Strategy	Description	Example
	Simple Subtraction	
Manipulatives Separating from	The value of the minuend is represented by objects. The number of objects represented by the subtrahend are then removed. The number of remaining objects represents the answer.	To solve 3 − 1, three blocks are counted out, and one block is then removed. The remaining blocks are counted, "1, 2," to get the answer.
Adding on	The value of the subtrahend is represented by objects. Blocks are added until the value of the minuend is reached. The answer is represented by the number of blocks added to the subtrahend.	To solve 3 − 1, one block is counted out. Two more blocks are added, as the child counts, "1, 2."
Matching	Two rows of objects are constructed with one-to-one correspondence, one row each for the minuend and subtrahend. The answer is represented by the number of unmatched objects.	To solve 3 − 1, one row consisting of one block is aligned with a second row consisting of three blocks. The two unmatched blocks are counted, "1, 2," to get the answer.
Counting fingers	The value of the minuend is represented by lifting a corresponding number of fingers. The number of fingers corresponding to the value of the subtrahend are folded down. The answer is represented by the number of remaining lifted fingers.	To solve 3 − 1, three fingers are first lifted. One finger is then folded down, and the remaining lifted fingers are counted, "1, 2," to get the answer.

continued

Table 2.2, continued

Strategy	Description	Example
	Simple Subtraction, *continued*	
Verbal counting		
Counting up	Counting up from the subtrahend until the value of the minuend is stated. The answer is represented by the number of counts.	To solve 3 − 1, the child counts "2, 3."
Counting down	Counting down from the minuend a number of times equal to the value of the subtrahend. The answer is represented by the number of counts.	Counting "2," to solve 3 − 1.
Retrieval	Direct retrieval of basic facts from long-term memory	Retrieving 2 to solve 3 − 1.
	Complex Subtraction	
Verbal counting		
Counting down	Same as above.	Counting "13," "12," "11" to solve 14 − 3.
Decomposition		
Down over the ten	Ten is subtracted from the minuend. The resulting difference is subtracted from the subtrahend, and this difference is subtracted from 10.	To solve 14 − 6: Step 1. 14 − 10 = 4 Step 2. 6 − 4 = 2 Step 3. 10 − 2 = 8
Take from the ten	The subtrahend is subtracted from 10. Ten is then subtracted from the minuend. The two provisional differences are added together to give the answer.	To solve 14 − 6: Step 1. 10 − 6 = 4 Step 2. 14 − 10 = 4 Step 3. 4 + 4 = 8

continued

Table 2.2, continued		
Strategy	**Description**	**Example**
	Complex Subtraction, *continued*	
Delete 10s rule	Increasing the value of the subtrahend to 10, then subtracting 10 from the minuend, and then adding the difference between 10 and the subtrahend to get the answer.	To solve $32 - 9$: Step 1. $9 + \underline{1} = 10$ Step 2. $32 - 10 = 22$ Step 3. $22 + \underline{1} = 23$
Columnar retrieval	Retrieving basic facts from long-term memory in a columnwise fashion.	To solve $32 - 9$: Step 1. $30 - 10 = 20$ (borrow) Step 2. represent 20 in working memory Step 3. $10 + 2 = 12$ Step 4. $12 - 9 = 3$ Step 5. $20 - 0 = 20$ Step 6. $20 + 3 = 23$

first number) the number of times that is indicated by the *multiplier* (second number) and then successively adding these numbers. To solve 3×4, the child would set up the problem as $3 + 3 + 3 + 3$ and then add $3 + 3 = 6, 6 + 3 = 9$, and $9 + 3 = 12$. The counting-by-n strategy is dependent on the child's ability to count by 2s, 3s, 5s, and so on. The counting-by-n strategy appears to be frequently used for problems that contain a 5. The solution to 3×5, for example, is based on the counting sequence 5, 10, 15. If paper and pencil are available, then children sometimes represent the multiplicand by sets of tally marks, one set for each value represented by the multiplier. The child then counts the tally marks. For instance, to solve 3×4, the child would mark four sets of three marks on the paper and then count the marks. A common error in the execution of the repeated-addition strategy involves adding the multiplicand too few or too many times, whereas undercounting and overcounting are common sources of errors with the count-by-n strategy (Siegler, 1988b).

Somewhat more sophisticated strategies involve the use of rules and derived facts (or decomposition). Rules are used primarily to solve problems containing a 0 or a 1, those in the form of $n \times 0 = 0$ and $n \times 1 = n$ (K. Miller et al., 1984). Although the use of rules to solve such problems is common, Cooney et al. (1988) found that not all children use these rules. When they are used, the most common error associated with the use of the $n \times 0$ and $n \times 1$ rules involves confusing addition and multiplication (K. F. Miller & Paredes, 1990). So, $n \times 0 = n$, and, for instance, $2 \times 1 = 3$, rather than 2. The derived-facts strategy is based on the child's ability to retrieve some multiplication facts from long-term memory, in particular, answers to doubles or tie problems (e.g., 2×2, 3×3). As with addition, doubles are memorized before other combinations (Siegler, 1988b). The retrieved answer to a double typically serves as the first step in the derived-facts strategy. For instance, to solve 5×6, a child using this strategy would first retrieve the answer to 5×5 and then add 5 to this provisional product to get 30. Derived-facts errors can occur if the child retrieves the wrong provisional product or incorrectly adds the remaining value to this provisional product.

Eventually, children are able to retrieve most basic multiplication facts from long-term memory. In fact, children seem to learn the basic multiplication table in less time than it takes to learn the addition table (Cooney et al., 1988; K. F. Miller & Paredes, 1990; Siegler, 1988b). Children appear to be highly motivated to learn the multiplication table, because the reconstructive strategies used for multiplication, such as repeated addition, are much more difficult than the reconstructive strategies used for addition. Even after basic facts have been committed to memory, problems differ in terms of the speed with which the associated answer can be retrieved and in terms of the chance of committing a retrieval error. Indeed, the same basic phenomena that have been found to be associated with addition-fact retrieval are also evident in multiplication-fact retrieval (Stazyk et al., 1982). As with addition, retrieval times and error rates increase as the size of the multiplicand and multiplier increases (the problem-size effect), although there are two exceptions to this finding. Specifically, answers for doubles and problems that include a 5 are retrieved rather quickly, independent of the value of the multiplicands (Siegler, 1988a; Stazyk et al., 1982). The problem-size effect in multiplication seems to result from differences in the frequency with which problems are presented in textbooks and by teachers and in the ease with which reconstructive strategies can be executed (Siegler, 1988b).

In keeping with findings in children's addition and subtraction, children's retrieval errors in multiplication tend to be very systematic (Campbell & Graham, 1985; Cooney et al., 1988) and fall into three general categories: *near misses, operation confusions,* and *table errors.* About half of all retrieval errors are table related (Campbell & Graham, 1985; Cooney et al., 1988; Siegler, 1988b). Table errors involve retrieving a multiplication fact that is incorrect for the presented problem but correct for problems with the same multiplicand or multiplier. For example, 24 is often retrieved for the problem 4 × 8, because both 4 and 8 are associated with the answer 24, as in 4 × 6 and 3 × 8 (Campbell & Graham, 1985). Operation confusions typically involve retrieving an answer that is correct for addition but incorrect for multiplication, such as retrieving 7 for the problem 3 × 4 (K. F. Miller & Paredes, 1990). Near misses occur less frequently than the table and operation-confusion errors and involve retrieving an answer that is 10% larger or smaller than the correct product (Siegler, 1988b).

There appear to be two general sources for these retrieval errors. The first involves confusing already-existing knowledge with the demands of the task at hand (K. F. Miller & Paredes, 1990). It is nicely illustrated by the operation-confusion errors. The second source of retrieval errors appears to be children's earlier errors in executing the reconstructive strategies (Siegler, 1988b). Early-strategy-execution errors account for many of the table and near-miss errors seen in children's retrieval of multiplication facts. As noted earlier, a common error associated with the repeated-addition strategy involves adding the multiplicand too many or too few times. For instance, to solve 4 × 5, a child might add 4 + 4 + 4 + 4, to get an answer of 16. If the child makes this same mistake on different occasions, then he or she will eventually retrieve 16 when presented with 4 × 5. Retrieving 16 for the problem 4 × 5 is, of course, a table error.

Complex Multiplication

As noted earlier, relatively little psychological research has been conducted on the strategies that are used to solve multicolumn multiplication problems, and much of the research that has been conducted has been with adults (Geary & Widaman, 1987, 1992; Geary et al., 1986). Nevertheless, as with children's addition, it is almost certain that children base their solution of complex multiplication problems, such as 46 × 9, on the strategies and knowledge acquired for simple multiplication. On the basis of the studies of adult performance, it appears that children eventually adopt formal algorithms for solving such problems. In the United States, a commonly taught algorithm involves solving the problems in the columnwise fashion, in much the same way that complex addition problems are solved. Indeed, Geary et al. (1986) found that American adults used nearly identical columnar algorithms for solving complex addition and complex multiplication problems.

Nevertheless, the columnwise algorithm is not the only way to solve such problems. So as not to become too culture-bound in our understanding of arithmetical problem solving, consider an alternative medieval method. In an intriguing work written in 952 and 953 A.D. (translated by Saidan, 1978), Al-Uqlidisi provided a comprehensive overview of the Arabic system of arithmetic. Among other things, the work describes a calculator for the blind as well as an Arabic method of multiplying: the *method of houses. House* in this context refers to a cell. The method of houses was apparently used before the now-common method of columnar multiplication was widely adopted. The use of this method is based on a spatial representation of place value. Figure 2.5 presents one such system, which, of course, can be expanded or contracted depending on the size of the multiplier and multiplicand. The display is read beginning with the left bottom cell—labeled *units* in the top portion of Figure 2.5— moving up each column and finally moving from left to right. During multiplication, each provisional product is placed in successive cells, beginning with the units cell; at the time of its use, provisional products were written in the sand. If the units cell is greater than 10, then the tens value is moved from the units cell to the tens cell above it. Once all of the provisional products have been completed, the values in the diagonal cells are added together. So the 2 tens cells are summed, then the hundreds cells, and so on. The sums represent the respective place values—that is, units, tens, and hundreds.

To better illustrate this method, consider the problem 36×72. The product of the first operation, 2×6, is written in the units cell, whereas the product of the next operation, 2×3, is written in the tens cell just above the units cell. Because the value in the units cell is greater than 10, the ten is shifted upward, leaving a 2 in the units cell and a 7 in the tens cell. We now shift to the next column of cells, to complete the multiplication of the tens-column information. The first product, 7×6, is written in the first cell of this column, and the second product, 7×3, is written in the second cell of this column. Next, the units value, 2, is brought down, the tens values are summed, $7 + 42$, and the hundreds value, 21, is brought down. Thus, there are 21 hundreds, 49 tens, and 2 units. The 4 from 49 is added to the 21, giving a total of 25 hundreds. Finally, the sequence is reversed, 25 9 2, giving the answer 2,592. The description of the method is much more laborious than the actual use of the method. In fact, with very little practice I found this method to be

Figure 2.5

Hundreds	Thousands	10 x Thousands
Tens	Hundreds	Thousands
Units	Tens	Hundreds

7 (6)	21	
2 (12)	42	
2	49	21

Schematic for using the method-of-houses approach to solving complex multiplication problems.

quicker than traditional columnar multiplication. Try it, and you will notice that the method reduces the working-memory demands of multiplying by making carrying unnecessary. Moreover, the basic units, tens, and hundreds value composition of the problem and answer is much more obvious with this method, as compared with traditional columnar multiplication.

Summary

In keeping with studies of children's addition and subtraction, children initially learn to multiply by relying on their knowledge of existing concepts and procedures (Siegler, 1983). In particular, initial multiplication strategies are based on the child's knowledge of addition and counting. These strategies include repeated addition and counting by n. Repeated addition involves representing the multiplicand a number of times represented by the multiplier and then successively adding these values; for example, adding $2 + 2 + 2$ to solve 2×3. The counting-by-n strategy is based on the child's ability to count by 2s, 3s, 5s, and so on. Here, a child might count 5, 10, 15, 20, to solve 5×4. Somewhat more sophisticated strategies involve the use of rules, such as $n \times 0 = 0$, and derived facts. Finally, most children appear to be able to retrieve most multiplication facts from long-term memory by the end of the elementary school years (K. F. Miller & Paredes, 1990). Little research has been conducted on children's strategies for solving complex multiplication problems. The research that has been conducted with adults, however, suggests that most children eventually adopt the formally taught columnwise procedure (Geary et al., 1986). The commonly used multiplication strategies are described in Table 2.3.

Finally, as with addition and subtraction, multiplication errors are systematic. These errors can result from the misexecution of a procedure, such as adding the multiplicand one too many or one too few times with the repeated-addition strategy, or are retrieval related. Multiplication-fact retrieval shows the same basic speed and accuracy pattern as is found in addition-fact retrieval (Stazyk et al., 1982). The larger the multiplicand and multiplier, the longer it takes to retrieve the fact from long-term memory and the more likely a retrieval error will occur. Retrieval errors most typically involve retrieving the answer to a related multiplication problem (Campbell & Graham, 1985), confusing addition for multiplication, or near misses. These retrieval errors, in turn, stem from the child's earlier-strategy-execution errors and from an interference from related knowledge about addition (K. F. Miller & Paredes, 1990; Siegler, 1988b).

Division

As compared with the three other arithmetical operations, even less research has been conducted on the psychological processes that govern

Table 2.3

Commonly Used Multiplication Strategies

Strategy	Description	Example
Simple Multiplication		
Repeated addition	The value of the multiplicand is written out a number of times indicated by the multiplier. These values are then added together to get the answer.	To solve 5×3, the problem is rerepresented as $5 + 5 + 5$. Next, the child adds $5 + 5 = 10$, and $10 + 5 = 15$.
Counting by n	The problem is solved by counting the value of the multiplicand or multiplier a number of times indicated by the other value.	To solve 5×3, the child counts, "5, 10, 15."
Rule	The answer is based on the following regularities in multiplication: $n \times 0 = 0$ and $n \times 1 = n$.	To solve 5×0, the child states 0, on the basis of the knowledge that $n \times 0 = 0$.
Derived facts (decomposition)	A related table fact is first retrieved from memory. The value of the multiplicand or multiplier is then added to this provisional product to get the answer.	To solve 5×6, the answer to 5×5 is retrieved. The remaining 5 is then added to this provisional product: $25 + 5 = 30$.
Fact retrieval	Direct retrieval of basic facts from long-term memory.	Retrieving 30 to solve 5×6.
Complex Multiplication		
Columnar retrieval	The problem is solved by retrieving columnwise products.	To solve 36×7: Step 1. $6 \times 7 = 42$ Step 2. Note trade (carry) value of 40 Step 3. $30 \times 7 = 210$ Step 4. $210 + 40$ (from trade) $= 250$ Step 5. $250 + 2 = 252$

children's solutions of division problems. Nevertheless, it is almost certain that many of the same regularities that have been found across addition, subtraction, and multiplication will also be found for children's division. Among these regularities are the tendency to use preexisting knowledge of related concepts when first learning a new concept (Siegler, 1983) and the tendency for problem-solving errors to be systematic (Ginsburg, 1977). Indeed, the research that has been conducted in children's division is consistent with this argument (Hughes, 1986; Ilg & Ames, 1951; Vergnaud, 1983).

Apparently, when first learning to solve division problems, children rely on their knowledge of addition and multiplication. The first of two classes of strategies that are used for solving division problems is based on the child's knowledge of multiplication (Ilg & Ames, 1951; Vergnaud, 1983). For example, the solving of 20/4 (20 is the *dividend*, and 4 is the *divisor*) is based on the child's knowledge that $5 \times 4 = 20$. To be consistent with terms used with the other arithmetic operations, we might call this strategy *multiplication reference*. For children who have not yet mastered the multiplication table, a derivative of this strategy is sometimes used. Here, the child multiplies the divisor by a succession of numbers until she or he finds the combination that equals the dividend. To solve 20/4, the sequence for this strategy might be $4 \times 2 = 8, 4 \times 3 = 12, 4 \times 4 = 16, 4 \times 5 = 20$. The second class of strategies described by Ilg and Ames and Vergnaud is based on the child's knowledge of addition. The first of these strategies involves a form of repeated addition. For example, to solve 20/4, the child would produce the sequence $4 + 4 + 4 + 4 + 4 = 20$, and then count the 4s. The number of counted 4s represents the quotient. Sometimes children try to solve division problems directly by means of their knowledge of addition facts. For instance, to solve the problem 12/2, the child would base her answer on the knowledge that $6 + 6 = 12$.

The use of multiplication-based and addition-based strategies to solve division problems suggests a rudimentary understanding that division involves determining the number of subsets of a smaller number, the divisor, that are contained within a larger number, the dividend. Even with this basic understanding, children's division performance declines when the problems are presented within even very simple word problems, such as "If Sue has $10 and wants to buy some books that cost $2 each, then how many books can she buy?" Even children who can easily solve 10/2, sometimes confuse the divisor and dividend in these problems, setting up the problem as 2/10 rather than 10/2 (Hughes, 1986). Other types of errors that occur with simple division involve retrieving the wrong multiplication fact, when using the multiplication-reference strategy, or having one too many or one too few divisors represented with the repeated-addition strategy.

In all, children's strategies and errors in simple division appear to follow the same general trends found for the other arithmetical operations. These trends include initial solution strategies that are based on other forms of arithmetical knowledge and errors that result from the misapplication of this knowledge. Nevertheless, our understanding of children's strategic and conceptual development for the domain of division is extremely rudimentary, almost nonexistent in fact. Normative studies on children's strategy and conceptual development, as well as on error patterns, in division are clearly needed.

Developmental Models

In this section, two general theoretical models of children's arithmetical development are presented. Both approaches seek to specify the cognitive changes associated with children's arithmetical development and how these changes relate to the observable behavior of the child, such as the child's problem-solving strategies and conceptual justifications about this behavior. The approaches differ, however, in the types of knowledge—and therefore the types of experiences—deemed most important for arithmetical development. The first of these approaches is represented by the *strategy-choice model,* which was developed by Siegler and his colleagues (Siegler, 1986; Siegler & Shrager, 1984). In preview, the strategy-choice model emphasizes children's overt strategy use and the factors that govern the child's choices among alternative strategies. The second model is based most generally on Piaget's (1950) work and focuses on the child's development of an interrelated framework of arithmetic knowledge. This latter emphasis is called the *schema-based model* (Anderson, 1984; Baroody, 1992; Steffe, 1992) and is described after the presentation of the strategy-choice model.

Strategy-Choice Model

Siegler (1986) developed the strategy-choice model to explain the fact that people have available to them many different ways to solve problems and tend to choose the best strategy available for solving each specific problem. The term *best* means that people tend to use the fastest and most accurate strategy available to them when presented with a particular problem to solve. For instance, young children tend to use the count-on-from-the-larger procedure to solve problems with one small and one large addend, such as 9 + 2, but the count-all procedure for problems with addends of about the same magnitude, such as 4 + 5 (Geary, 1990; Siegler & Jenkins, 1989). For such problems, the difficulties associated with determining which is the larger number and keeping track of the counting from the appropriate value (i.e., 5) makes the counting-all procedure more accurate, although slower, than the counting-on procedure for some children.

The tendency to make adaptive strategy choices is found for lower-middle- and upper-middle-class children (Siegler, 1993), in young and elderly adults (Geary, Frensch, & Wiley, 1993; Geary & Wiley, 1991), and in groups of Chinese as well as American children (Geary, Fan, & Bow-Thomas, 1992). In fact, even though the Chinese first-grade children assessed in this study outperformed their American peers on just about all measures of addition skills, the groups did not differ in the tendency to choose adaptively among the strategies available to them. Adaptive strategy choices are also evident for children's spelling, reading, time telling, and reasoning (Siegler, 1986, 1988b; Siegler & Taraban, 1986) and are strongly correlated with performance on traditional achievement and ability measures (see the Cognitive Perspective section of chapter 4; Geary & Burlingham-Dubree, 1989).

The strategy-choice model includes three general mechanisms that influence both children's strategy choices and changes in the mix of existing strategies: *memory representations, processes that act on these representations,* and a *learning mechanism* (Siegler & Jenkins, 1989). Memory representations include associations between problems, such as 1 + 2, and potential answers, such as 3, as well as representations between specific problems and different methods for solving the problem, such as counting or retrieval. These associations vary in strength. For instance, consider the first panel shown in Figure 2.6. The information shown in the figure was constructed from the retrieval efforts of preschool and kindergarten children for solving simple addition problems (Geary & Burlingham-Dubree, 1989). Associative strength represents the proportion of times these children retrieved the associated answer (the confidence criterion is discussed below). Returning to the first panel, we see that the answer of 3 was retrieved about 80% of the time for the problem 1 + 2, whereas the answers of 2 and 4 were each retrieved about 10% of the time. Thus,

Figure 2.6

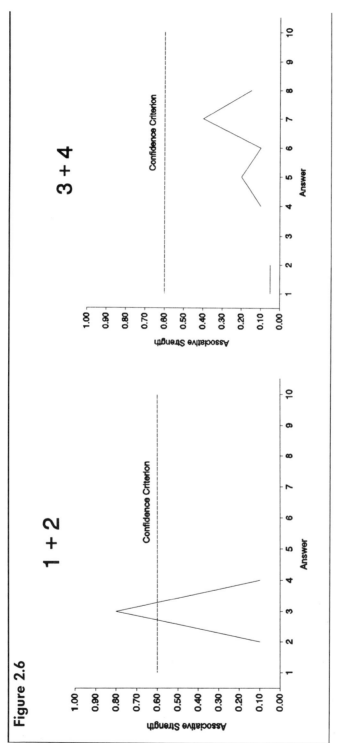

Associative strength and confidence criterion values for two simple addition problems.

when preschool and kindergarten children are presented with the problem 1 + 2, the number 3 is retrieved almost all of the time; 3 is retrieved, in part, because it is the counting-string associate of 1, 2 (Siegler & Shrager, 1984).

For the problem 3 + 4, on the other hand, the correct answer, 7, was retrieved only about 40% of the time, and the counting-string associate, 5, was retrieved about 20% of the time. Because the strength of the association between 1 + 2 and its correct answer is much stronger than the association between 3 + 4 and its correct answer, retrieval is a much more adaptive choice for solving 1 + 2 than for solving 3 + 4. Clearly, children sometimes retrieve the answer to problems such as 3 + 4, but these answers are often wrong. This is where the confidence criterion influences problem solving. The confidence criterion represents an internal standard for gauging one's confidence in the accuracy of the retrieved answer. Children with a high confidence criterion state only answers that they are very sure are accurate, whereas those with a low confidence criterion state just about anything that comes to mind, accurate or not (Siegler, 1988a). A retrieved answer is only stated if the associative strength between the answer and the problem exceeds the value of the child's confidence criterion. If the retrieved answer is not stated, then the problem has to be solved by means of some other strategy, such as counting. The confidence criterion shown in Figure 2.6 would lead to most children's retrieving and stating 3 for the problem 1 + 2 but not stating any answer that might be retrieved, even 7, for the problem 3 + 4. This is because the child would be too uncertain, in relation to his or her confidence criterion, about the accuracy of any answer that might be retrieved for 3 + 4.

During problem solving, the child not only retrieves potential answers but also retrieves procedures that have been used in the past to solve the problem or similar problems (Ashcraft, 1987; Siegler & Jenkins, 1989). Thus, if the child typically counts on to solve 3 + 4, then the counting-on procedure and potential answers will be retrieved from memory simultaneously and will compete for expression in a horse race fashion. One child might retrieve and state 7, whereas another child might automatically begin to count before any potential answer comes to mind. For arithmetical problem solving, then, one important determinant of whether a child will use one strategy or another depends on the strength of the association between the problem and potential answer, combined with the child's confidence criterion, as well as the strength of the association between the problem and procedures that have been used in the past for solving the presented problem or similar problems.

All of these memory representations develop through the *process* of problem solving (Siegler & Jenkins, 1989). For instance, the first time a child counts "4, 5, 6, 7" to solve the problem 3 + 4, an associative representation between the problem and the generated answer is developed, and the scheme governing the use of this counting-on procedure also becomes associated with the problem. Each time the child correctly counts to solve 3 + 4, the associative strength between the problem and the answer increases. Eventually, the child will quickly retrieve 7 when presented with the problem 3 + 4, before the scheme for counting on can be activated and used. At this point, the child no longer counts to solve the problem, because the correct answer is quickly and accurately retrieved. Note that if the child makes a lot of counting errors, then she or he will begin to retrieve wrong answers. Either way, the use of counting and other types of reconstructive procedures results in their own extinction, because their use leads to the learning and retrieval of basic facts (Siegler, 1989b).

From this scenario, we can see that the *learning mechanism* involves a combination of procedural use and the development of memory representations. The child learns basic facts through the execution of counting and other procedures and also learns about the efficiency and accuracy of these procedures. Adaptive strategy choices arise not from conscious choice, but rather from the pattern of memory representations the child develops between specific problems and answers and procedures associated with these problems, along with the confidence criterion. Adaptive strategy choices are a natural, unconscious result of these memory representations. Although the child's eventual reliance on memory retrieval to solve most arithmetic problems follows easily from this learning mechanism, this mechanism cannot explain how children discover novel strategies. This is a problem for the model, because it is clear that most children abandon the counting-all procedure, for instance, in favor of the more efficient counting-on procedure without formal instruction (Groen & Resnick, 1977).

Indeed, Siegler and his colleagues (Siegler & Crowley, 1991; Siegler & Jenkins, 1989) have argued that the strategy-choice model needs to be expanded so as to include mechanisms that allow for strategy discovery. At this point, it is not clear exactly how children discover new strategies, although it is almost certain that the construction of novel strategies results from the child's goal-directed behavior and from related knowledge (Siegler & Crowley, in press). For example, the construction of the counting-on procedure for solving addition problems is constrained by the child's general knowledge of addition, that both numbers have to be combined and—for example—that to achieve the goal of adding one can count forward but not backward. Moreover, newly discovered strategies do not appear out of a vacuum but rather are constructed from bits and pieces of existing procedures (Siegler & Jenkins, 1989). The counting-on procedure is constructed from the counting-all procedure and the child's ability to start counting from numbers other than 1. The child's understanding that stating 4, for the problem 3 + 4, is a shortcut to counting "1, 2, 3, 4" also seems to be an important precursor to the discovery of the counting-on procedure (Fuson, 1982).

Summary

For the domain of arithmetic, the goal of the strategy-choice model is to explain why and how children make adaptive strategy choices during problem solving, why development results in a greater and greater reliance on direct retrieval, and how children discover new problem-solving strategies. The emphasis is on children's observable strategies and on the cognitive processes that govern when one strategy or another strategy is used for problem solving. The adaptive nature of children's strategy choices, according to Siegler (1986), is not the result of conscious decisions, but rather flows naturally from the pattern of associations of problems with answers and procedures associated with these problems. The execution of reconstructive strategies, such as counting, results in the child's learning about these strategies and learning basic arithmetic facts. Eventually, basic facts are retrieved quickly and accurately, which makes the use of reconstructive strategies unnecessary. Finally, the discovery and construction of novel problem-solving strategies are based on the child's problem-solving goals, as well as bits and pieces taken from already-existing procedures. These bits and pieces are reconstructed into a new strategy, which is based, in part, on the child's knowledge of related concepts.

Schema-Based Model

A schema represents a coordinated body of knowledge and mental operations (and sometimes actions, as in pointing and counting; Piaget, 1950) and provides the structure for the child's observable behavior (e.g., strategy choices) and justifications about that behavior. The schema-based approach to arithmetical and mathematical development rests on very different assumptions about knowledge acquisition than does Siegler's and other information-processing researchers' approach to arithmetical development (e.g., Ashcraft, 1992). In brief, the basic assumption is that arithmetical knowledge is constructed by the individual child through her or his social discussions about arithmetic and related topics (e.g., Belmont, 1989; Steffe & Wood, 1990). Moreover, these arithmetical schemas are seen as more or less unique to each individual. For instance, two children who are presented with an identical lecture on addition will, because of differences in their current arithmetic knowledge, hear the teacher say different things about addition and will come to develop a somewhat different understanding of the presented material (Steffe, 1992). Observable behaviors, such as strategy choices, are not especially important in and of themselves, but rather are useful only to the extent that they provide a window that allows the teacher or researcher to glimpse the child's developing arithmetical schemas.

The grist of development is not specific cognitive skills, such as counting, or specific cognitive representations, such as problem-answer associations, but rather the child's abstract conceptual understanding of arithmetic. This is not to say that cognitive skills are not important but rather that these skills will emerge as the child develops appropriate abstract schemas (Baroody, 1992; Case & Sowder, 1990). With the strategy-choice model, on the other hand, the child's knowledge of arithmetic develops through the execution of procedures, not the other way around. To better illustrate this issue, consider my 7-year-old daughter Corie's schema of addition facts and procedures. To get an understanding of her addition knowledge, I followed the general strategy used by Chi and Koeske (1983). During five 20-min sessions over the course of about a 2-week period, I obtained information on the simple addition facts that she knew, as well as her knowledge of addition procedures, by asking her to tell me all of the addition facts that she could think of, as they came to mind, and all of the different ways to solve simple addition problems. The addition facts that Corie recalled were very consistent across sessions and clustered into groups of double problems (e.g., $3 + 3 = 6$, or $30 + 30 = 60$) and nondouble problems ($1 + 2 = 3$). In keeping with research in this area, across all five sessions, clusters of doubles were recalled first and with relatively short pauses between problems. Clusters of nondouble problems were less frequent and tended to differ across sessions.

More germane to the current topic are Corie's descriptions of the "different ways to solve simple addition problems." I had asked her to describe ways to solve simple problems only, such as $5 + 4$, but she

Table 2.4

Corie's Knowledge of Addition Procedures

Procedure	Number of Sessions	Comments
	Simple Addition	
Remember	5	"Easy." "Get to know it by counting over and over again." "Do it over and over again and you know it stronger."
Counting	5	"For 5 + 5, you can count 6, 7, 8, 9, 10 or 1, 2, 3, . . . 10." "The first way is easier. The second way you could get the wrong answer." "Usually mess up."
Derived facts	3	She couldn't label this strategy but demonstrated it. "7 + 5 = 5 + 5 = 10, and 10 + 2 = 12." "Easier than counting." "Easy to remember doubles."
Subtraction	2	"5 + 5 = 10, because 10 − 5 = 5."
Counting objects	2	She represented a problem by tally marks and then counted them.
Guess	1	"If you don't know, you can just guess."
Word problems	1	"Like there were 5 pigs and 2 more came, how many were there altogether?"
Write it out	1	She wrote out 6 + 5 on paper, retrieved the answer, and then wrote it down.
	Complex Addition	
Columns	4	"50 + 87 = 137." "The 1 equals hundreds, the 3 equals tens, and the 7 equals ones."
Trades	2	"Ten units makes a trade."

occasionally provided information on ways to solve more complex problems, such as 50 + 87. Her descriptions of these procedures and the number of sessions in which she described them are presented in Table 2.4. Across sessions, Corie described eight different methods that could be used to solve simple addition problems. *Remember* and *counting* were described in all five sessions; *guessing*, *word problems*, and *write it out* were described only once. Examination of some of the comments that she provided during these sessions showed that she had a rather insightful understanding of addition procedures and their interrelationships. She knew which procedures were easy (*remember*) or hard (*counting all*), which factors influenced her knowledge ("Do it [*remember*] over and over again and you know it stronger"), and how procedures and knowledge interact ("Get to know it [*remember*] by counting over and over again").

The schema-based position holds that abstract knowledge of arithmetic procedures, as reflected in Corie's descriptions, is the basis for her problem-solving skills. That is, she uses this knowledge to construct, for example, the derived-facts strategy. Thus, arithmetical development involves, for the most part, a greater and greater elaboration of these schemas. The schemas, in turn, are elaborated primarily through the child's assimilation and accommodation of related knowledge during interactions in social groups (e.g., group discussions of arithmetic concepts). The alternative view is that much of Corie's arithmetic knowledge has been induced through her noticing how different procedures, such as counting, influence quantity. With this alternative view, strategy choices occur more or less automatically, on the basis of the pattern of memory representations for facts and procedures (Siegler, 1986), and conceptual knowledge is constructed as the child notices how the procedures work and their outcomes. In fact, it is more likely that both types of factors, schema driven and memory driven, influence arithmetical development. For instance, Geary, Bow-Thomas, and Yao (1992) found that the adoption of the counting-on procedure in favor of the counting-all procedure for simple addition was dependent on the child's counting knowledge. Counting procedures might be executed automatically, but their modification appears to be dependent on, at least to some extent, counting schemas (Ohlsson & Rees, 1991).

On the other hand, the position that children's mathematical schemas are unique to each child and therefore can never be fully understood by others is overstated (Steffe, 1992). The finding that humans have an innate sense of number and basic arithmetic argues that there is a common ground for understanding each other's number and arithmetic schemas. Moreover, with formal education, children's arithmetic skills (e.g., types of strategies) become increasingly similar across cultures (e.g., Ginsburg, 1982; Saxe, 1985)—which suggests that formal education leads to the construction of very similar conceptual schemas for children throughout the world. This is not to say that we should not attempt to understand how individual children understand arithmetic. We should. The argument is, there are going to be more similarities than differences in children's developing arithmetic knowledge. Nevertheless, one additional feature of this approach does merit further consideration: social and contextual influences on arithmetical learning (Greeno, 1993). It is very likely that one of the more natural methods for learning is through social discourse (Eibl-Eibesfeldt, 1989; Geary, in press). In fact, much of children's early arithmetical development occurs through the course of social interactions in natural settings (e.g., Saxe et al., 1987), assuming, of course, that the topic of conversation is numerical.

It does not necessarily follow, however, that all arithmetical learning can and does occur in a social context, nor should this approach lead to a de-emphasis on teaching procedural skills and on practice (see Educational Philosophy section of chapter 8). Researchers in this area often make the assumption that procedural skills are learned without conceptual knowledge and that memorized facts are "inert": essentially useless knowledge. Children, however, develop arithmetic schemas, at least in part, by inducing how procedures work and what procedures do. These inductions occur during the course of practice. Granted, memorized facts are not conceptual knowledge. Nevertheless, the ability to automatically retrieve basic facts from memory makes the solving of more complex word problems and reasoning tasks less effortful and quicker (Geary & Burlingham-Dubree, 1989; Geary & Widaman, 1992). For example, consider this very simple word problem: "Bill has three apples, John has two apples, and Karen has four apples. How many do they have altogether?" The ability to retrieve the answer to 3 + 2 makes solving the next step (adding 4) much less cumbersome and error-prone than if the child had to count to solve 3 + 2. In all, arithmetical development needs to proceed on at least two general and interrelated dimensions. The first involves the development of procedures and memory representations so that they are executed and retrieved automatically by the child. The second involves a conceptual understanding of these procedures (i.e., the associated schema) and more general problem-solving schemas, as well as the child's understanding of the contexts within which the schemas are most appropriately used.

Summary

The schema-based approach to children's arithmetical development focuses on the child's conceptual understanding of arithmetic rather than on directly observable strategies and procedures. The web of conceptual knowledge is the schema. From this perspective, the development of schemas is the result of the dynamics of social discourse. Children come to spin their own webs of arithmetical knowledge through active exploration and discussion of mathematical materials. The observable skills of the child, such as specific problem-solving strategies, emerge from these schemas. The focus on arithmetical development within a social context adds another dimension to our understanding of children's cognitive growth. Nevertheless, the focus on the social construction of arithmetical knowledge at the expense of more mechanical skills (e.g., learning to use procedures or memorize basic facts) will probably be detrimental to children in the long run. This is not to say that the schema-based approach is not useful. It is. I am arguing that a more balanced approach to children's arithmetical and mathematical development is needed, an approach that acknowledges the importance of procedural as well as conceptual knowledge (see Geary, in press).

Conclusion

Arithmetic is an activity that is seen in cultures throughout the world (Crump, 1990; Ginsburg, 1982; Saxe, 1982b; Zaslavsky, 1973). More important, the general features of early arithmetic calculation and development appear to be universal, at least in terms of their deep structure (Starkey et al., 1991). Children throughout the world initially perform arithmetic calculations through the use of their culture's counting system and show similar development features, such as an initial reliance on body parts—such as fingers—to aid in counting. The findings that infants and young preschool children have an implicit sense that addition increases quantity and that subtraction decreases quantity strongly support the argument that arithmetic is a natural domain (Starkey, 1992; Wynn, 1992a). However, this early intuitive understanding of arithmetic appears to be limited to quantities of 2 (as in $1 + 1$) for infants and to quantities of less than 4 (e.g., $3 - 1$) for preschool children. Whether this basic intuitive understanding of arithmetic gradually increases with age or whether the child's understanding of arithmetic with larger numbers is dependent on an understanding of the meaning of conventional number words (Fuson, 1988) remains to be seen. Either way, this early implicit knowledge of arithmetic almost certainly provides the seed for later arithmetical development—in particular, the structure around which the child's early use of counting as an arithmetic tool develops.

Culture also greatly influences the child's developing arithmetical knowledge and skill. Some cultural influences are subtle but pervasive. These influences would include how supportive the culture's language is for number and arithmetical development (Fuson & Kwon, 1991; Saxe, 1982b), as well as the nature and extent of early parent-child activities that are centered around numerical tasks (Saxe et al., 1987; Stevenson, Lee, Chen, Stigler, et al., 1990). Many of the early cultural differences in the expression of arithmetical knowledge, such as counting on fingers or counting body parts as the Oksapmin do, represent relatively unimportant differences in the surface structure of number and arithmetic knowledge. Nevertheless, these arguments should in no way be taken to mean that all basic arithmetic knowledge is natural (Geary, in press). Studies of children's understanding of fractions clearly illustrate that there is little inherent understanding of this important domain of arithmetic (Clements & Del Campo, 1990). In fact, it is very possible that the implicit understanding of addition and subtraction is limited to small quantities and that arithmetic knowledge for larger quantities develops more slowly— as the child learns the culture's number words, associates these words with sets of objects (e.g., associating *five* with all of the fingers on one hand), and then learns to manipulate these quantities through arithmetic.

Even with inherent influences on arithmetic knowledge, the strongest influence on arithmetical development is formal education (Ginsburg et al., 1981b). The studies of Ginsburg et al. and Saxe (1982a) have shown that formal instruction leads to the development of skills that almost certainly would not have emerged in a more natural environment, as well as to a convergence in the surface structure of arithmetic skills across cultures. To a large extent, differences in educational practices and in general expectations for academic achievement appear to underlie cultural variations in the rate of arithmetic learning and in the ultimate level of developed skill (see chapter 7), although language differences for number words seem to also, though probably to a lesser extent, influence cultural variations in arithmetical development (Fuson & Kwon, 1992a, 1992b). Despite cultural differences in rate and level of arithmetical development, there are many similarities in schoolchildren's arithmetic skills.

Throughout the world, it appears that arithmetical development does not occur in a linear, step-by-step fashion, but rather is more like a wave (Siegler & Jenkins, 1989). For instance, children do not first solve addition problems exclusively by means of finger counting, then through verbal counting, and finally through fact retrieval. Rather, children have available to them a mix of problem-solving strategies (Siegler, 1986). Across arithmetic operations, "development involves changes in the mix of existing strategies as well as construction of new ones and abandonment of old ones" (Siegler & Jenkins, 1989, p. 27). The crest of the wave that represents this mix of strategies gradually changes from a general reliance on the use of manipulatives and fingers, to verbal counting, and finally to retrieval-based strategies. Arithmetical development also involves changes in the child's schemas about numbers and arithmetic—which, in turn, influence the child's observable skills (e.g., problem-solving strategies; Steffe, 1992).

There is much that remains to be learned about children's arithmetical development. Some of the issues that must be addressed include a better understanding of the limits of infants' and children's intuitive knowledge of number and arithmetic and how and to what extent this knowledge influences later development. We must come to understand more fully the relationship between the child's counting knowledge and early arithmetic knowledge and skill (Geary, Bow-Thomas, & Yao, 1992). It is essential to understand how, and the extent to which, early parent-child interactions (e.g., arithmetic games) and other ecological factors influence arithmetic knowledge (Greeno, 1993; Saxe et al., 1987). It is equally important to know the limits of the influence of social discourse on early mathematical learning. Finally, we have to realize that arithmetical development proceeds on many dimensions—including observable problem-solving strategies, underlying memory representations, and abstract schema-based knowledge. It is imperative that all of these dimensions be recognized as not only important but essential. We need to understand better how these different types of skills interact and what mix of instructional methods and ecological experiences is needed for the optimal development of each of these dimensions of arithmetic skill (see Geary, in press).

3

Learning Mathematical Problem Solving

Learning to problem solve is an essential feature of children's mathematical development, but unfortunately it is a skill that most children in the United States apparently do not master (Mullis, Dossey, Owen, & Phillips, 1991). The focus of this chapter is on the processes underlying children's mathematical problem solving and the factors that appear to influence the development of problem-solving skills. For the most part, the strategies described in chapter 2 represent only the last phase of problem solving and only the strategies that are associated with solving relatively simple problems. In this chapter, we consider more fully the processes that *precede* the actual execution of a problem-solving strategy, in particular how these processes are manifested during the solving of more complex word problems. These issues are considered in two basic sections: The first section presents developmental research on children's ability to solve arithmetic word problems, whereas the second presents research on the skills needed to solve algebraic word problems. The chapter closes with a summary of the basic findings in these two areas and a discussion of the relationship between the basic skills described in chapter 2 and those introduced herein.

Arithmetic Word Problems

Arithmetic word problems are introduced in the United States during the first grade, although many children can solve simple word problems

---

Summaries are provided at the end of each of the main sections: Arithmetic Word Problems (p. 115) and Algebra (p. 127).

before this point (Stigler, Fuson, Ham, & Kim, 1986). As was found for basic arithmetic (Fuson et al., 1988), national differences exist in the frequency and grade with which various types of word problems are introduced to children. Stigler et al., for example, found that textbooks in the United States tend to present much easier problems than do comparable textbooks in the former Soviet Union. Moreover, many of the more difficult types of word problems (described below) are introduced in first grade in the former Soviet Union, but if they are introduced at all in American textbooks, it is often not until third grade. Even when complex problems are presented in American textbooks, they are not presented frequently enough to allow American children to become skilled at solving such problems. The implication of this and similar analyses (e.g., Fuson et al., 1988) is that the development of the problem-solving skills of American children most likely lags behind that of children in many other nations. In other words, because much of the research described in this section is based on American children, the extent to which the same developmental patterns or sources of problem-solving difficulties will be evident with children in other cultures cannot be stated with any certainty.

Arithmetic word problems are thought to represent an important bridge between the child's developing computational skills, described in chapter 2, and the application of these skills in real-world contexts (Briars & Larkin, 1984). It is therefore important to understand how children develop problem-solving skills and to identify the sources of problem-solving difficulty. Research on the solution of word problems in and of themselves is important because children make more errors when solving word problems than when solving comparable number problems (Carpenter, Kepner, Corbitt, Lindquist, & Reys, 1980). For instance, children can often solve the problem, "How much is 3 + 2?" more accurately than a comparable word problem: "Mary had three candies. Amy had two candies. How many candies did they have altogether?"

Early research on children's solving of arithmetic word problems focused on identifying the features of the problems themselves that made word problems more difficult than number problems (Jerman & Rees, 1972). More recent studies, on the other hand, focus on how the child understands and conceptually represents word problems and how this affects the child's problem-solving processes. The overall goal is to better understand how problem features interact with the child's knowledge in determining problem-solving accuracy and the sources of skill development. The discussion of children's skills in solving arithmetic word problems is presented in four sections. In the first section, the influence of general structural features of the problem, such as its length, on problem-solving accuracy is presented. In the second section, the influence of the semantic structure of the problem on the types of strategies that children use for problem solving is discussed. The effect of conceptual knowledge on problem solving is presented in the third section. A discussion of developmental considerations and sources of children's problem-solving errors is presented in the final section.

Problem Features

Jerman and his colleagues conducted a series of exhaustive studies of the relationship between structural features of arithmetic word problems and problem-solving accuracy (Jerman & Mirman, 1974; Jerman & Rees, 1972). The features they examined were extensive and included factors such as average word length; number of arithmetic operations in the problem; number of sentences in the problem; average number of words in each sentence; and frequency of nouns, verbs, and conjunctions. In the largest of these studies, Jerman and Mirman examined 73 linguistic features of word problems as they were related to problem-solving accuracy for groups of fourth- to ninth-grade children. In short, the studies were designed to determine which features of the problems influenced whether a problem was relatively difficult or relatively easy to solve.

Overall, Jerman found that several features of the problem accounted for a substantial proportion of the variability in problem difficulty (i.e., accuracy rates). Jerman and Rees (1972), for instance, found that a combination of linguistic features (e.g., the number of words in the problem) and computational demands (e.g., the number of required arithmetic operations) accounted for nearly 87% of the variability in problem difficulty for a group of fifth graders. Difficult problems were relatively long and required multiple arithmetic operations. Note that working-memory demands were probably important contributors to the length effect found in this study. Jerman and Mirman (1974) found that different features predicted problem difficulty for elementary school (Grades 4 to 6) and junior high school (Grades 7 to 9) children. For elementary school children, the most important factors that contributed to problem difficulty were computational demands and the number of mathematical terms, such as *diameter* or *remainder*, in the problem. The most difficult computational features of the problems were the presence of trades and the presence of more than one arithmetic operation.

Neither the computational nor the math vocabulary variables proved to be important factors in predicting problem difficulty for the junior high school students. Several linguistic features of the problems, however, did influence problem difficult for these children. In particular, the number of words between the first number presented in the problem and the last number presented in the problem accounted for nearly 30% of the variability in problem difficulty. The significance of this variable suggests that the working-memory demands of the problem might have been the primary determinants of problem difficulty for junior high school children. As the number of words between the numerical features of the problem increases, the likelihood of forgetting the first number before reading the second number also increases (Kintsch & Greeno, 1985). In all, these studies show that certain features of arithmetic word problems play an important role in the relative difficulty of these problems. During the elementary school years, the child's basic computational skills and math vocabulary, along with the memory demands of the problem, appear to be important determinants of the child's success in solving arithmetic word problems. As computational skill and vocabulary improve, other problem features such as memory demands become more salient influences on the child's ability to solve arithmetic word problems.

Semantic Structure

In addition to problem features, such as length, the semantic structure of arithmetic word problems also influences the child's ability to solve these problems (Carpenter, Hiebert, & Moser, 1981; Carpenter & Moser, 1983; De Corte & Verschaffel, 1987; Grouws, 1972; Riley, Greeno, & Heller, 1983). *Semantic structure* refers to the meaning of the statements in the problem and their interrelationships. Riley et al. and Carpenter and Moser have classified addition and subtraction word problems on the basis of semantic structure. Most word problems can be classified into four general categories: *change, combine, compare,* and *equalize.* Examples of these different types of arithmetic word problems are presented in Exhibit 3.1 for addition and subtraction. Comparable analyses for multiplication and division problems suggest similar general categories (e.g., Hardiman & Mestre, 1989; A. B. Lewis, 1989; A. B. Lewis & Mayer, 1987). The focus, however, is on addition and subtraction, because most of the research on children's skills has been conducted using addition and subtraction problems, not multiplication and division problems.

Change problems imply that some type of action is performed by

Exhibit 3.1

Classification of Arithmetic Word Problems

Change

1. Amy had two candies. Mary gave her three more candies. How many candies does Amy have now?
2. Amy had five candies. Then she gave three candies to Mary. How many candies does Amy have now?
3. Amy had two candies. Mary gave her some more candies. Now Amy has five candies. How many candies did Mary give her?
4. Mary had some candies. Then she gave two candies to Amy. Now Mary has three candies. How many candies did Mary have in the beginning?

Combine

1. Amy has two candies. Mary has three candies. How many candies do they have altogether?
2. Amy has five candies. Three are chocolate stars and the rest are chocolate hearts. How many chocolate hearts does Amy have?

Compare

1. Mary has three candies. Amy has two candies. How many fewer candies does Amy have than Mary?
2. Mary has five candies. Amy has two candies. How many more candies does Mary have than Amy?
3. Amy has two candies. Mary has one more candy than Amy. How many candies does Mary have?
4. Amy has two candies. She has one candy less than Mary. How many candies does Mary have?

Equalize

1. Mary has five candies. Amy has two candies. How many candies does Amy have to buy to have as many candies as Mary?
2. Mary has five candies. Amy has two candies. How many candies does Mary have to eat to have as many candies as Amy?
3. Mary has five candies. If she eats three candies, then she will have as many candies as Amy. How many candies does Amy have?
4. Amy has two candies. If she buys one more candy, then she will have the same number of candies as Mary. How many candies does Mary have?
5. Amy has two candies. If Mary eats one of her candies, then she will have the same number of candies as Amy. How many candies does Mary have?

the child. For instance, most kindergarten and first-grade children can easily solve the first type of change problem shown in Exhibit 3.1. Children typically represent the meaning of the first term, "Amy had two candies," through the use of two blocks. The meaning of the next term, "Mary gave her three more candies," is represented by three blocks. Some children then answer the "how many" question by literally moving the two sets of blocks together and then counting them (e.g., Carpenter & Moser, 1984; Riley & Greeno, 1988). This example reflects what Carpenter and Moser termed a *change-join problem*, in that the change involves adding to the initially defined set (Amy's two candies). There are also *change-separate problems*, in which the action involves removing a subset from the originally defined set. The second example presented under the Change category in Exhibit 3.1 illustrates this type of problem.

The change problems and the combine problems are conceptually different, even though the basic arithmetic involved in solving these problems is the same (Briars & Larkin, 1984; Carpenter & Moser, 1983; Riley et al., 1983). Combine problems involve a static relationship rather than the implied action found with change problems. For instance, consider the first examples under both the Change and Combine categories presented in Exhibit 3.1. Both problems require the child to add 2 + 3. In the change problem, the quantities in Amy's and Mary's sets differ after the action (i.e., addition) has been performed. In the combine example, however, the status of Amy's set of candies and Mary's set of candies is not altered by the addition. The combination does not change what each individual has. Rather, the operation involves developing a *superordinate*, or more general, set that represents the total quantity in the two subsets. From the child's perspective, these are very different scenarios. For the change problem, the wealth—so to speak—of Amy increases whereas the wealth of Mary decreases. No such change in wealth occurs with the combine problem. Thus, although the change and combine problems are exactly the same in terms of computational demands, differences in the language within which the problems are presented give them very different meanings to children. These differences in meaning can influence how children represent and interpret the problems, which in turn can influence the child's conceptual understanding of what is being asked as well as the types of strategies used to solve the problem (De Corte & Verschaffel, 1987). The second type of combine problem shown in Exhibit 3.1 involves the initial presentation of the superordinate set and the value of one of the subsets. The child must then determine the quantity of the other subset.

Compare problems involve the same types of static relationships as are evident in combine problems. The quantity of the sets does not change. Rather, the arithmetic operation results in determining the exact quantity of one of the sets by reference to the other set, as shown with the third and fourth compare examples in Exhibit 3.1. The first and second compare examples, on the other hand, involve a more straightforward *greater than/less than* comparison. As described in the Developmental Patterns section, the compare problems shown in Exhibit 3.1 tend to be relatively difficult for most children to solve. However, compare problems that help the child to develop a mental model of the comparison appear to be easier to solve than other forms of compare problems. For instance, Hudson (1980; cited in Riley et al., 1983) showed that even kindergarten children can easily solve compare problems in the following form: "There are five birds and three worms. If the birds run over to get the worms, how many birds won't get a worm?" This form of problem is often referred to as a *won't get* compare problem.

Equalize problems have not been as extensively studied as the other categories of problems. Nevertheless, examination of the equalize problems presented in Exhibit 3.1 reveals that they are conceptually similar to change problems. The action, or arithmetic operation, that is performed results in a change in the quantity of one of the sets. The change is constrained, so the result is that both sets are equal once the action has been completed, whereas there is no such constraint with change problems.

Research on the relationship between the semantic structure of word problems and children's solving of these problems has been conducted on two levels. The work of Carpenter and Moser (e.g., Carpenter & Moser, 1983, 1984) and others (e.g., De Corte & Verschaffel, 1987) has focused on the influence of the problem's semantic structure on observable problem-solving strategies. The work of Riley and Greeno (Kintsch & Greeno, 1985; Riley & Greeno, 1988; Riley et al., 1983); Briars and Larkin (1984); and A. B. Lewis and Mayer (1987) has focused on how problem semantics influence the child's or adult's conceptual representation of the problem. We consider first the work of Carpenter and Moser. Research on the conceptual representation of word problems is presented in the section on Problem-Solving Processes.

The basic assumption underlying Carpenter and Moser's research is that when solving arithmetic word problems, children "may match their strategies to a given problem's structure by modeling the implied actions or relationships in the problem" (Carpenter et al., 1981, p. 29). To test this hypothesis, Carpenter et al. administered various forms of addition and subtraction word problems to a group of first-grade children. To reduce the working-memory demands of the task, the problems were reread if the child forgot any part of the problem before it could be solved. Manipulatives were available, in the form of cubes, if the child needed these to aid his or her problem solving. In this research, the focus was on the use of strategies that involved physically representing the values to be added or subtracted (i.e., the use of manipulatives and finger-counting strategies). The strategy that the child used to solve the problem was recorded on a trial-by-trial basis. There was not a lot of variability in the children's solution strategies for the addition problems, as most of the problems were solved by means of the counting-all strategy. As a result, for addition there was little relationship between the semantic structure of the problem and the solution strategy.

Nevertheless, there was a strong relationship between the semantic structure of subtraction word problems and problem-solving strategies. For instance, the second change problem shown in Exhibit 3.1 is similar in implied action to the separating-from strategy, which is used with manipulatives to solve subtraction problems. Recall that the separating-from strategy involves representing the value of the minuend by a corresponding number of objects and then removing a number of objects represented by the value of the subtrahend (see Table 2.2). In this study, the separating-from strategy was used on 76% of the change-problem trials of the following type: "Amy had five candies. Then she gave three candies to Mary. How many candies does Amy have now?" In other words, to solve this problem, most of the children first gathered five cubes, to represent Amy's five candies, and then removed three cubes, one at a time, to represent the three that were given to Mary. The final answer was obtained by counting the remaining blocks. The second most frequently used strategy was matching, which was used on only 11% of the change-problem trials.

On the other hand, the compare problems are conceptually similar to the matching strategy. The matching strategy involves separately representing the values of the minuend and subtrahend by rows of objects (see Table 2.2). The number of unmatched objects represents the difference. Carpenter et al. (1981) used a problem similar to the first compare problem shown in Exhibit 3.1: "Mary has three candies. Amy has two candies. How many fewer candies does Amy have than Mary?" In keeping with their hypothesis, Carpenter et al. found that the matching strategy was used on 49% of the compare-problem trials and that the separating-from strategy was used on 23% of the compare-problem trials. With the matching strategy, the children formed one row of three cubes, to represent Mary's three candies, and an adjacent row of two cubes, to represent Amy's two candies. The single unmatched block represented the difference.

De Corte and Verschaffel (1987) argued that Carpenter and Moser (Carpenter et al., 1981; Carpenter & Moser, 1984) did not find a clear relationship between the semantic structure of addition word problems and problem-solving strategies because they had not scored the addition strategies in enough detail to detect a relationship. For instance, there are variations of the counting-all strategy. With the use of manipulatives, one method involves the construction of a set of objects to represent the value of the augend, then the child adds to this set a number of objects equal to the value of the addend. She or he then counts all of the objects. This method is called the *adding method*. A second, *joining*, method involves representing the value of the augend and addend with separate sets of blocks, physically moving these sets together, and then counting all of the blocks. The third method differs from the second method in that the sets of blocks are not physically moved together and is therefore called the *no-move method*.

The adding method corresponds nicely to the first change problem shown in Exhibit 3.1, whereas the no-move method corresponds quite closely to the first combine problem. When manipulatives were used, De Corte and Verschaffel (1987) found that the adding method was used on 75% of the change-problem trials, whereas the no-move method was used on 68% of the combine-problem trials. An equally important finding was that the verbal-counting strategies used by the children also appeared to be influenced by the semantic structure of the problem. For instance, the counting-on-from-the-larger strategy was used much more frequently to solve combine problems than to solve change or compare problems. For subtraction problems, De Corte and Verschaffel replicated the findings of Carpenter and Moser (Carpenter et al., 1981; Carpenter & Moser, 1984) and showed that semantic structure influenced verbal-counting strategies, as well as strategies associated with finger counting and the use of manipulatives.

In all, this line of research indicates that the semantic structure of word problems, in particular the action implied in the problem and the manner in which the sets to be added or subtracted are presented, has a considerable influence on the types of strategies that young elementary school children use in problem solving. These results should not be taken to mean that the influence of problem features, such as memory demands, described in the preceding section are not important (e.g., Jerman & Mirman, 1974). In the Carpenter and Moser (1984) and De Corte and Verschaffel (1987) studies, many of the important problem features identified by Jerman and his colleagues were held constant across the semantically different word problems. For instance, the problems were re-read if the child forgot features of the problem. As a result, memory demands should not have influenced the child's problem-solving performance. By taking this type of precaution, these researchers were able to determine the influence of semantic features of word problems on problem-solving strategies above and beyond any other problem features, such as length, which might have influenced the children's performance.

Problem-Solving Processes

Research described in the two preceding sections focused on how features associated with the problem influence children's error rates and solution strategies. In this section, research that has focused on problem-solving processes that occur within the child is presented (Briars & Larkin, 1984; Kintsch & Greeno, 1985; Riley & Greeno, 1988; Riley et al., 1983). Mayer (1985) has argued that problem solving in this area occurs in four stages: *problem translation, problem integration, solution planning*, and *solution execution*. Problem translation and problem integration are features that determine the child's representation of the problem. The focus of this section is on the problem-representation and solution-planning processes associated with compare problems. This is because these problems are especially difficult for children, and for many adults, to solve. *Solution execution* refers to the specific strategies used in problem solving. These strategies were described in detail in chapter 2 and will not be repeated here.

The child's ability to mentally represent the meaning of an arithmetic word problem strongly influences the types of problems that the child can successfully solve. The building of a problem representation must first start with an understanding of the text within which the problem is embedded (Kintsch & Greeno, 1985). Text comprehension involves understanding not only the meaning and mathematical implication of specific words (e.g., *more* implies addition) but also the structure of the entire problem. The order and manner with which the information is presented can make the problem more or less difficult to comprehend and represent. For instance, consider the following compare problems from Exhibit 3.1:

> Amy has two candies. Mary has one more candy than Amy. How many candies does Mary have?
>
> Amy has two candies. She has one candy less than Mary. How many candies does Mary have?

The first of these problems is much easier to understand and therefore to solve than the second problem. This is at least in part because the key word *less* in the second problem suggests that the problem should be solved by means of subtraction rather than by means of addition (Huttenlocher & Strauss, 1968; A. B. Lewis & Mayer, 1987; Mayer, Lewis, & Hegarty, 1992).

More important, however, are differences in the structure of the second sentence. In the first problem, Mary is the subject of the relational sentence, whereas Mary is the object of the relational sentence in the second problem. To better understand why this is important, consider how a child—or adult—might process the first problem. When the first sentence is read, its meaning is organized by a *schema*, or a general format for extracting and representing or translating the basic meaning of the statement (Kintsch & Greeno, 1985; Mayer, 1985; Riley et al., 1983). The organizing feature of the schema that represents the basic meaning of the first sentence is quantity. Quantity is then elaborated by reference to who, what, and how many. This schema is concretely represented in Figure 3.1. In the top portion of Figure 3.1, we see the important features of the first sentence represented in a manner that is easily comprehended. It is assumed that with experience, most children develop schemas of this sort and simply plug in, so to speak, the appropriate *who, what,* and *how many* for different problems (Stigler et al., 1986). This same schema is also used to represent the meaning of Sentence 2: "Mary has one more candy than Amy." The only difference between the representation of the meaning of Sentence 1 and the meaning of Sentence 2 is that the amount of candy is, as yet, unknown for Sentence 2.

Now consider the child's ability to represent the meaning of the second compare problem stated above. As noted in Figure 3.2, the first sentence is represented in exactly the same way as the first sentence in Problem 1. The representation of the second sentence, "She has one candy less than Mary," is much more difficult. If the meaning of the problem statements are represented on the basis of the "who, what, and how many" schema, then when Amy (i.e., she) is the subject of the second sentence, it initially leads to the schematic representation shown in the middle

Figure 3.1

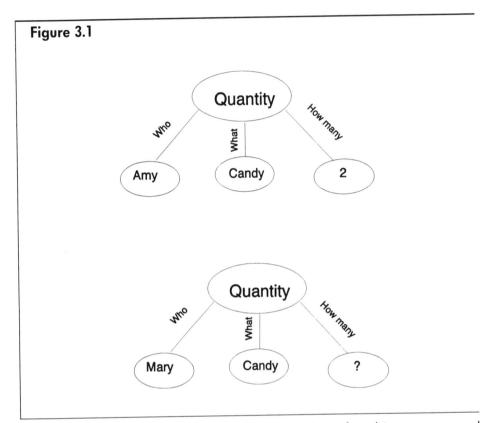

A representation of the *who, what,* and *how many* schema for solving compare word problems.

section of Figure 3.2. The initial subject of the representation is Amy, not Mary. At this point, the representation of the information presented in the problem does not allow the child to solve the problem, because there is no schematic representation for Mary.

According to A. B. Lewis and Mayer (1987), one way that this type of problem can be solved is to mentally reverse the statement so that Mary is the subject rather than the object of the second sentence. The reversal involves setting up a third representational schema, as shown in the bottom portion of Figure 3.2, as well as reversing the implied action of the relational term from *less* to *more* (Stern, 1993; Verschaffel, De Corte, & Pauwels, 1992). Both setting up a third representational schema and reversing the meaning of the relational term increase the chances that an error will be committed when solving this type of problem. The increased error rate for this type of problem is likely to be related to the increased working-memory demands that are associated with these rep-

Figure 3.2

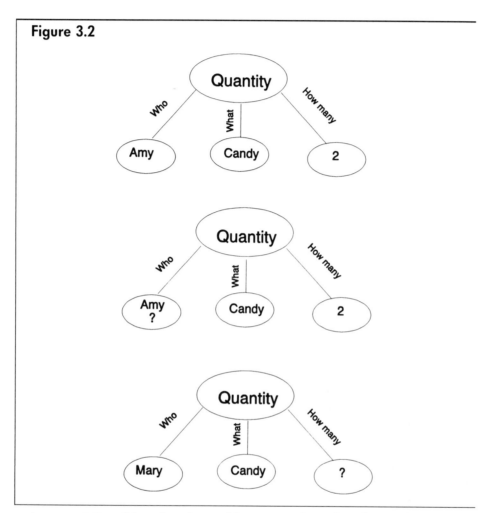

A representation of the *who, what,* and *how many* schema for solving difficult compare word problems.

resentational/reversal operations, as well as an inability to set up the appropriate representation. The most important point is that the language structure of the problem influences the child's (and adult's) ability to mentally or concretely (as with manipulatives) represent the meaning of the word problem, which in turn influences problem-solving skill.

Once the basic meaning of the problem statements has been mentally or concretely represented, a second type of schema has to be constructed. This time, however, the schema must represent the *relationship* between the important features of the two problem statements rather than representing the meaning of each individual sentence. In Mayer's (1985)

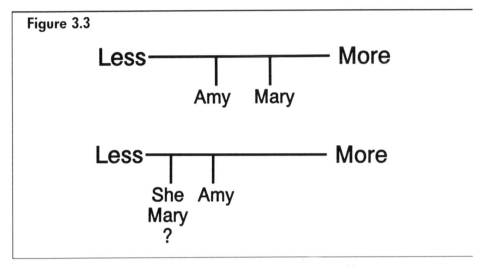

Figure 3.3

A representation of relational schemas for solving compare problems.

process model, this refers to problem integration. One way to understand relational schemas is shown in Figure 3.3 (A. B. Lewis, 1989). The top portion of the figure represents the relationship between the representations depicted in Figure 3.2. The individual schemas are, of course, defined by quantity. This dimension, quantity, depicts the most important feature of each individual sentence and is represented as the horizontal line in Figure 3.3. The relationship between the amount of candy that Amy has in comparison with Mary is shown by the relative positions along the horizontal line. Because the number of candies that Amy has is directly given in the first problem statement, her position on the line is the reference point for the relationship. When the next sentence is processed, "Mary has one more than Amy," the owner of the next set, Mary, is noted to the right of the notation for Amy. The placement of the notation for Mary is based on the key word *more*.

Now consider the initial relational schema for the second compare problem, which is shown in the bottom portion of Figure 3.3. Understanding and representing the relationship between the amounts of candies owned by Amy and Mary are rather difficult when the relationship is embedded in statements such as "She has one candy less than Mary." Here, the term *less* is highly salient and results in many people's placing the relative position of Mary to the left of Amy instead of to the right of Amy (A. B. Lewis, 1989; A. B. Lewis & Mayer, 1987; Stern, 1993). A. B. Lewis showed that graphically representing relationships in compare problems in a manner similar to that depicted in the top portion of Figure

3.3 greatly reduced the number of errors that college students made when trying to solve such problems. This result suggests that skilled problem solvers probably develop relational schemas that are analogous to that shown in the top portion of the figure and base their solution plan on this schema.

Once the meaning of the problem has been represented, the next step is solution planning (Mayer, 1985; Riley et al., 1983). Solution planning involves choosing the most appropriate strategy, such as counting all, for solving the problem. Riley et al. argued that a third type of schema, called an *action* schema, provided the link between the relational schema and the actual strategy that the child uses to finally solve the problem. Action schemas provide implicit knowledge about the results that the various strategies produce, such as combining sets for addition strategies, and the contexts in which they are most typically used. For instance, the adding method, which is sometimes seen with the counting-all strategy, is typically used in contexts where objects are added to the initially defined set, as with the first change problem shown in Exhibit 3.1 (De Corte & Verschaffel, 1987). The actual strategy chosen to solve the presented word problem is based on the best fit between the actions implied in the problem representation and the actions associated with the schemas that represent the outcome of each of the strategies available to the child (Riley et al., 1983). In other words, solution planning involves building a bridge between the child's representation of the important relationship in the problem and the most appropriate strategy for acting on this relationship.

To summarize, the essential features of children's problem solving for arithmetic word problems involve developing an appropriate representation of the problem (i.e., problem translation and integration) and then linking this representation to the best strategy for actually solving the problem (i.e., solution planning). Problem representation appears to occur in several steps. First, the meaning and implications of the text within which the problem is embedded must be understood by the child (Kintsch & Greeno, 1985). Text comprehension is influenced by specific words with mathematical implications—key words such as *more* or *less*— as well as by the overall structure of the problem (A. B. Lewis & Mayer, 1987). Next, the problem needs to be translated into a representation that highlights the quantitative features of the problem (Mayer, 1985). Once all of the statements in the problem have been represented, the next step is problem integration. Problem integration involves developing a way to understand the important quantitative relationships among the individual statements within the problem (A. B. Lewis, 1989). The final step in problem solving, before the child actually executes a strategy, involves solution planning. Here, the relationships developed during the problem representation phase are mapped onto the implied actions and outcomes that are associated with the various problem-solving strategies available to the child. The actual strategy chosen for problem solving is based on the best match between the actions implied in the problem representation and the actions associated with the strategies available to the child.

Developmental Patterns

In this section, the relative difficulty of the different types of arithmetic word problems that are summarized in Exhibit 3.1 is discussed, followed by a discussion of the factors that appear to govern developmental improvements in problem-solving skills. The arrangement of problems within each category, such as Change or Combine, in Exhibit 3.1 is roughly from easiest to most difficult, on the basis of the proportion of children that solve the problems correctly across all grades (e.g., Riley et al., 1983). Considering the change, combine, and compare problems, the first two change problems and the first combine problem shown in Exhibit 3.1 tend to be the easiest to solve. The compare problems, with the exception of won't-get problems, tend to be the most difficult, with the remaining problems of intermediate difficulty. There has not been much research on children's skill at solving equalize problems. Nevertheless, the work that has been conducted suggests that the first type of equalize problem shown in Exhibit 3.1 is relatively easy, whereas the remaining problems appear to be of intermediate difficulty (Morales, Shute, & Pellegrino, 1985).

Most kindergarten children in the United States can solve the first two types of change problems and the first type of combine problem, shown in Exhibit 3.1, with little difficulty (Riley & Greeno, 1988). Riley and Greeno found that when kindergarten children used manipulatives, they rarely made errors for these types of problems and that when manipulatives were not available, the children correctly solved about 70% of the problems. The actions implied in these problems are easy to understand and conform nicely to the types of problem-solving strategies already available to the typical kindergarten child. For instance, consider the second change problem shown in Exhibit 3.1. "Amy had five candies. Then she gave three candies to Mary. How many candies does Amy have now?" This problem conforms to the frequently used separating-from strategy (see Table 2.2). In contrast, kindergartners' performance on the other types of change and combine problems, as well as on compare problems, was very poor; Riley and Greeno did not administer equalize problems.

Most first-grade children are able to solve the easier forms of change and combine problems (e.g., the first two change problems and the first combine problem in Exhibit 3.1) without error, even without the use of manipulatives (Riley & Greeno, 1988; Riley et al., 1983). The performance of first graders for more difficult change and combine problems is better than that of kindergarten children but is still not at the level found for the easier problems in these categories. Compare problems are still rather difficult for most first-grade children (Stern, 1993). It is not until second or third grade that performance on compare problems improves appreciably for most children (Morales et al., 1985; Riley & Greeno, 1988). Moreover, Morales et al. found that third-grade children were able to solve the first type of equalize problem shown in Exhibit 3.1 more than 90% of the time but had considerable difficulty in solving the second type of equalize problem shown in the exhibit. By fifth or sixth grade, equalize problems were solved with little difficulty by most children. In fact, most of the problems shown in Exhibit 3.1 are easily solved by most children by the end of the elementary school years (Riley & Greeno, 1988).

The just-described findings do not mean that arithmetic-word-problem-solving skills are fully developed by the end of elementary school. All of the problems shown in Exhibit 3.1 and most of the research on children's skill at solving word problems have been based on one-step problems (Briars & Larkin, 1984; Carpenter & Moser, 1984; Riley et al., 1983). Two-step word problems are considerably more difficult than one-step problems, even for many college students (Hegarty, Mayer, & Green, 1992; A. B. Lewis & Mayer, 1987; Verschaffel et al., 1992). A difficult two-step problem from A. B. Lewis and Mayer (1987) is the following: "At ARCO gas sells for $1.13 per gallon. This is 5 cents less per gallon than gas at Chevron. How much do 5 gallons of gas cost at Chevron?" (p. 366). The results of A. B. Lewis and Mayer and others (e.g., Verschaffel et al., 1992), combined with Stiger et al.'s (1986) findings that mathematics textbooks in the United States do not present very many two-step problems, suggest that the performance of most sixth-grade American children would be poor even for moderately complex two-step problems. American textbooks do not present many difficult problems. When difficult problems are presented, it is at a later grade than for comparable textbooks used in many other nations (Fuson et al., 1988; Stigler et al., 1986). This pattern suggests that the development of problem-solving skills might be accelerated for children in many other nations as compared with American children (see chapter 7).

A variety of factors appear to influence developmental change from one grade to the next in the child's skill at solving arithmetic word problems. These factors include reading ability (Moyer, Sowder, Threadgill-Sowder, & Moyer, 1984; Muth, 1984), basic conceptual knowledge and problem-solving strategies (Carpenter & Moser, 1984; Wolters, 1983), working memory (Case, 1985), and the ability to form problem representations (Dean & Malik, 1986; De Corte, Verschaffel, & De Win, 1985; Morales et al., 1985). Not surprisingly, poor readers are not as skilled as good readers at solving arithmetic word problems. Moyer et al. (1984), for instance, found that good readers performed substantially better than poor readers on a measure that assessed their conceptual understanding of arithmetic word problems. The same pattern was found for third- to seventh-grade children. When the same types of problems were presented visually rather than verbally, across grades the performance of the poor readers improved an average of 14%, as compared with an average improvement of 3% for the good readers. Even with this improvement, the poor readers still did not perform at the same level as the good readers. Nevertheless, the results of this study suggest that reading skills directly contribute to skill at comprehending the meaning of arithmetic word problems, because eliminating the need to read the problem by presenting it visually leads to improved performance in poor readers but not in good readers. Important reading abilities most likely include general comprehension skills, as well as mathematics vocabulary (Jerman & Mirman, 1974; Kintsch & Greeno, 1985).

Wolters (1983) argued that even with adequate reading skills, before children can successfully solve any type of word problem, they need to understand basic counting and arithmetic concepts. For instance, consider the first type of combine problem shown in Exhibit 3.1: "Amy has two candies. Mary has three candies. How many candies do they have *altogether*?" Even though this type of combine problem is easily solved by most kindergarten children, the ability to solve this problem rests on certain basic conceptual knowledge. In particular, children need to have a basic understanding of part-whole relationships: Amy and Mary's candies are subsets (parts) of the more general set (whole) implied by "how many altogether." Thus, children's early number, counting, and arithmetic concepts provide the foundation for their initial problem-solving skills.

Working memory can be defined as the ability to keep important information in mind while mentally performing some type of operation on this information. Working-memory capacity improves gradually through the elementary school years and into adulthood (Case, 1985; Kail, 1992) and is almost certainly one source of developmental change in children's ability to solve arithmetic word problems. As the amount of information presented in the problem and the number of required arithmetic operations increase, the probability that the child will commit a problem-solving error also increases (e.g., Jerman & Mirman, 1974). This type of result implicates working memory as one factor that influences problem-solving skill. Indeed, Geary and Widaman (1992) found that both computational skills and working-memory capacity influenced adults' skill at solving complex arithmetic word problems: The larger the working-memory capacity, the better the performance on the word problems.

Baddeley (1986) has shown that working-memory capacity is, at least in part, related to how quickly information can be rehearsed. For instance, the more quickly words can be verbally stated, the more words that can be remembered. The implication of this finding is that faster readers are going to remember more information than slower readers. So as children become faster readers, with practice and maturation, their ability to solve increasingly complex (i.e., longer) word problems should improve, in part because faster reading should result in the ability to remember larger pieces of information. Thus, reading speed might be one factor that influences the relationship between reading skills and problem-solving accuracy (Moyer et al., 1984). Regardless of why capacity increases, improvements in working-memory capacity almost certainly contribute to developmental improvements in children's ability to solve arithmetic word problems. Future studies of developmental change in children's skill at solving arithmetic word problems would benefit greatly by the inclusion of working-memory measures or by the direct assessment of factors that might influence both working-memory capacity and problem-solving skill, such as reading speed.

Improvements in the mix of strategies available to the child for problem solving, as well as improvements in the child's ability to represent word problems, also appear to contribute to developmental change in problem-solving skill (Carpenter & Moser, 1984; Dean & Malik, 1986; Morales et al., 1985). As described in the Semantic Structure section, there is a close relationship between the child's ability to solve arithmetic word problems and the different types of strategies available to the child (e.g., Carpenter et al., 1981; De Corte & Verschaffel, 1987). For instance, consider the first change problem shown in Exhibit 3.1: "Amy had two candies. Mary gave her three more candies. How many candies does Amy have now?" This problem might be directly translated into the counting-on-from-the-larger strategy (see Table 2.1). This counting-on strategy will be much more efficient than the counting-all strategy for solving this type of problem. To take advantage of the relationship between the semantic structure of this problem and the counting-on strategy, the child must, of course, know how to count on (De Corte & Verschaffel, 1987). Dean and Malik (1986) argued that the relationship between strategies and conceptual knowledge was bidirectional. The child's conceptual knowledge influences the types of strategies that he or she can use in problem solving, and at the same time the use of these strategies leads to a greater elaboration of conceptual knowledge. Conceptual knowledge is elaborated as the child notices the outcomes produced by different strategies (Geary, Bow-Thomas, & Yao, 1992). The development of conceptual knowledge and the development of strategic skills are interwoven and influence one another in a leapfrog fashion.

Finally, improvements in the ability to represent and conceptually understand the meaning of arithmetic word problems contribute to developmental change in problem-solving skill. In a nicely done study, Morales et al. (1985) examined the relationship between representation of arithmetic word problems and problem-solving accuracy for groups of third- and fifth/sixth-grade children. Problem representation was determined for change, combine, compare, and equalize problems by having the children sort various examples of these types of problems into more or less similar groups. This is a method that is often used to distinguish how experts and novices understand a domain. Novices tend to sort problems on the basis of the surface structure of the problems, whereas experts tend to sort problems on the basis of the underlying conceptual similarities. For novices, problems that use the same types of phrases tend to be classified together, even if they are conceptually different. The surface structure of the problem does not typically influence the classifications of experts (Larkin, McDermott, Simon, & Simon, 1980).

For instance, the first combine problem and the first compare problem shown in Exhibit 3.1 start with the same two statements, although in a different order: "Amy has two candies. Mary has three candies." A child who sorts problems on the basis of surface structure will place these two problems together. On the other hand, a child who understands the conceptual meaning of these same problems will place them in different categories. Morales et al. (1985) found that fifth/sixth-grade children were more accurate at solving all four categories of problems than were the third graders. For both grade levels, more than two thirds of the errors were conceptually based rather than due to arithmetic computational errors. More important, they found that the categorizations of the third-grade children were more strongly influenced by the surface structure of the problems than by any underlying conceptual similarities.

Conversely, the categorizations of the fifth/sixth-grade children were more strongly influenced by conceptual similarities than by similarities in surface structure. In fact, the categorizations of these children were similar in many respects to the categorizations developed by experts in the area, that is, categorizations developed by Riley et al. (1983) and Carpenter and Moser (1984). For the fifth/sixth-grade children, the problems were generally sorted into the change, combine, compare, and equalize categories shown in Exhibit 3.1. Finally, for the third-grade group, children who tended to organize the problems on the basis of conceptual categories rather than surface structure were much more accurate at actually solving the problems than were their peers who focused more on the surface structure of the problems. The fifth/sixth graders did not make enough problem-solving errors to conduct this type of analysis. In all, this study shows that the child's ability to conceptually represent the meaning of arithmetic word problems influences skill at solving such problems, and developmental change in conceptual skills most likely contributes to the improvements in performance that are seen from one grade to the next. Improvements in conceptual understanding, in turn, are more than likely influenced by changes in a variety of areas—such as maturation, working memory (Case, 1985), and overall practice at solving word problems (Bransford & the Cognition and Technology Group of Vanderbilt University, 1993; Mayer, 1981; Stigler et al., 1986).

Summary

Identifying the skills that are associated with children's ability to solve arithmetic word problems is the focus of ongoing research by mathematics educators and by cognitive and developmental psychologists, as well as by cognitive scientists. The intense interest in children's problem-solving skills is the result of the assumption that these skills provide the link between the observable strategies described in chapter 2 and the use of these strategies in real-world settings. The focus on problem solving has provided a wealth of information on this facet of children's mathematical growth. We now know that structural features of the problems themselves, such as length and complexity, make problems more or less difficult for children to solve (Jerman & Mirman, 1974).

Another intriguing outcome of this research is the knowledge that the actions implied in a problem are reflected in the strategies used to solve the problem (Carpenter et al., 1981). For instance, young children tend to solve problems that include "Mary gave her three more candies" by counting out three blocks and then adding them to an already-defined set of blocks. The ability to understand word problems is also influenced by key words, such as *more* or *less*, as well as by the building of "who, what, and how many" schemas. The development of these and other schemas is important for representing the quantitative meaning of individual statements, for integrating or pulling together these statements, and for selecting appropriate problem-solving strategies (Briars & Larkin, 1984). Finally, developmental changes in problem-solving skill appear to be influenced by a variety of factors—including improvements in reading skills and increased working-memory capacity, as well as improvements in the ability to understand and represent the deeper meaning of word problems (i.e., schema development) as compared with focusing on surface features.

Algebra

To be conceptually continuous with the discussion of arithmetic problem solving, the focus of this section is on the solution of algebraic word problems, rather than on simple algebraic equations such as "Solve for y: $3y = 150$." Much of the research presented in this section has, of course, been conducted with junior high school, high school, and college students. This is because most people, at least in the United States, receive most of their exposure to algebra in junior high school and high school. Nevertheless, many of the same features that influence the ability of elementary school children to solve arithmetic word problems also appear to influence the ability of older children to solve algebraic word problems. These features include the acquisition of schemas to guide problem solving and the ability to appropriately translate important features of word problems into corresponding equations (Clement, 1982; Mayer, 1982). Algebraic problem solving, however, differs from the solving of arithmetic word problems on several important dimensions, which are the foci of this section.

The first of these dimensions is that the translation of algebraic word problems into equations tends to be more difficult than the translation of arithmetic word problems. Second, the greater complexity of algebraic problems results in a larger problem space (C. Lewis, 1981; Wenger, 1987). *Problem space* refers to all of the procedures and rules that the person knows about a particular type of problem, as well as all of the different ways that the problem can be solved. The complexity of many algebra problems means that there is almost always more than one method that can be used to solve the problem. In this circumstance, problem solving requires not only using appropriate procedures but also choosing the best route (i.e., sequence of procedures) to the final goal (Schoenfeld, 1987). The processes involved in choosing the best overall problem-solving strategy are referred to as *metacognition*. The first of three Algebra sections therefore focuses on how people translate algebraic word problems into equations. The influence of problem space and metacognition on algebraic problem solving is considered in the second section. In the final section, factors that appear to influence the development of algebraic skills are discussed. This is not to say that features described for arithmetical problem solving, such as problem length, are not going to influence algebraic problem solving—they almost certainly will. Rather, factors that affect algebraic problem solving above and beyond those described for arithmetic word problems are presented in this section.

Problem Translation and Algebraic Bugs

Mayer (1982) argued that the solution of algebraic word problems requires two general sets of processes: *problem translation* and *problem solution*. Problem translation involves transforming the meaning of the problem statements into a set of algebraic equations and includes the problem-translation as well as the problem-integration and solution-planning processes described for arithmetic word problems. *Problem solution* simply refers to the actual use of algebraic or arithmetical procedures to solve the resulting equations. An inappropriate use of arithmetical and algebraic procedures is a common source of problem-solving errors and is called an *algebraic bug* (Schoenfeld, 1985). Of the two sets of processes (i.e., translation and bugs), problem translation appears to be the primary source of problem-solving errors in algebra (Clement, 1982; Hinsley, Hayes, & Simon, 1977; Mayer, 1982; Mayer et al., 1992). Thus, problem translation is considered first, followed by a brief description of algebraic bugs.

Hinsley et al. (1977) showed that the translation of algebraic word problems is guided by schemas. These schemas—that is, mental representations of the similarities among categories of problems—work in much the same way as the schemas described for the solution of arithmetic word problems. For instance, problems that ask students to determine the relative ages of two individuals, as shown below, can often be solved by using similar ways to represent the problem and by using similar algebraic procedures to solve the resulting equations. These representations and procedures, however, differ from those used to solve other types of problems. The memory for how to represent and solve these problems forms the schema, or script, for age problems, which differs in some respects from the schema for other categories of problems (Mayer, 1981). Even though the schemas for different types of word problems differ from one another, most algebraic problems have a similar structure: Schemas appear to be constructed around this basic structure.

Mayer (1981) showed that most algebra problems presented in high school textbooks are centered around four types of statements: *assignment statements, relational statements, questions*, and *relevant facts*. Assignment statements, not surprisingly, involve assigning a particular numerical value to some variable, such as "M&Ms cost $1.00 per pound." A relational statement specifies a single relationship between two variables: "Chocolate balls cost twice as much as M&Ms." Questions are just that: "How much do 2 lb of chocolate balls cost?" Relevant facts are information presented in the problem that might be needed to solve the problem. For instance, "Amy has $3.00. How many pounds of chocolate balls can she buy?"

Problem translation involves taking each of these different types of information and using them to develop corresponding algebraic equations. The translation of assignment statements, questions, and relevant facts does not pose much of a problem for most high school students (Mayer, 1982; Wenger, 1987). However, translation errors frequently occur during the processing of relational statements. For this reason, the discussion of problem translation focuses on the processing of these statements. For instance, consider a simple problem: "There are six times as many students as professors at this university." Clement (1982) presented this problem to freshman engineering students and asked them to write an equation that represented the relationship between the number of students and the number of professors, using S for the number of students and P for the number of professors. Of the engineering students, 37% committed an error on this problem (liberal arts majors, thereafter?)—typically $6S = P$. This type of error is not due to inattention and is, in fact, fairly common (Hinsley et al., 1977).

There appear to be two reasons for this type of translation error. The first is that the *syntax*, or structure, of the relational statement is literally translated into an algebraic expression. So, "six times . . . students" is literally translated into 6S.

Second, many students appear to interpret relational statements as requesting static comparisons. In this example, 6S is used to represent the group of students and P to represent the group of professors. In other words, the equivalence symbol, =, does not represent a mathematical relationship for these students; it simply separates the two groups. Students who correctly translate this relational statement understand that S and P represent quantities, not static groups, and that the quantities represented by S and P are unequal. To make the number of professors equal to the number of students, some type of operation has to be performed. Specifically, the smaller quantity, P, has to be increased so as to make it equal to the larger quantity, S. These students correctly translate the relational statement as $S = 6P$ (Clement, 1982). The same result can, of course, be achieved by reducing the number of students, $S/6 = P$, although most people do not translate the problem in this way.

Consider another problem, from Mayer (1982): "Laura is 3 times as old as Maria was when Laura was as old as Maria is now. In 2 years Laura will be twice as old as Maria was 2 years ago. Find their present ages" (p. 202; hint: Maria's past age is 6). This type of problem is fairly common in high school algebra textbooks and can be used to illustrate several important features of algebraic problem solving. First, the problem includes more complex relational statements than the student/professor problem and therefore more opportunity for translation errors. Second, once the problem has been correctly translated, there are several different ways that the resulting algebraic expressions can be solved. Thus, the problem space is more complex than that of the student/professor problem, a point that we consider in the next section.

Once the student understands the goal of the problem—find Laura and Maria's present ages—then the next process is the assignment of variables to these ages. If the student simply reads the statements and sets up the relational information on the basis of the syntax of these statements, then he or she might assume that the problem has only two variables: Laura's present age, Z, and Maria's present age, Y. Once the relevant variables are assigned, then the relations are translated into algebraic expressions. So "Laura is three times as old as Maria" might be translated as $3Z = Y$, or $3Y = Z$. Either way, the translation is incorrect. The next sentence, "In 2 years Laura will be twice as old as Maria was 2 years ago," might be translated as $Z + 2 = 2(Y - 2)$ or as $2(Z + 2) = Y - 2$. The first of these translations is, of course, correct, but the second is not. The second translation represents the same type of error as described for the student/professor problem. The point is, the translation of this problem is not that straightforward and therefore provides the student with many opportunities to commit errors. At this point, no matter which combination of translations the student has performed, no amount of algebraic effort expended on manipulating the resulting equations will satisfy the goal of finding Laura's and Maria's present ages. This is because a more careful reading of the relational statements shows that there are in fact three different ages presented in the problem, not two.

One way to better understand these relationships and to reduce the working-memory demands of the problem is to diagram the age relationships (A. B. Lewis, 1989). The top portion of Figure 3.4 shows the relationships among Laura and Maria's past and present ages. Once the relationships are drawn graphically, it becomes clear that this is a three-, not a two-, variable problem. Here, X represents Maria's past age, and Y represents Maria's present age as well as Laura's past age. Now that the relational information is properly represented, the translation of the problem can be completed rather easily. The first statement, "Laura is 3 times as old as Maria was when Laura was as old as Maria is now," translates into the first algebraic expression shown in Figure 3.4: $3X = Z$. The second statement translates into the second expression, $Z + 2 = 2(Y - 2)$. At this point, both relational statements have been translated into appropriate algebraic expressions. However, manipulating these two expressions will not achieve the final goal, because there are three unknowns and only two equations. Nevertheless, once this type of problem has been translated, many students immediately attempt to solve the problem without a consideration of whether enough information is available to complete problem solving or what the most efficient solution strategy might be (Schoenfeld, 1985). Frustration and a lot of wasted effort are the typical result of this "jump in without thinking" approach to problem solving (Schoenfeld, 1987).

Skilled problem solvers, in contrast, will typically take much more time to analyze the problem than the typical high school or college student (Schoenfeld, 1987). In this example, further analysis of the problem reveals that a third algebraic expression is needed before the solution process can be completed. Specifically, the relations among the three variables, X, Y, and Z, need to be expressed. Because the relationships are

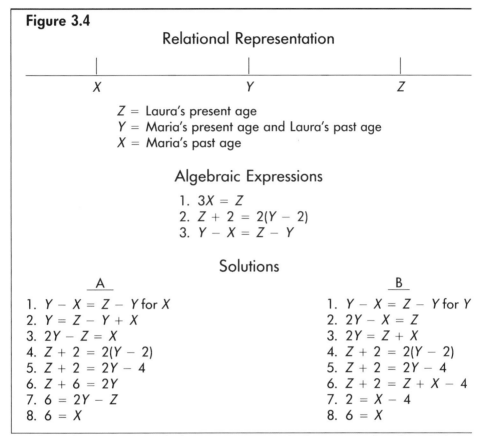

Figure 3.4

Relational Representation

Z = Laura's present age
Y = Maria's present age and Laura's past age
X = Maria's past age

Algebraic Expressions

1. $3X = Z$
2. $Z + 2 = 2(Y - 2)$
3. $Y - X = Z - Y$

Solutions

A

1. $Y - X = Z - Y$ for X
2. $Y = Z - Y + X$
3. $2Y - Z = X$
4. $Z + 2 = 2(Y - 2)$
5. $Z + 2 = 2Y - 4$
6. $Z + 6 = 2Y$
7. $6 = 2Y - Z$
8. $6 = X$

B

1. $Y - X = Z - Y$ for Y
2. $2Y - X = Z$
3. $2Y = Z + X$
4. $Z + 2 = 2(Y - 2)$
5. $Z + 2 = 2Y - 4$
6. $Z + 2 = Z + X - 4$
7. $2 = X - 4$
8. $6 = X$

Relational representations, algebraic expressions, and solution options for solving the Laura and Maria age problem.

linear, $Z > Y$ and $Y > X$, and because the relative age difference between Laura and Maria does not change as they get older, these relationships can be represented by the third algebraic expression shown in Figure 3.4, $Y - X = Z - Y$. Note that translating these variables too quickly, without considering what they mean in the real world, can result in a conceptual error. Specifically, although $X - Y = Y - Z$ is correct algebraically, it does not make sense for this particular problem, because it results in a negative age difference. This is yet another opportunity for the student to commit a translation error. It should be clear at this point that translating story problems into algebraic expressions is a complex and difficult task for most people. To make matters worse, once the algebra problem has finally been successfully translated, unlike arithmetic word problems, the next step is not always obvious. This is where the issues of problem space and metacognition become important. I finish the solving of the

Table 3.1	
Algebraic Bugs	
Expression	Incorrect Transformation
$(X + Y)^2$	$X^2 + Y^2$
$\sqrt{X + Y}$	$\sqrt{X} + \sqrt{Y}$
$\dfrac{X}{Y + Z}$	$\dfrac{X}{Y} + \dfrac{X}{Z}$
3^{XY}	$3^X + 3^Y$
$X(YZ)$	$XY \cdot XZ$

Laura/Maria age problem with the discussion of problem space and metacognition.

However, before the importance of problem space and metacognition is considered, the issue of simple algebraic errors that occur during the solution process should be discussed. Not all errors that occur during the solution of algebraic word problems result from difficulties in representing and translating problem statements. Sometimes students perform incorrect algebraic procedures on correctly translated expressions (Schoenfeld, 1985). As was described in chapter 2 for subtraction errors (VanLehn, 1990), many of these types of algebraic errors appear to reflect bugs (Schoenfeld, 1985; Wenger, 1987). Schoenfeld described a number of common algebra bugs, some of which are shown in Table 3.1. Recall that bugs are procedures that are correct for some problems, or for simple expressions, but are incorrectly applied to the present expression. With algebra, bugs appear to reflect incorrect assumptions about how radicals and exponents can be manipulated. For instance, consider the first bug in Table 3.1. Even though it is correct that $(X + Y)^1 = X^1 + Y^1$, this rule does not apply with an exponent other than 1. The last bug represents a confusion between the rules used for addition and those used for multiplication. Thus, although it is correct that $X(Y + Z) = XY + XZ$, the same separation of terms is incorrect when the parenthetical values are multiplied.

Moreover, errors can occur in performing basic operations, such as moving a term from the left side of the expression to the right side of the expression (G. Cooper & Sweller, 1987; C. Lewis, 1981). For instance, consider the problem, $[C(A + D)]/F = G$, solve for A. One type of error might involve moving $/F$ from the left to the right rather than multiplying both sides of the equation by F. In this case, the right side of the equation reads G/F, rather than GF. Other errors are even more basic, such as dividing by 0 (C. Lewis, 1981). C. Lewis found that even experts committed this type of error, although experts usually checked their work and therefore almost always caught and corrected these basic procedural errors, whereas novices were less likely to do so.

Problem Space and Metacognition

In comparison with arithmetic, there are many more ways to solve algebra problems and therefore many more decisions to be made about the most efficient sequence of operations to be performed during problem solving. As noted earlier, all of the potential operations that the student can perform to solve a particular problem represent the problem space. An important issue in algebraic problem solving then becomes determining the most efficient search through the problem space (Lester, 1983; Schoenfeld, 1987). Skilled problem solving involves planning the sequence of moves that will achieve the goal with the fewest operations. *Metacognition* refers to the planning and decision-making processes that guide this search through the problem space. Schoenfeld (1985) argued that metacognition was an essential feature of mathematical problem solving. The importance of metacognitive skills can be illustrated by comparing the activities of novice and expert mathematical problem solvers.

One difference between the problem solving of novices and that of experts is the amount of time spent on analyzing the problem and developing a strategy for solving the problem (Schoenfeld, 1987). Novices almost invariably read the problem and then spend most of their problem-solving time executing what they believe are the most appropriate computations. Experts, on the other hand, spend most of their problem-solving time analyzing the problem and developing potential solution strategies. Thus, even if novices and experts have the same basic computational skills needed to successfully solve a mathematics problem, their behavior during problem solving is very different (Schoenfeld, 1987). These differences in problem-solving activities reflect metacognitive differences between novices and experts. To illustrate, let us return to the algebraic expressions shown in Figure 3.4. All three expressions can be manipulated by means of algebraic procedures. Knowledge and execution of these procedures represent the basic cognitive processes governing the algebraic computations.

Metacognitive processes come into play in determining which of these expressions to work with first and in determining the most efficient way to manipulate the expressions. For instance, one way to begin the solution process for the Laura/Maria age problem is with the third expression shown in Figure 3.4. Once this decision has been made, the problem space has been reduced from all of the potential operations that can be performed on all three expressions to only those that can be performed on Expression 3. Here, one can solve for X, Y, or Z. The first three steps under Solutions in Figure 3.4 show the operations needed to solve for X and Y and the sequence for solving for X and Y. Next, for Solution A, the fourth expression, $Z + 2 = 2(Y - 2)$, is transformed so that the two unknowns are on the right side of the equation, Steps 4 to 7. Step 8 results when X is substituted for $2Y - Z$ (see Step 3). For Solution B, $Z + X$ is substituted for $2Y$ at Step 5, giving the result noted in Step 6. From there it is relatively simple to solve for X and then solve for the remaining unknowns by substituting the value of X (i.e., 6) into the original expressions. Solutions A and B represent two of many different ways to solve the Laura/Maria age problem and illustrate different solution routes through the problem space.

Developmental Patterns

In comparison with research on skill development for solving arithmetic word problems, research on algebraic problem solving has not been especially systematic, at least from a developmental perspective. For instance, much of the research on skill acquisition has focused on comparisons of adult novices and experts in algebra rather than on longitudinal studies of skill development in children. As a result, the developmental processes presented in this section should be viewed as a personal best guess, given the research that has been conducted, rather than a description of well-established developmental trends. Nevertheless, it seems reasonable to assume that many of the same factors that contribute to developmental improvements for arithmetical problem solving also contribute to developmental change, from junior high school to college, in algebraic problem solving. These factors would include improvements in reading skills, working memory, basic computational skills, and so on. Moreover, comparisons of experts and novices, as well as practice-related effects, suggest that the development of problem-solving skills in the domain of algebra involves at least three additional dimensions. These include schema acquisition, rule automation (i.e., learning basic algebraic procedures), and metacognitive skills.

G. Cooper and Sweller (1987) argued that the development of problem-solving skills in algebra involve schema acquisition and rule automation. The processes underlying the acquisition of these two skills, as well as the amount of experience needed for skill development, appear to be independent (Adams et al., 1988; Novick, 1992; Sweller, Mawer, & Ward, 1983). Rule automation concerns the ease of use of basic algebraic procedures, such as subtracting or adding variables to each side of the equation. More specifically, *automation* refers to the automatic execution of a procedure without having to think about the rules governing the use of that procedure. One of the benefits of rule automation is a reduction in the working-memory demands associated with using the procedure. The freeing of working-memory resources makes the processing of other features of the problem easier and less error-prone. In a series of studies conducted with junior high school and high school students, G. Cooper and Sweller showed that rule automation occurs gradually and only with the extensive use of the rule, or procedure. Once the rules have been memorized and automatically executed for one set of problems, they are readily used to solve different types of algebra problems. Before this point, however, rules learned in the context of one type of algebra problem are not readily transferred—that is, these rules are not automatically used to solve other types of algebra problems. In all, rule automation not only reduces working-memory requirements of the problem but also leads to the use of these rules for solving other forms of algebra problems. Rule automation, however, appears to occur only with extensive practice.

Schema acquisition, on the other hand, appears to occur rather quickly, that is, with very little practice (G. Cooper & Sweller, 1987; Hinsley et al., 1977; Novick & Holyoak, 1991). This is not to say that solving a few problems in and of itself will lead to the development of a problem-solving schema. Rather, schema acquisition appears to be dependent on the way in which related problems are solved (Adams et al., 1988; Novick, 1992; Wenger, 1987). Novick, for instance, has argued that when two problems of a similar category (e.g., age or speed) are presented in sequence, then problem solving occurs by analogy, at least in the rare case when the student recognizes that the problems can be solved by using similar procedures. The relationships used for problem translation and the associated procedures for problem solution for the first problem are mapped onto and adapted for use on the second problem (Novick & Holyoak,

1991). Apparently during the process of mapping and adaptation, a more general schema for this particular class of problem is developed.

At this point, the schema includes the features common to both problems. For instance, after solving several age problems, such as the Laura/Maria problem discussed earlier, a general schema for solving similar problems is acquired. This schema might involve first graphically depicting the age relationships, as I did in Figure 3.4, and then translating the representation into algebraic expressions, being careful not to commit earlier syntax errors. Finally, alternative ways of solving the problem might be considered, and then what appears to be the most efficient method is executed. This sequence of actions then provides the framework for solving problems that are in the same category. Other age-category problems do not lead to the formation of new schemas but can lead to the elaboration and refinement of the general problem-solving script for this class of problem (Sweller et al., 1983).

As noted earlier, in addition to rule automation and schema acquisition, successful problem solving in algebra appears to require metacognitive skills (Schoenfeld, 1987). Metacognitive skills become increasingly important as the problem space for a particular domain increases. The more options that are available for problem solving, the more important it becomes to choose effectively among these options. The importance of the role of metacognition in algebraic problem solving has been assessed primarily by means of comparing the performance of experts and novices (Schoenfeld, 1985). As a result, it is not clear how developmental factors influence this type of skill in the domain of algebra. Nevertheless, it is clear that metacognition in mathematical problem solving requires extensive experience in solving similar types of problems (Bransford & the Cognition and Technology Group of Vanderbilt University, 1993). With experience, people presumably figure out which approaches work and which do not work for solving problems in the same category (e.g., age problems).

Another factor that appears to influence general problem-solving skills is the *goal* of problem solving. Sweller et al. (1983) found that asking students to simply solve for X, as is done in most classrooms, did not greatly influence their problem-solving approaches. Even after extensive practice, they still used problem-solving approaches that are commonly used by novices. However, more general goals, such as asking students to find different ways to solve the same problem, did lead to the use of problem-solving approaches typically used by experts. Thus, the development of the metacognitive skills associated with mathematical problem solving is probably influenced by more general changes in cognitive development (Case, 1992), as well as by solving the same types of problems in many different ways and in many different contexts.

Summary

Most people find the solving of algebraic word problems a cumbersome task. The primary source of difficulty in solving algebraic word problems is translating the story into appropriate algebraic expressions. Translation involves assigning variables, noting constants, and representing relationships among variables. Of these, relational aspects of the problem are particularly difficult to translate correctly (Mayer, 1982). The specifics of algebraic translation errors have not been examined as closely as the translation errors associated with arithmetic word problems. Nevertheless, it seems reasonable to assume that algebraic translation errors result from some of the same factors that contribute to arithmetic translation errors, such as the semantic structure and memory demands of the problem. Once the problem has been translated, problem-solving errors can and do still occur and are often due to bugs (C. Lewis, 1981; Schoenfeld, 1985). In other words, errors that occur during the manipulation of algebraic expressions (i.e., during problem solution) typically involve the inappropriate use, or misapplication, of an algebraic or arithmetical procedure.

Once fundamental algebraic skills have been acquired, such as knowing how to correctly execute basic rules and procedures and knowing how to represent the problem, skilled problem solving appears to require metacognitive processes (Schoenfeld, 1985). These processes involve knowing which problem-solving options are available, evaluating the potential usefulness of these options, and then choosing the most efficient route to the goal (i.e., search through the problem space). Metacognitive skills also involve knowing one's own skills and limitations within the domain, as well as knowing when to give up on a chosen path through the problem space and try something else.

Finally, more developmental research on algebraic problem solving is clearly needed. Much of the research that has been conducted thus far has not been from a developmental perspective. For instance, although comparisons of novice and expert problem solvers provide some insights into how children might problem solve, they do not necessarily inform us about how developmental change might influence skill acquisition. Even though many features of skill acquisition will be the same for adults and adolescents, there will almost certainly be some differences. One aspect of development involves improvements in the ability to allocate attention and to control working-memory resources, as well as improvements in other metacognitive skills (e.g., Welsh, Pennington, & Groisser, 1991). We therefore need to better understand, for instance, how developmental change in basic metacognitive, or executive, skills influences schema acquisition and rule automation.

Conclusion

An essential aspect of children's mathematical development involves learning how to problem solve. Children's problem solving has been studied primarily with the use of arithmetic and algebra word problems. The solving of these problems is not as straightforward as the solving of the simple arithmetic equations described in chapter 2. Not only does the child need to have a solid grasp of basic number and counting concepts, as well as basic arithmetic skills, the child must also acquire an array of other skills to become proficient in mathematical problem solving. These include reading comprehension (Kintsch & Greeno, 1985), the ability to translate problem texts into arithmetical and algebraic equations (A. B. Lewis & Mayer, 1987), and metacognitive skills (Schoenfeld, 1987). Of fundamental importance are good reading skills (Moyer et al., 1984; Muth, 1984). Essential reading skills appear to include a basic mathematical vocabulary (Jerman & Rees, 1972), as well as the ability to understand the meaning of the problem (Kintsch & Greeno, 1985). To understand the basic meaning of the problem, the student must know that some words have mathematical implications (e.g., *more* implies addition), and—more important—the student must be able to comprehend the meaning of the relationships described within the problem (A. B. Lewis & Mayer, 1987).

Comprehending relational information, such as "Laura is three times as old as Maria was when Laura was as old as Maria is now," and translating that information into an appropriate equation are primary sources of problem-solving errors for both arithmetical and algebraic word problems (A. B. Lewis & Mayer, 1987; Mayer, 1982). Relational statements appear to be relatively difficult to translate, because the translations often involve a number of mental operations that do not appear to come naturally for most children and adults. For instance, consider again the fourth compare problem shown in Exhibit 3.1: "Amy has two candies. She has one candy less than Mary. How many candies does Mary have?" To solve this problem, the child must use addition. Yet, the key word *less* implies subtraction. A. B. Lewis and Mayer found that overriding the tendency to subtract when terms such as *less* are stated in the problem is difficult (see also Huttenlocher & Strauss, 1968). Thus, the failure to override this tendency is common and often results in the use of the wrong operation. Moreover, relational statements often involve mentally manipulating information. For most people to understand the relational meaning of the second sentence above, "She has one candy less than Mary," the sentence has to be recast to read, "Mary has one more candy than Amy." Recasting relational statements appears to be a common source of problem-solving errors. In combination, all of these factors appear to make the representation and translation of relational information a potent roadblock to skilled problem solving.

The difficulties associated with algebraic problem solving are compounded by the fact that as the complexity of the problem increases, the number of ways that the problem can be solved also increases. As the number of solution options increases, metacognitive skills become especially important for skilled problem solving (Schoenfeld, 1985, 1987). Metacognitive skills include making decisions about the best ways to solve a problem as well as deciding when to abandon one strategy in favor of another. This aspect of mathematical problem solving is something that is not obvious to most high school and college students, who often assume that there is only one correct way to solve each problem. Thus, the first basic task in developing metacognitive skills in mathematics is to understand that there are many routes to the same answer. Fortunately, Sweller et al. (1983) showed that requiring students to solve problems in different ways improved their approaches to mathematical problem solving (i.e., their approaches became more similar to those used by experts). It follows from this finding that improvements in metacognitive skills might be achieved with slight modifications in the classroom curriculum for algebra.

To complicate the situation further, acquiring problem-solving skills as an adult is likely to be different in many respects than skill acquisition in developing children. This is not to say that there will not be similarities but rather that understanding developmental change in translational and metacognitive skills, for instance, will require a consideration of many more factors than is required to understand skill acquisition in adults. Developmental change on a variety of dimensions will likely influence the changes in problem-solving skills seen from one grade to the next. These factors include improvements in reading comprehension, reading speed, working-memory resources, and basic computational skills. Each of these skills may have different developmental trajectories for different children and may differ in their relative importance to translational skills, for example, for children of different ages. Even though comparisons of novices and experts or children of different ages provide important insights into mathematical problem solving, additional research strategies are needed. Specifically, to fully understand the development of problem-solving skills (Carpenter & Moser, 1984), longitudinal research that considers all of these factors in concert, as related to children's problem solving, needs to be conducted.

Finally, the relationship between the skills described in chapter 2 and those skills described in this chapter needs to be considered. Basically, the skills described in chapter 2 represent an essential component and building block for the development of problem-solving skills, just as number and counting concepts are essential for the child's developing arithmetic skills (Fuson, 1982; Geary & Burlingham-Dubree, 1989; Kaye, 1986). It is important not to lose sight of the intimate relationship between basic skills and the acquisition of more complex problem-solving skills. Thus, skilled problem solving must rest on a solid foundation of basic skills: Attempts to improve the mathematical problem solving of children should not be done instead of teaching basic skills but rather in addition to teaching these basic skills. Future research, in addition to examining the factors (e.g., working memory) that influence the development of problem-solving skills, should also consider more carefully the relationship between the child's developing computational and problem-solving abilities (Dean & Malik, 1986).

4 Individual Differences in Mathematical Ability

The most general goal of this chapter is to describe *why* people differ in mathematical ability. Research described in the first three chapters essentially depicted what children or adolescents do to solve mathematical problems, but this type of research does not typically provide insights as to why some individuals are better than others in mathematics. In this chapter, individual differences in arithmetical and mathematical problem-solving skills are approached from three perspectives: psychometric, cognitive, and behavioral genetic. The psychometric studies are typically concerned with individual differences in performance on traditional paper-and-pencil ability tests and are useful for determining, among other things, whether different types of mathematical abilities exist (Thurstone, 1938). The cognitive approach to individual differences, on the other hand, is concerned with identifying the mental processes that underlie individual differences on traditional psychometric tests (Hunt, 1978). In combination, these two perspectives provide unique insights into the different domains of mathematical ability, as well as into the cognitive processes that contribute to ability differences within these domains.

The behavioral genetic approach is concerned with understanding the extent to which individual differences in ability, among other traits, are due to individual differences in the constellation of genes that support the associated skills (Plomin, DeFries, & McClearn, 1990). With this in mind, behavioral genetic studies of numerical and mathematical abilities were reviewed. The goal of this review was to determine the extent to which ability differences, identified in the psychometric and cognitive

---

Summaries are provided at the end of the main sections: Psychometric Perspective (p. 139); Cognitive Perspective (p. 146); and Behavioral Genetic Perspective (p. 149).

literatures, appear to be influenced by genetic factors. Research in each of these respective areas is presented in turn, followed by an integrative Conclusion.

Psychometric Perspective

The psychometric approach to the study of cognition is concerned with understanding individual differences in human abilities and intelligence. Historically, the approach has involved the administration of large numbers of paper-and-pencil tests to groups of individuals and then, by means of factor-analytic and related techniques, identifying clusters of tests that share some common property (e.g., Spearman, 1927; Thurstone, 1938). Cognitive tests that cluster together define an *ability factor*. The factor is thought to represent a system of conceptual knowledge or cognitive processes that all of the tests that define the factor have in common. However, before it can be argued with any certainty that ability tests share a set of cognitive processes, it must be demonstrated that the same tests cluster together across many studies. This criterion is called *factor stability* or *factor invariance* (Thurstone, 1938). The demonstration of factor stability, in turn, suggests that the factor represents a basic domain of human ability, implying that the skills represented by the factor are distinct from other cognitive abilities for educational, cultural, or biological reasons.

In this section, psychometric studies that have focused on or included numerical and mathematical tests are summarized, with the goal of identifying basic domains of mathematical ability and developmental changes within these domains. In preview, the studies suggest two general mathematical domains: numerical facility and mathematical reasoning. The existence of several other types of relatively distinct mathematical abilities, such as estimation skills, has been identified in some but not all psychometric studies. The basic findings in each of the apparent mathematical domains are presented below. Within each section, the stability of the associated factor, developmental changes in the types of tests that define the factor, and the cognitive abilities that appear to be represented by the factor are discussed.

Numerical Facility

In the psychometric literature, the numerical facility construct is measured by using arithmetic tests (Thurstone, 1938). The factor-analytic studies of this construct are important for two reasons. First, if it can be demonstrated that the Numerical Facility factor is stable across studies, then this finding would be consistent with the argument that arithmetic represents a fundamental domain of human ability (Gelman & Gallistel, 1978; Wynn, 1992a). Second, once the stability of the factor has been established, then factor-analytic studies can be used to study the development of the associated skills. In particular, these techniques can be used to determine the age at which arithmetical skills emerge as a distinct ability and how the relationship between arithmetical skills and, for instance, mathematical reasoning skills changes with the age of the child.

With regard to factor stability, the Number, or Numerical Facility, factor has been identified in dozens of psychometric studies (e.g., Canisia, 1962; Chein, 1939; Coombs, 1941; French, 1951; Goodman, 1943; Thurstone, 1938; Thurstone & Thurstone, 1941). A distinct Numerical Facility factor has been found with studies of American, Chinese, and Filipino students (Guthrie, 1963; Vandenberg, 1959), as well as with separate analyses of groups of males and females at varying ages (Dye & Very, 1968). In fact, Pawlik (1966) argued that "Verbal Comprehension (V) and Numerical Facility (N) are best confirmed of all aptitude factors known" (pp. 546–547). French (1951), in an exhaustive review of ability factors, stated that "the number factor is the clearest of them all" (p. 225). Even Spearman (1927), a staunch supporter of the position that individual differences in human abilities are best explained by a single factor, called General Intelligence (g), stated that basic arithmetical skills "have much in common over and above . . . g" (p. 251).

Across studies, the Numerical Facility factor is most strongly defined by arithmetic computation tests (e.g., tests that require the solving of complex multiplication problems) and by tests that involve a conceptual understanding of number relationships and arithmetical concepts, but not by tests that simply contain numbers as stimuli (Thurstone, 1938; Thurstone & Thurstone, 1941). Basically, the Numerical Facility factor appears to encompass all, or nearly all, of the basic arithmetical skills described in chapter 2 (Geary & Widaman, 1987). From the psychometric perspective, then, it can be argued that arithmetic, or numerical facility, "represents a unique ability" (Thurstone, 1938, p. 83). In other words, the psychometric studies suggest that arithmetical skills cluster together and define a single domain, which is distinct from other domains, such as spatial ability. This is not to say that individual differences in general intelligence do not influence performance on arithmetical and other ability measures but rather that there appears to be a set of processes and concepts that influence individual differences on arithmetic tests above and beyond the influence of general intelligence.

The next question to be addressed is, when in development does the Numerical Facility factor begin to emerge, that is, at what age are arithmetical skills identifiable as a distinct ability? Osborne and Lindsey (1967), in a longitudinal study of the factor structure of the Wechsler Intelligence Scale for Children (WISC; Wechsler, 1949), found support for the existence of a more or less distinct Numerical Facility factor for a sample of kindergarten children. The factor was defined by the Arithmetic and Digit Span subtests of the WISC, as well as by quantitative items from the Information subtest (e.g., How many pennies make a nickel?). This result suggests that for kindergarten children, numerical facility encompasses counting, simple arithmetic, working memory for numbers, and general knowledge about quantitative relationships. Meyers and Dingman (1960), in a review of the basic abilities of preschool children, also argued that numerical skills are identifiable as a relatively distinct ability by 5 to 7 years of age. This is not to say that distinct numerical and arithmetical skills are not evident before this age but rather that given data-gathering constraints (e.g., children's ability to take paper-and-pencil tests), distinct number skills are measurable using standard psychometric tests by the time children are in kindergarten. The Numerical Facility factor has also been identified for samples of children and adolescents in elementary school, junior high school, high school, and college (e.g., Dye & Very, 1968; Osborne & Lindsey, 1967; Thurstone & Thurstone, 1941). A distinct Numerical Facility factor is also evident in samples of elderly adults (Schaie, 1983).

Even though the Numerical Facility factor has been identified across the life span, there appears to be two developmental changes in the types of tests that define this factor. First, the factor becomes more exclusively arithmetical in nature with development. For instance, for third-grade children the Numerical Facility factor encompasses tests of arithmetic and working memory for numbers (Osborne & Lindsey, 1967), but for older adolescents and young adults the tests that define this factor are almost exclusively arithmetical in nature (e.g., Thurstone & Thurstone, 1941). The finding that tests for working memory for numbers and tests for arithmetic cluster together for preschool and elementary school children (Osborne & Lindsey, 1967) is probably due to the working-memory requirements of executing arithmetic procedures, such as the counting-on procedure described in chapter 2 (Widaman & Little, 1992). For elementary school children, Little and Widaman (in press) found that working memory for numbers directly contributed to performance on numerical facility tests, whereas Geary and Widaman (1992) found no such direct relationship for adults. In other words, tests of numerical memory span cluster with arithmetic tests for groups of elementary school and younger children because numerical memory span is important for early arithmetic skills, not because it is a basic feature of the arithmetic domain per se. As arithmetical processes become more automatic, that is, effortlessly executed, the numerical facility construct becomes almost exclusively arithmetical in nature.

The second developmental trend involves the relationship between performance on numerical facility tests and tests of mathematical reasoning, described below. Basically, for relatively unskilled samples, such as junior high school children, the numerical facility and mathematical reasoning tests are moderately correlated: The better the basic arithmetic skills, the better the performance on mathematical reasoning tests, which typically include arithmetic word problems (e.g., Chein, 1939; Dye & Very, 1968; Murray, 1949; Thurstone & Thurstone, 1941). For samples that include older adolescents or young adults, where—at least for studies 25 years or more old—nearly all of the subjects have mastered basic arithmetic, there is no relationship between the Numerical Facility and Mathematical Reasoning factors (e.g., Dye & Very, 1968).

Mathematical Reasoning

In addition to Numerical Facility, a Mathematical Reasoning factor has been consistently identified in factor-analytic studies (Canisia, 1962; Dye & Very, 1968; Ekstrom, French, & Harman, 1976; French, 1951; Goodman, 1943; Thurstone, 1938; Vandenberg, 1959; Very, 1967). Across studies, this factor has been given many different labels, including Arithmetical Reasoning, General Reasoning, and Mathematical Reasoning. For the sake of consistency, the factor will be called Mathematical Reasoning hereafter. Tests that define the Mathematical Reasoning factor typically require the ability to find and evaluate quantitative relationships and to draw conclusions on the basis of quantitative information. In fact, the mathematical reasoning tests used in the associated factor-analytic studies include, for the most part, problems that are essentially the same as those described in chapter 3. Thus, the descriptions of mathematical problem-solving skills in chapter 3 appear to describe mathematical reasoning ability as defined by psychometric studies. The psychometric studies can, however, add to the understanding of mathematical problem-solving skills in several ways. First, psychometric studies can be used to determine whether, and if so when, Mathematical Reasoning emerges as a distinct factor. Second, the psychometric approach can be used to assess the relationship between mathematical reasoning ability and other number and reasoning skills.

Thurstone and Thurstone (1941), in a large-scale ($N = 1,154$) study of eighth-grade children, found that arithmetical reasoning tests clustered with the traditional numerical facility tests. This result, as well as results from related studies (e.g., Guthrie, 1963), suggests that the Numerical Facility and Mathematical Reasoning factors do not necessarily represent distinct sets of skills for most junior high school children. In a second large-scale study, Dye and Very (1968) administered a battery of psychometric tests to 358 high school students in Grades 9 and 11 and to 198 college students. The tests included measures of numerical facility, perceptual speed, verbal reasoning, deductive reasoning, mathematical reasoning (called *arithmetical reasoning* in their study), and estimative ability, among others. A distinct Mathematical Reasoning factor was found for both males and females at all three grade levels.

Even though distinct Numerical Facility and Mathematical Reasoning factors were identified in the Dye and Very (1968) study, performance on numerical facility tests was moderately correlated with performance on mathematical reasoning tests for the ninth-grade students. However, the Numerical Facility and Mathematical Reasoning factors were uncorrelated for the college students. The relationship between the Numerical Facility and Mathematical Reasoning factors for ninth graders was probably found because many of the mathematical reasoning tests involved solving arithmetical word problems. Because good arithmetic skills will facilitate the solving of such problems, tests of numerical facility and mathematical reasoning should be modestly correlated for groups of subjects who have not yet mastered basic arithmetic (Geary & Widaman, 1992). The Dye and Very study suggests that the development of mathematical reasoning skills is initially dependent on, at least to some extent, numerical facility but that for college-age students, mathematical reasoning and numerical facility are distinct skills. In all, these studies suggest that the cognitive systems underlying numerical facility, or arithmetic, and mathematical reasoning begin to diverge during high school.

The development of mathematical reasoning skills that are distinct from arithmetical knowledge leaves unanswered the question of whether mathematical reasoning is distinct from other forms of reasoning (e.g., deductive and inductive reasoning). Dye and Very (1968) did not find a strong relationship between performance on the verbal and deductive reasoning tests and performance on mathematical reasoning tests. This result suggests that for some adolescents and young adults, the ability to reason in a quantitative domain is distinct from the ability to reason in nonquantitative domains. Nevertheless, other studies have found that tests of arithmetical and algebraic problem solving sometimes cluster with verbal and other nonquantitative reasoning tests (e.g., Ekstrom, French, & Harman, 1979; French, 1951; Vandenberg, 1959). In combination, this pattern of results suggests that mathematical reasoning initially develops from the skills represented by the Numerical Facility factor and by more general reasoning abilities. A distinct system of mathematical reasoning skills appears to emerge only with older adolescents (i.e., end of high school or early college) who have had a lot of experience with mathematics (e.g., Very, 1967).

Other Mathematical Abilities

The Numerical Facility and Mathematical Reasoning factors represent the only systems of mathematical abilities that have been consistently found across many factor-analytic studies. However, other mathematics-related factors have been identified in a few studies. It is not clear why these other potential mathematical abilities have not been found as consistently as numerical facility and mathematical reasoning, but one reason appears to be the type of tests that have been used across studies. Thurstone and Thurstone (1941), for instance, found evidence for a Dot Counting factor, which was defined by items that assessed the ability to quickly and accurately count arrays of dots. Dot counting tests have not typically been included in other psychometric studies. As a result, the stability of this factor has not actually been assessed. In addition to Dot Counting, several studies have also identified separate Digit Flexibility and Estimation factors (e.g., Canisia, 1962; Very, 1967).

The Dot Counting factor, identified by Thurstone and Thurstone (1941) for a group of eighth-grade children, is of interest because the tests that define this factor are very similar to the tasks that are used in subitizing experiments (see the Numerical Competencies in Human Infants section of chapter 1). In this study, performance on numerical facility tests was unrelated to performance on the dot counting tests. This result suggests that for junior high school children, arithmetic skills are not related to the skills that are associated with dot counting, that is, subitizing, counting, and estimating (Mandler & Shebo, 1982). Children, of course, sometimes count to solve arithmetic problems, which is why counting and arithmetic tests tend to cluster together for preschool children (Osborne & Lindsey, 1967), but counting was probably not used very often by these junior high school students to solve arithmetic problems (Ilg & Ames, 1951). Thus, counting skills would not have contributed to arithmetic skills for the children assessed in the Thurstone and Thurstone (1941) study. The finding of no correlation between the Dot Counting and Numerical Facility factors suggests that basic subitizing skills might be distinct from arithmetic skills, contrary to recent arguments that subitizing and arithmetic emerge from the same innate preverbal counting system (Gallistel & Gelman, 1992). This is not to say that subitizing is not used with the early development of arithmetic skills. Rather, even if subitizing can be used in counting and arithmetical development, this result suggests that the cognitive systems underlying subitizing and arithmetic might be different, at least by early adolescence.

The Digit Flexibility factor, which was identified by Very (1967) and Canisia (1962), involves the "ability to manipulate, to arrange, and to compare (numbers), without performing arithmetic operations" (Very, 1967, p. 187). Tests associated with this factor involve, for instance, making rapid *greater than* or *less than* comparisons of sets of numbers, writing as many numbers as possible that satisfy certain conditions (e.g., odd or even), or completing number series (e.g., 2, 4, 6, __). Canisia found that the Digit Flexibility factor was not correlated with the Numerical Facility and Mathematical Reasoning factors for a group of high school students. An Estimation factor was identified by Very (1967) and involves the ability to make quantitative estimations. Thurstone (1938) found that a quantitative estimation test did not cluster with the traditional numerical facility tests, though it did cluster with mathematical reasoning tests. Indeed, Very (1967) argued that the Estimation factor involved "the ability to create and evaluate new hypotheses quickly in order to draw proper conclusions" (p. 187), implying that estimation required reasoning skills. Thus, the *speed* of making quantitative comparisons appears to be distinct from arithmetical and mathematical reasoning skills. The ability to make good quantitative estimations might not differ, however, from mathematical reasoning ability.

In conclusion, although the evidence is not as consistent as in the studies of the numerical facility and mathematical reasoning constructs, some psychometric studies suggest that more minor mathematical domains might exist. These domains would appear to include the skills associated with subitizing (i.e., dot counting; Thurstone & Thurstone,

1941), digit flexibility (i.e., the ability to rapidly make numerical comparisons), and skill at making quantitative estimations. Of these three potential mathematical domains, evidence for the distinctiveness of the skills associated with the Dot Counting and Digit Flexibility factors is stronger than the evidence associated with the Estimation factor (Thurstone & Thurstone, 1941; Very, 1967). Dot counting and digit flexibility appear to be distinct from numerical facility and mathematical reasoning skills, at least for skilled individuals. The processes associated with the Estimation factor, on the other hand, might simply reflect mathematical reasoning ability (Petitto, 1990).

Summary

Factor-analytic studies are often not appreciated by contemporary developmental, cognitive, and mathematics education researchers. Yet these studies, when carefully conducted, can provide unique and useful information about human abilities. In particular, factor-analytic studies are useful for identifying potentially unique ability domains. Once it has been shown that a particular domain, such as verbal comprehension, is consistently found across studies and across samples drawn from different populations (e.g., China and the United States), then factor-analytic techniques can also be used to follow trends in the development of the domain. The techniques are especially useful for determining when the associated skills emerge as a separate domain and how the relationships among various domains might change with development.

For mathematics, factor-analytic studies have consistently found two basic domains. The first is numerical facility, which essentially encompasses the arithmetic skills described in chapter 2. The second domain is mathematical reasoning, which essentially encompasses the mathematical problem-solving skills described in chapter 3. Of these two domains, numerical facility is the most stable. In fact, the Numerical Facility factor is among the clearest and most stable factors identified across decades of psychometric research (French, 1951; Thurstone, 1938). The Numerical Facility factor is initially found for children between the ages of 5 and 7 years (e.g., Meyers & Dingman, 1960) and is always, or almost always, found for samples of elementary school children and adolescents in junior high school and high school, as well as for groups of younger and older adults. The most important developmental trend for this factor is that as children become more mathematically skilled, the factor becomes almost exclusively arithmetical in nature. In all, the stability of this factor is quite impressive and strongly supports the argument made in chapter 2: Arithmetic involves a fundamental domain of human ability.

Although a Mathematical Reasoning factor has been found in most studies that have included tests of arithmetical and algebraic problem solving, it is not clear to what extent mathematical reasoning is distinct from arithmetical and general reasoning skills (Dye & Very, 1968; Thurstone & Thurstone, 1941). Overall, it appears that the skills associated with the Mathematical Reasoning factor develop with experience in solving mathematics problems. That is, for most children a distinct set of mathematical problem-solving skills more than likely emerges during the high school years from basic arithmetic and general reasoning abilities. Nevertheless, once these skills have been developed, mathematical reasoning appears to be distinct from other numerical and reasoning abilities (e.g., Dye & Very, 1968). Finally, this does not preclude the possibility that some children and adolescents are especially skilled at reasoning about quantitative relationships (e.g., Benbow, 1988). The development of mathematical reasoning abilities for these mathematically gifted children might differ from that for children in general (e.g., O'Boyle, Alexander, & Benbow, 1991; see the section of chapter 6 entitled, The Mathematically Gifted).

Cognitive Perspective

Beginning in the 1970s, cognitive psychologists became interested in identifying the cognitive processes that underlie performance on traditional psychometric tests (e.g., Hunt, Lunneborg, & Lewis, 1975). For instance, Hunt et al. found that good performance on tests of verbal ability was associated, among other things, with the speed with which verbal information could be retrieved from long-term memory. The faster verbal information could be accessed from long-term memory, the better the performance on tests of verbal ability. The overall effort to link the psychometric and cognitive approaches to understanding individual differences in human abilities has led to mixed results (e.g., Keating, List, & Merriman, 1985). Nevertheless, research that has attempted to link cognitive studies of arithmetic (Ashcraft, 1992) with the psychometrically defined numerical facility and mathematical reasoning constructs has been relatively successful (e.g., Geary & Widaman, 1987; Siegler, 1988a). These studies have provided insights into the cognitive and strategic processes that underlie individual differences on numerical facility tests, as well as some information on the processes that contribute to individual differences in mathematical reasoning abilities. The results and implications of these studies are summarized in this section.

Numerical Facility

There have been several studies that have directly examined the relationship between the cognitive processes associated with solving arithmetic problems, such as retrieving arithmetic facts from long-term memory, and performance on numerical facility tests (Geary & Widaman, 1987, 1992; Widaman et al., 1992). In the first of these studies, Geary and Widaman administered a battery of paper-and-pencil tests to 100 college students and asked the students to solve a series of computer-administered arithmetic problems. The paper-and-pencil battery included numerical facility, perceptual speed (i.e., speed of encoding, or reading, symbols such as numbers), and spatial ability tests. The data from the computer task, along with some statistics, allowed us to break down the solving of arithmetic problems into component processes. The component processes included number encoding (i.e., reading numbers off the computer screen), retrieving arithmetic facts from long-term memory, and carrying (or trading). Moreover, these techniques provided information on the speed with which each subject could execute each of these processes. For instance, when solving complex addition problems, such as 47 + 79, one subject required 0.9 s to carry (or trade) from one column to the next, whereas a second subject required only 0.5 s to perform the same process.

The results showed a very strong relationship between the speed with which basic arithmetical processes could be executed and performance on the traditional numerical facility tests. Specifically, the faster arithmetic facts could be retrieved from long-term memory and the faster the carry (or trade) operation could be executed, the better the performance on the paper-and-pencil numerical facility tests. Moreover, the speed of executing these arithmetical processes was not directly related to performance on the perceptual speed and spatial tests. This pattern of results is important for several reasons. First, at least in the mathematical domain, the results show that contemporary cognitive researchers (e.g., Ashcraft, 1992) identified the same processes that underlie individual differences in basic arithmetical skills as did the early psychometric researchers (e.g., Thurstone, 1938). Second, the results of the study showed that the numerical facility construct is clearly arithmetical in nature and that the processes that contribute to arithmetical skills appear to be distinct from those processes that are associated with other mental abilities, in particular, spatial ability in this study (Geary & Widaman, 1987).

There have also been several studies that have examined the relationship between arithmetical processes and performance on traditional arithmetic tests for groups of children (Geary & Brown, 1991; Geary & Burlingham-Dubree, 1989; Siegler, 1988a; Widaman et al., 1992). In one of these studies (Geary & Burlingham-Dubree, 1989), preschool and kindergarten children were administered the Wechsler Preschool and Primary Scale of Intelligence (WPPSI; Wechsler, 1967) and the Arithmetic subtest of the Wide Range Achievement Test (WRAT; Jastak & Jastak, 1978) and were videotaped as they solved a series of simple addition problems. The videotapes were scored to determine the types of strategies the children used to solve the addition problems, such as counting on fingers, and the speed with which they executed the different strategies. Next, two cognitive variables were constructed. The first provided an estimate of how quickly the children could retrieve addition facts from long-term memory for those problems for which they correctly remembered an answer.

The second variable provided information on the child's skill at using alternative strategies. For instance, if the child could correctly retrieve the answer to an addition problem, then a skilled strategy choice would involve simply stating the retrieved answer. On the other hand, if an answer could not be correctly retrieved from memory, then a better strategy choice would involve solving the problem by, for example, counting on fingers (see the Strategy-Choice Model section of chapter 2). For children of this age, finger counting is often preferable to memory retrieval, because finger counting is typically more accurate than memory retrieval. A poor choice would involve stating any answer that came to mind, whether or not the answer was likely to be correct. In this study, as well as in several studies conducted by Siegler (1988a, 1993), it was found that children vary in the adaptiveness of their strategy choices (Geary & Burlingham-Dubree, 1989). Some children almost always choose the best problem-solving strategy, whereas other children consistently make poor strategy choices. Skilled strategy choices involve using the fastest and most accurate strategy available for solving each individual problem: choosing the strategy that provides a good balance between the amount of time required to execute the strategy and the likelihood that the strategy will produce the correct answer. Poor strategy choices often involve stating any answer that comes to mind, whether or not the retrieved answer is likely to be correct, or using inefficient counting strategies.

The primary purpose of the Geary and Burlingham-Dubree (1989) study was to determine the relationship of these speed-of-fact-retrieval and strategy-choice variables with performance on the ability and achievement tests. The results showed a strong to moderate relationship between the strategy-choice variable and performance on the Arithmetic subtests of the WPPSI and WRAT. The ability to make good strategy choices was associated with better performance on both arithmetic tests. Strategy choices were also moderately related to performance on several spatial subtests of the WPPSI but were unrelated to the verbal subtests. The relationship between strategy choices for solving addition problems and spatial ability was likely due to young children's use of spatial information, as in representing addition problems with their fingers, to solve arithmetic problems (e.g., Carpenter & Moser, 1984). The speed of addition-fact retrieval was also related to performance on the arithmetic tests but not to performance on the verbal or spatial subtests of the WPPSI. As was found by Geary and Widaman (1987) for college students and Widaman et al. (1992) for elementary school children, the faster basic arithmetic facts can be retrieved from memory, the better the performance on arithmetic tests for preschool children.

This is not to say, from a cognitive perspective, that individual differences in arithmetical skill are not related to differences in conceptual knowledge. In fact, Geary, Bow-Thomas, and Yao (1992), in a study of first-grade children, found that counting knowledge strongly influenced individual differences in making skilled strategy choices for solving addition problems. In particular, the child's use of the efficient counting-on procedure appeared to be dependent on a mature understanding of counting concepts. Both conceptual knowledge and skill at making good strategy choices were related to performance on a traditional mathematics achievement test. Individual differences in the understanding of counting concepts contributed to individual differences in the use of developmentally sophisticated counting strategies, such as the counting-on procedure, to solve addition problems and directly contributed to individual differences in mathematical achievement scores. In a more recent study, Geary, Bow-Thomas, et al. (1993) found that working memory also influenced kindergarten children's use of counting strategies. In particular, the use of verbal counting strategies, as opposed to finger counting, was related to good working-memory skills.

These studies show a clear continuity between the psychometric studies described in the previous section and contemporary studies of individual differences in basic arithmetical abilities. Psychometric and cognitive researchers have identified the same sources of individual differences in basic arithmetical skills. Both approaches also suggest that the source of individual differences in basic skills changes somewhat with development. For preschool and elementary school children, individual differences in skill at making good strategy choices and factors that affect those choices, such as conceptual knowledge and working memory, appear to be a primary source of individual differences in arithmetical ability. The speed with which basic operations, such as fact retrieval, can be executed also influences ability differences in children but appears to be a minor factor as compared with individual differences in strategic skills and perhaps conceptual knowledge. As basic facts become committed to memory and as the use of procedures becomes more or less automatic, strategy differences across individuals largely disappear. At this point, the primary source of individual differences in arithmetical abilities is the speed with which basic arithmetical processes, such as carrying or trading, can be executed.

In conclusion, children who excel in arithmetic appear to have a good understanding of basic number and arithmetic concepts and to have a good numerical working memory (Geary, Bow-Thomas, et al., 1993; Geary, Bow-Thomas, & Yao, 1992). The well-developed conceptual and working-memory skills of these children result in an ability to make effective strategy choices during problem solving. The effective use of alternative problem-solving strategies, in turn, appears to be the primary source of individual differences in early arithmetical abilities, although the speed of executing basic arithmetical processes is also important (Geary & Burlingham-Dubree, 1989; Siegler, 1988a).

Mathematical Reasoning

Although there have been several cognitive studies of individual differences in reasoning skills in general (e.g., Sternberg, 1977; Sternberg & Gardner, 1983), there has been relatively little research, from a cognitive perspective, on individual differences in mathematical reasoning ability. In one study that was conducted in this area, Geary and Widaman (1992) examined the relationships among cognitive measures of the speed of executing arithmetical operations, such as carrying; the ability to perform arithmetical operations in working memory; and performance on a battery of psychometric tests for 112 Air Force recruits. The psychometric tests defined the Numerical Facility, Perceptual Speed, Mathematical Reasoning (called General Reasoning in this study), and Memory Span factors. The Mathematical Reasoning factor was defined by two sets of arithmetic-word-problem tests. One of the sets of tests required the recruits to actually solve word problems, whereas the second required the recruits to determine the sequence of arithmetic operations, such as addition and multiplication, that were necessary to solve the problem, without actually solving the problem. The Memory Span factor was defined by tests that required the recruits to remember visually and verbally presented strings of numbers for a short period of time.

One goal of this study (Geary & Widaman, 1992) was to determine the extent to which individual differences in mathematical reasoning abilities were related to the speed of executing basic arithmetical processes and skill at mentally performing arithmetic operations in working memory. The results replicated the findings of our first study; the speed of executing basic arithmetical processes was strongly related to performance on the numerical facility tests. In fact, across both studies, the speed of executing basic processes accounted for roughly 80% of the individual differences in the basic arithmetical abilities of young adults. Although the speed of executing basic arithmetical operations also contributed to performance on the tests of mathematical reasoning skills, they were not a primary source of individual differences in mathematical reasoning ability, in keeping with the findings of the earlier-described psychometric studies. The ability to mentally perform arithmetic operations in working memory, however, was an important source of individual differences in mathematical reasoning ability, explaining about 25% of the differences across adults in this ability.

On the basis of this study, in combination with individual-differences research on general reasoning abilities (Sternberg & Gardner, 1983) and research described in chapter 3, it can be tentatively argued that individual differences in mathematical reasoning ability are probably related to four sets of cognitive skills. The first two involve the skills associated with mathematical problem solving described in chapter 3 (A. B. Lewis & Mayer, 1987). Although these skills have not been explicitly assessed within an individual-differences framework, the studies described in chapter 3 do allow inferences to be made about potential sources of individual differences in mathematical problem-solving skills. First, the ability to mentally set up representations of arithmetic word problems is an important source of problem-solving errors for college students. This result suggests that individual differences in representational skills will likely be an important source of individual differences in mathematical reasoning ability. Second, the ease with which problem-solving schemas are developed also appears to be a likely source of individual differences in mathematical reasoning skills (Mayer, 1981).

The third source of individual differences in mathematical reasoning appears to be working memory, or the ability to keep important information in mind while performing mathematical operations. Finally, the speed with which basic mathematical processes, such as carrying or trading, can be executed also contributes to differences in mathematical reasoning ability. In all, individuals who are skilled at mathematical reasoning appear to execute basic computations quickly and automatically, are able to keep important information in mind while performing other operations, and have developed schemas to aid in the representation, translation, and solution of mathematical problems.

Summary

The cognitive studies of individual differences in basic mathematical abilities provide an important supplement to the earlier-described psychometric studies. In particular, the cognitive research has provided insights into the specific mental processes that underlie performance on tests that define the Numerical Facility and Mathematical Reasoning factors. For preschool and elementary school children, individual differences on tests of numerical facility, or arithmetic, appear to be primarily related to the child's sophistication and skill at using alternative problem-solving strategies. Skilled strategy choices are reflected in the ability to balance speed and accuracy demands when choosing to use one problem-solving strategy over another (Geary & Burlingham-Dubree, 1989; Siegler, 1988a) and appear to be dependent on a good understanding of the underlying concepts (Geary, Bow-Thomas, & Yao, 1992) and working memory (Geary, Bow-Thomas, et al., 1993).

More precisely, a skilled strategy choice involves using the strategy that is most easily executed and will most likely provide the correct answer for each problem. Children who are skilled in this area might retrieve the answer to one problem, count verbally to solve the next problem, and count on their fingers to solve yet other problems. Poor strategy choices in children are associated with using inefficient counting strategies and with the overuse of memory retrieval to solve arithmetic problems. That is, many of these children state retrieved answers whether or not they are likely to be correct. Finally, the speed with which elementary processes can be executed also contributes, though to a lesser extent, to individual differences among children in basic arithmetical abilities (Geary & Brown, 1991; Widaman et al., 1992). With more skilled samples, where individual differences in strategy choices are less evident, performance differences on numerical facility tests are determined by the speed with which basic arithmetical processes, such as carrying or trading, can be executed. The more quickly and accurately the processes can be executed, the better the performance on paper-and-pencil tests of numerical facility (Geary & Widaman, 1987, 1992).

The cognitive factors that contribute to individual differences in mathematical reasoning ability are not as well understood as those associated with the numerical facility construct. Nevertheless, it appears that the speed with which basic processes can be executed and the ability to keep information in mind while mentally performing mathematical operations are important sources of individual differences in mathematical reasoning ability (Geary & Widaman, 1992). Moreover, it is very likely that individual differences in the ability to mentally translate and represent the meaning of arithmetical and algebraic word problems (e.g., A. B. Lewis & Mayer, 1987), along with the ease with which the associated schemas develop, are important sources of performance differences in mathematical reasoning. Individual-difference studies that explicitly examine these skills, in concert with arithmetical processing and working-memory skills, for their relation to performance on mathematical reasoning tests are needed to fill in the gaps in our understanding of this area.

Behavioral Genetic Perspective

Behavioral genetic research is concerned with determining the extent to which genetic factors contribute to individual differences in observable traits in a variety of areas (e.g., intelligence, physical development, and personality; Plomin et al., 1990). The primary measure used to express this relationship is heritability, or h^2. Heritability values provide an estimate of the amount of variability in a particular trait for a given population that is due to individual differences in the constellation of genes that support the trait; h^2 cannot be used to make judgments about specific persons, only populations. For instance, if it is determined that the h^2 estimate for intelligence is .50, then it can be argued that 50% of the variability in IQ scores across people appears to be due to variability in the genes that support intelligence (cf. Bouchard, Lykken, McGue, Segal, & Tellegen, 1990). It cannot be stated, however, that 50% of one person's IQ score is due to genes and that the remaining 50% is due to environmental factors or to some interaction between genetic and environmental factors.

In the behavioral genetic studies of the heritability of mathematical ability, two commonly used methods for estimating h^2 have been used (Plomin et al., 1990). The first involves comparing the degree of trait similarity between sets of identical or monozygotic (MZ) twins with the degree of trait similarity between sets of fraternal or dizygotic (DZ) twins. Monozygotic twins are genetically identical, whereas DZ twins share, on average, about 50% of their genes. One type of heritability estimate can be obtained by doubling the difference between the trait correlation between MZ and DZ twin pairs (Plomin et al., 1990). So if MZ and DZ twin pairs correlated .80 and .60, respectively, on some test, then the h^2 estimate would be .40, that is, 2(.80 − .60). The other commonly used method for making judgments about potential genetic, as well as environmental, influences on traits is to examine the extent to which parents and children are similar on a given trait; these are called *familial resemblance studies*. The focus of this section is on twin and familial resemblance studies of the numerical facility and mathematical reasoning constructs described throughout this chapter. The goal is to assess the extent to which arithmetical and mathematical problem-solving abilities might be heritable. Information on the heritability of arithmetic and mathematical reasoning abilities will provide insights into the relative contributions of biological and environmental factors on the development of these abilities.

Numerical Facility

Twin studies of the numerical facility construct consistently indicate moderate heritability for the associated arithmetical abilities (Husén, 1959; Nichols, 1978; Vandenberg, 1962, 1966). For instance, in a review of twin studies of psychological traits, Vandenberg (1966) found the median h^2 estimate, across several studies, for performance on the Numerical Facility factor (i.e., overall performance summed across several numerical facility tests) to be .45. Similarly, in another study, Vandenberg (1962) found h^2 estimates for individual numerical facility tests, such as an addition test, to range between .43 and .58. Husén, in a large-scale study of twin-pair Swedish military inductees, found an h^2 estimate of .66 for elementary school grades in arithmetic. DeFries and his colleagues showed that the arithmetical skills of parents and children are positively correlated, suggesting that both genetic and environmental factors contribute to good arithmetical skills (DeFries et al., 1976, 1979). In all, these studies suggest that roughly one half of the variability in arithmetical ability is due to genetic factors.

Mathematical Reasoning

Because many of the studies noted below did not include specific tests of mathematical reasoning, the results from studies of the heritability of mathematical reasoning abilities are less certain than the results associated with studies of the numerical facility construct. Nevertheless, Vandenberg (1966) found that performance on general reasoning tests, which often include arithmetic word problems, showed, across several studies, a median h^2 estimate of .27. Loehlin and Nichols (1976) studied 850 twin pairs, who were selected from a pool of high school juniors who had taken the National Merit Scholarship Qualifying Test (NMSQT). Among other things, they examined the differences between MZ and DZ correlations for five NMSQT subtests, including Mathematics. Although the NMSQT Mathematics subtest assesses a variety of mathematical abilities, given the relatively skilled sample assessed in this study, it is likely that individual differences in performance on this measure were largely due to mathematical reasoning skills and not due to basic arithmetic skills. The h^2 estimates for this subtest, across two cohorts and gender, ranged between .36 and .66.

In all, these studies suggest that a small to moderate proportion of individual differences in mathematical reasoning ability are due to genetic factors. It is not clear, however, whether these results reflect general reasoning skills or specific mathematical reasoning abilities. Behavioral genetic studies of performance on traditional psychometric tests of mathematical reasoning (e.g., Ekstrom et al., 1976), in combination with numerical facility and nonquantitative reasoning tests, are necessary before more definitive conclusions can be drawn about the heritability of mathematical reasoning abilities above and beyond the heritability of arithmetic and general reasoning abilities.

Summary

Behavioral genetic studies of the heritability of the numerical facility construct suggest that roughly one half of the variability in arithmetical abilities is due to genetic differences across people (e.g., Vandenberg, 1966). As noted in the previous section and in chapter 2, arithmetical ability involves a variety of component skills (e.g., Geary & Burlingham-Dubree, 1989). Behavioral genetic studies of these different component skills, such as the speed of arithmetic fact retrieval and computational skills, would be very useful, as would studies of potential developmental changes in the associated h^2 estimates. At this point, it can only be stated that overall arithmetical ability appears to have a substantial heritable component. It cannot be stated to what extent individual differences in strategic, conceptual, or speed-of-processing skills in arithmetic are heritable. As has been done in other domains (Plomin et al., 1990), behavioral genetic studies that explicitly assessed the heritability of these separate component skills would be a logical next step in the genetic studies of the numerical facility construct.

The results of the heritability studies of mathematical reasoning are less conclusive than the studies of the numerical facility construct. This is because the mathematics and reasoning tests used in these studies were not typically the same tests that were used in the psychometric studies of the Mathematical Reasoning factor. Nevertheless, studies of performance on complex mathematical tests do suggest that at least some proportion of individual differences in mathematical reasoning ability might be heritable (e.g., Loehlin & Nichols, 1976; Nichols, 1978). Without behavioral genetic studies that directly use traditional mathematical reasoning tests, however, it is premature to make h^2 estimates for mathematical reasoning. Moreover, on the basis of findings that suggest that mathematical reasoning skills emerge in most individuals only after extensive mathematical experience (e.g., Very, 1967), it seems likely that the reported h^2 estimates for complex mathematics tests reflect the heritability of general reasoning abilities rather than mathematical reasoning abilities per se. If mathematical reasoning were a biologically primary and therefore heritable trait, then the Mathematical Reasoning factor would presumably be distinct from the Numerical Facility factor and other reasoning skills without extensive mathematics course work in high school.

Conclusion

The study of individual differences in numerical and mathematical ability is often ignored or misunderstood by contemporary developmental, cognitive, and mathematics education researchers (Schoenfeld, 1985). Yet the individual-differences perspective has much to offer, in terms of identifying basic domains of mathematical ability and understanding why people differ in the skills associated with these domains. The psychometric research, for instance, strongly supports the position that two relatively distinct types of mathematical abilities exist: numerical facility and mathematical reasoning (French, 1951; Thurstone, 1938). Indeed, it is almost certainly not a coincidence that researchers in the mathematical development and mathematics education areas have implicitly identified the same two domains. These domains were, of course, described in chapter 2 and chapter 3 and are termed *arithmetic* and *mathematical problem solving*, respectively, by contemporary researchers. At this point, it is not clear whether the number skills described in chapter 1 represent a mathematical domain distinct from arithmetic and mathematical problem solving.

Within the psychometric tradition, computational skills and conceptual knowledge of arithmetic have been consistently identified with and represented by the Numerical Facility factor (Thurstone, 1938). The numerical facility construct has emerged as one of the clearest and most stable human abilities identified through decades of psychometric research (e.g., French, 1951). The psychometric, cognitive, and behavioral genetic studies described in this chapter, along with some of the infancy studies (e.g., Starkey, 1992; Wynn, 1992a), strongly support the argument that many aspects of arithmetic, or numerical facility, represent a fundamental, perhaps a biologically primary (i.e., inborn), domain of human ability. This is not to say that culture and experience do not influence arithmetical development. They surely do. Rather, the results from the psychometric and behavioral genetic studies are consistent with the conclusion drawn in chapter 2: An intuitive understanding of many, though not all, features of arithmetic might be an inherent human ability.

From a developmental perspective, a more or less distinct Numerical Facility factor is identifiable in children by 5 years of age (Meyers & Dingman, 1960). At this age and for elementary school children, numerical facility encompasses arithmetic, as well as related skills such as working memory (Osborne & Lindsey, 1967). The clustering of arithmetical and memory span tests under the numerical facility umbrella most likely reflects children's reliance on working-memory skills to solve arithmetic problems (Widaman & Little, 1992) instead of reflecting working memory as a part of the arithmetic domain per se. In support of this view is the finding that with more skilled samples, such as older adolescents, the Numerical Facility factor is almost exclusively arithmetical in nature (Thurstone, 1938). This is most likely because as basic arithmetical processes become more automatic, working-memory resources become unnecessary for basic arithmetic. At this point, arithmetic and working-memory tests should not, and do not, cluster together.

The cognitive studies also show that the nature of the numerical facility construct changes somewhat with development. For preschool and elementary school children, individual differences in arithmetical abilities appear to be primarily related to differences in children's working memory; differences in the conceptual understanding of arithmetic and related domains, such as counting; and differences in the ability to effectively use alternative problem-solving strategies. Children who excel in arithmetic are able to implicitly choose the best problem-solving strategy available to them to solve each individual arithmetic problem. These children implicitly balance the time and effort required to execute the strategy with the likelihood that the strategy will produce the correct answer (Geary & Burlingham-Dubree, 1989; Siegler, 1988a). Less skilled children, on the other hand, tend to use the same strategy, such as memory retrieval, to solve most problems, whether or not it is the most efficient and accurate strategy for a particular problem.

For older adolescents and adults, strategy choices become less necessary because most problems can be solved by retrieving the answer from long-term memory and through executing overlearned procedures. At this point, individual differences in arithmetical ability are strongly related to the speed with which basic arithmetical processes can be executed (Geary & Widaman, 1987, 1992). Finally, the behavioral genetic studies suggest that roughly one half of the individual differences in arithmetical abilities are due to genetic differences (Vandenberg, 1966), although it is not clear whether the heritability of arithmetical abilities varies with the age of the individual. Moreover, all of the behavioral genetic studies of arithmetical abilities have been conducted with traditional ability and achievement tests. So at this point, it is not clear to what extent the individual components of arithmetic ability, such as conceptual knowledge, strategy choices, or speed of processing, are heritable. Future behavioral genetic studies that use the cognitive measures of arithmetic skills for children and adolescents of varying ages should help to clarify these issues.

The psychometric studies have also consistently identified a Mathematical Reasoning factor (e.g., Dye & Very, 1968; Thurstone, 1938). The mathematical reasoning construct essentially encompasses the mathematical problem-solving skills that were described in chapter 3. A distinct set of mathematical problem-solving abilities appears to emerge, for most people, during the high school years (Dye & Very, 1968). Before this point, tests of mathematical reasoning tend to cluster with the traditional numerical facility tests or with nonquantitative reasoning tests (Thurstone & Thurstone, 1941). This developmental pattern suggests that for the typical adolescent, mathematical problem-solving skills are to a large extent learned, as opposed to being a biologically primary ability. In other words, for most adolescents, mathematical problem-solving strategies are probably constructed, or learned in school, from their understanding of the arithmetical domain combined with general problem-solving skills. Either way, it appears that people who are skilled at mathematical reasoning can execute basic mathematical operations quickly and automatically; are able to manipulate many mathematical operations simultaneously in working memory; and have well-developed schemas that can be used to represent, translate, and solve mathematical problems.

Even though the behavioral genetic studies suggest a genetic contribution to individual differences in mathematical reasoning ability (e.g., Nichols, 1978), the nature of these particular studies makes it unclear whether the associated heritability estimates reflect general reasoning skills or mathematical reasoning abilities in particular. For the most part, the finding that the emergence of a distinct Mathematical Reasoning factor does not occur, for most individuals, except after a lot of mathematics courses in high school suggests that these skills are primarily learned and not inherited. Nevertheless, the research of Benbow (1988) and her colleagues does suggest that a few adolescents are especially skilled in reasoning about quantitative relationships before much exposure to complex mathematics. On the basis of these findings, it seems likely that there is more than one route to strong mathematical problem-solving skills (see the section of chapter 6, entitled, The Mathematically Gifted). Clearly, there is much to be learned about developmental and individual differences in mathematical reasoning.

5 Mathematical Disabilities

I t has been estimated that about 6% of school-age children have some type of learning disorder in mathematics (Badian, 1983; Kosc, 1974). About 70% of these children are boys (Badian, 1983). Many, though not all, mathematically disabled (MD) children also have problems in learning how to read, whereas many reading disabled (RD) children have problems with mathematics. Clearly, learning problems in mathematics, in one form or another, affect many children (Sutaria, 1985). The goal of this chapter is to integrate cognitive, neuropsychological, and behavioral genetic research on MD and related disorders to provide a better understanding of the different forms of MD (see Geary, 1993, for a more technical presentation); suggestions for treating the various forms of MD discussed in this chapter are provided in the Remediation section of chapter 8. The overall focus of this chapter is on arithmetic rather than mathematical problem solving, because most of the research that has been conducted with MD children is in the arithmetic area. Nevertheless, a brief consideration of the relationship between arithmetic-related forms of MD and specific difficulties in mathematical problem solving follows the presentation of the arithmetic-related disabilities.

In all, the chapter includes five major sections plus a conclusion. In the first, cognitive research that has contrasted normal and MD children on basic arithmetic skills is presented. In preview, these studies suggest two potentially distinct forms of cognitive deficit in MD children. The first involves difficulties in representing arithmetic facts in, or retrieving facts from, long-term memory. In other words, many MD children have

---

Summaries are provided at the end of the following main sections: Cognitive Component (p. 168) and Neuropsychological Component (p. 174).

problems remembering basic arithmetic facts, such as $5 + 9 = 14$, even with extensive drilling (Geary et al., 1991; Howell, Sidorenko, & Jurica, 1987). The second deficit involves difficulties in executing arithmetical procedures, such as carrying or trading in complex addition, or in executing counting procedures to solve simple addition problems. Many MD children appear to outgrow the procedural deficit, but the memory-retrieval problems tend to be persistent (Geary et al., 1991).

The second section presents clinical studies of acquired and developmental dyscalculia (e.g., Temple, 1991). *Acquired dyscalculia* refers to number and arithmetic problems that result from some form of brain injury. *Developmental dyscalculia* refers to children's problems in understanding numerical concepts or arithmetic learning. These children have not necessarily suffered from some form of overt brain injury, although it is assumed that a neuropsychological deficit underlies the learning problems. The research presented in this section suggests that three types of arithmetic-related learning disorders exist. In keeping with the cognitive studies, the neuropsychological research suggests that there are distinct fact-retrieval and procedural deficits in the dyscalculias. The third deficit involves the disruption of the ability to spatially represent numerical information (e.g., columnar alignment in complex problems, such as $35 + 74$) and the conceptual understanding of the representation (e.g., place value). Because cognitive researchers have not typically assessed the visuospatial skills of MD children, it is not surprising that they have not identified a spatial-related deficit in their studies.

Unfortunately, there have been no behavioral genetic studies specifically of MD. Nevertheless, it is argued in the third section that certain forms of MD might be heritable. The rationale for this position is based by analogy on two lines of evidence for genetic contributions to RD. First, behavioral genetic studies suggest that certain forms of RD are heritable (Olson, Wise, Conners, Rack, & Fulker, 1989; Pennington, Gilger, et al., 1991). Some scientists argue that RD is etiologically distinct from normal reading skills, whereas others argue that RD might simply represent the lower end of the distribution of normal reading ability (e.g., Gilger, Borecki, Smith, DeFries, & Pennington, 1993; Shaywitz, Escobar, Shaywitz, Fletcher, & Makuch, 1992). If RD represents the lower end of the distribution of normal reading skills, then the genes that influence normal variability in reading skills might be the same genes that underlie certain forms of RD.

By analogy, then, the evidence for the heritability of arithmetic skills (presented in the Behavioral Genetic Perspective section of chapter 4) would suggest that certain forms of MD might simply represent the lower end of normal variability in arithmetic skills and therefore show a heritable component. Second, certain forms of RD and MD occur together (Richman, 1983). Because there is a substantial heritable component to RD, if the same cognitive or neuropsychological systems contribute to RD and MD in these children, then certain forms of MD might also be heritable (Gilger, Pennington, & DeFries, 1992). The relationship between RD and MD is explored further in the fourth section. Here, potential cognitive, neuropsychological, and genetic commonalities in RD and MD are discussed. As noted above, the relationship between learning disorders in arithmetic and mathematical problem solving is presented in the fifth section.

Before beginning, note that children who have been identified as having some form of MD are almost certainly a heterogeneous group (Strang & Rourke, 1985). In other words, different children have difficulties in learning basic mathematics for different reasons. No doubt, in many of the studies that are described in this chapter, children have been identified as MD for different reasons. It is also likely that some children who perform poorly in mathematics have *no* underlying cognitive or neuropsychological deficit at all (Geary, 1990). In our studies, roughly half of the children who had been identified as having a learning problem in mathematics did not show any form of cognitive deficit (e.g., Geary, 1990; Geary, Bow-Thomas, & Yao, 1992). These children probably perform poorly in mathematics because of a lack of experience, poor motivation, or anxiety (M. D. Levine, 1987). Despite these problems, both the cognitive and neuropsychological studies converge on the same conclusion: Many children have difficulty in acquiring even basic mathematical skills because of one or more underlying cognitive or neuropsychological deficits, certain forms of which might be heritable.

Cognitive Component

As noted in the Cognitive Perspective section of chapter 4, the use of cognitive methods provides much more precise information on individual differences in component skills than does the use of paper-and-pencil tests. Thus, the focus of this section is on cognitive studies rather than studies that have compared MD and normal children by means of traditional paper-and-pencil achievement measures (e.g., Cawley & Miller, 1989). As noted earlier, the focus is on the basic cognitive processes that are associated with solving arithmetic problems. Cognitive measures of arithmetic are well suited for the study of the basic skills of MD children, because much is known about the development of arithmetic skills (see chapter 2). In all, the review of the cognitive studies of MD is presented in three sections: Arithmetic Development in MD Children, Review of Arithmetic Development in Normal Children, and Specific Cognitive Deficits of MD Children.

Arithmetic Development in MD Children

In this section, a brief description of the arithmetic skills of MD children is provided. After a brief review of arithmetic development, the potential deficits that might account for this pattern are discussed. Basically, in comparison with their normal peers, when MD children solve simple arithmetic problems, they tend to use immature problem-solving strategies, for instance, they use the counting-all rather than the counting-on procedure; they have rather long solution times; they frequently commit procedural and memory-retrieval errors; and they make poor strategy choices (Fleischner, Garnett, & Shepherd, 1982; Garnett & Fleischner, 1983; Geary, 1990; Geary & Brown, 1991; Geary et al., 1991; Geary, Widaman, Little, & Cormier, 1987; Goldman, Pellegrino, & Mertz, 1988; Kirby & Becker, 1988; Svenson & Broquist, 1975).

In regard to their development, MD children use the immature counting-all procedure for several years beyond the point when most academically normal children have abandoned it in favor of counting on. Whether they use the counting-all or counting-on procedure, younger MD children make many more errors than normal children. Also, MD children do not remember as many arithmetic facts as their normal peers, and when they do retrieve a fact, it is often wrong. This pattern is evident for many MD children, even at the end of the elementary school years (Geary et al., 1987). Finally, MD children require much more time, on average, to solve each problem than do normal children (Kirby & Becker, 1988).

Review of Arithmetic Development in Normal Children

Even though the components that contribute to the development of arithmetic skills have been described in chapter 1 and chapter 2, they are briefly reviewed here. A review of this information should make the cognitive assessment of the arithmetic skills of MD children easier to follow. This section contains a review of children's counting knowledge, strategy choices, and the development of representations of arithmetic facts in long-term memory.

Counting Knowledge

Counting knowledge and counting skills provide the framework for early arithmetical development (Kaye, 1986). It is therefore possible that MD children's difficulties in basic arithmetic are related to a poor understanding of counting concepts (Geary, Bow-Thomas, & Yao, 1992; Ohlsson & Rees, 1991). Recall that there are two basic approaches to understanding children's counting knowledge (see the Number and Counting section of chapter 1). In the principles-first model, it is assumed that children have an implicit understanding of counting and that this knowledge governs their counting behavior (Gelman & Gallistel, 1978). Of particular importance are three how-to-count principles: one-one correspondence, stable order, and cardinality. *One-one correspondence* refers to the implicit understanding that one and only one word tag (e.g., *one* or *two*) can be assigned to each counted object, whereas knowledge of the *stable-order* principle is reflected in the use of the same order of word tags across counted sets. Finally, *cardinality* refers to the child's understanding that the value of the final word tag represents the number of counted items.

With the alternative procedures-first model, counting is initially done by rote, that is, without an understanding of counting concepts (Briars & Siegler, 1984). Gradually, with practice, children induce the essential and nonessential features of counting. Of particular importance is the child's understanding of nonessential but common features of counting. One of these features is adjacency. This feature reflects the fact that when items are counted, they are typically counted one after the other rather than skipping around. Children who have a relatively mature understanding of counting know that items can be counted in any order. Children who have an immature conceptual understanding of counting, on the other hand, believe that counting that violates the adjacency feature is wrong.

Strategic and Memory Development

As described in chapter 2, children use many different types of strategies when solving arithmetic problems. If they cannot retrieve the correct answer from long-term memory, then they might guess or resort to some type of backup strategy to complete problem solving. Recall that backup strategies, for children in the United States, typically involve either counting on fingers or counting verbally (Carpenter & Moser, 1984; Siegler, 1987). For addition, if counting, either on fingers or verbally, is required to solve the problem, then children typically use either the counting-on procedure or the counting-all procedure. Counting on involves starting the count with the cardinal value of the larger addend and then counting the number of times indicated by the value of the smaller addend until a sum is obtained (Fuson, 1982). Counting all is a developmentally less mature procedure that involves counting both addends to get an answer. Across all arithmetic operations, strategy development involves a gradual shift away from using counting procedures to the direct retrieval of facts from long-term memory.

Changes in the child's mix of strategies appear to be related to several factors, including counting knowledge, working memory, and the development of memory representations of basic facts (Geary, Bow-Thomas, et al., 1993; Siegler & Jenkins, 1989). Counting knowledge appears to be important for the child's abandonment of the immature counting-all procedure, in favor of the more sophisticated counting-on procedure (Geary, Bow-Thomas, & Yao, 1992). *Working memory* refers to the child's ability to keep information in mind while performing other mental operations. Working-memory skills are related to several factors, the details of which are beyond the scope of this presentation (Woltz, 1988). Basically, working memory appears to influence a child's skill at executing arithmetical procedures and the development of long-term memory representations of basic facts (see Geary, 1993, for details).

Recall that the development of memory representations appears to be related to the use of counting procedures (see the Strategy-Choice Model section of chapter 2). When counting is used to solve an arithmetic problem, the answer generated by the count and the problem appear to become associated in memory. So if the child counts "5, 6, 7," to solve the problem 3 + 4, then the last number stated, 7, becomes associated with 3 + 4. After many such counts, the child eventually retrieves 7 when given the problem 3 + 4 to solve. However, for counting to lead to the development of memory representations, it appears that the child must have the augend, addend, and generated answer simultaneously in mind or in working memory (Geary et al., 1991). If the child uses the counting-all procedure to solve 3 + 4, for instance, then the child states "1, 2, 3, 4, 5, 6, 7." With such a long count, many children probably forget what the first number was (i.e., 3) before finishing the count. Thus, the child might get the right answer but forget all or parts of the problem in the process. In this circumstance, the problem and generated answer are less likely to become associated in long-term memory.

In other words, with slow counting the child might not develop associations between problems and answers and, as a result, have trouble memorizing the basic arithmetic tables. From this perspective, the development of memory representations for basic arithmetic facts should be related to how quickly the child can count. Slow counting should lead to a slow development of memory representations of basic facts, whereas faster counting should lead to the development of memory representations with many fewer counting trials (Geary, Bow-Thomas, et al., 1993). The accuracy of the child's counting is also important, because if the child makes a lot of counting errors, then these incorrect answers would become associated with the problem. As a result, the child is more likely to retrieve wrong answers during subsequent attempts to solve the problem (Geary, 1990; Siegler, 1986).

Specific Cognitive Deficits of Mathematically Disabled Children

On the basis of the above-described developmental factors, the basic cognitive deficits of MD children potentially involve five component skills: counting or other types of arithmetical procedures, fact retrieval, conceptual knowledge, working memory, and speed of processing (especially counting speed). Of these, the procedural and fact-retrieval components are functional skills. That is, they are used directly to solve arithmetic problems. The three remaining components, on the other hand, represent skills that contribute to the procedural and fact-retrieval components. For example, working memory is important for the accurate use of counting procedures, but it is the counting that produces an answer to the problem, not the child's working-memory skills. To make this a little clearer, the potential relationships among these component skills are presented in Figure 5.1. A consideration of each of these components as a potential cause of arithmetic-related forms of MD is presented in the following sections.

Procedural and Fact-Retrieval Skills

One of the more detailed cognitive studies of MD children was conducted by Geary (1990). In this study, groups of first-grade MD and normal children were administered a cognitive addition task. The cognitive addition task provided information on the types of strategies the children used to solve simple addition problems and on how quickly they executed

Figure 5.1

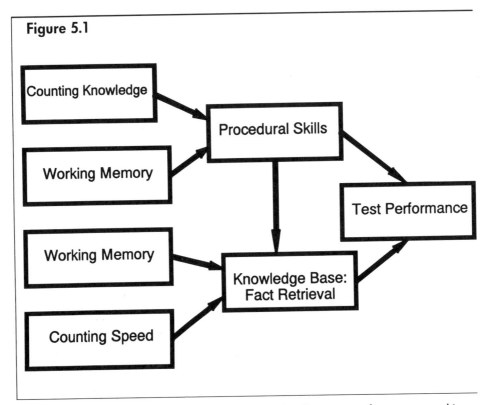

Schematic representation of the proposed relationships between performance on achievement and ability tests, contributing cognitive skills (i.e., procedural and fact retrieval), and underlying cognitive components (i.e., counting knowledge, working memory, and counting speed). (Adapted from "Mathematical Disabilities: Cognitive, Neuropsychological, and Genetic Components," by D. C. Geary, 1993, *Psychological Bulletin, 114*, p. 349. Copyright 1993 by the American Psychological Association.)

the strategies. Selection of the MD group was based on placement in the Chapter 1 remedial education program in mathematics (a federally funded program for children with low achievement scores in mathematics). The MD children were placed in this program on the basis of performance on achievement tests that were administered at the end of the kindergarten year. At the end of first grade, about half of these children tested out of remedial education. Thus, there were two groups of MD children: those who showed improved mathematic skills from the end of kindergarten to the end of first grade and those who showed no relative change in mathematic skills. The resulting three groups of children—normal, MD-improved, and MD-no change—were administered the cognitive addition task at the end of first grade.

On the addition task, children in all three groups used the same types of problem-solving strategies (i.e., retrieval, verbal counting, and counting fingers) but differed in the skill and speed of executing the strategies. The MD-improved and normal children did not differ substantively in the skill or speed of executing any of the strategies. This result suggests that the MD-improved children were cognitively normal and apparently were misidentified at the end of kindergarten as having a learning problem. The performance of the MD-no change group, in comparison with the two other groups, was characterized by a high frequency of procedural (i.e., counting procedures) and fact-retrieval errors, the frequent use of the immature counting-all procedure, and a lot of variability in the speed of executing the counting and retrieval strategies. When they counted to solve an addition problem, these children sometimes counted as fast or faster than the normal and MD-improved children, but at other times they counted rather slowly. The solution-time patterns for fact retrieval were highly unusual (i.e., unsystematic) in relation to the pattern that is found with academically normal children. In all, this study suggests that first-grade MD children have poor procedural skills (specifically, counting procedures in this study) and an atypical representation of arithmetic facts in long-term memory.

A follow-up study of these children was conducted about 10 months later, at the end of second grade (Geary et al., 1991). Of particular interest were developmental changes in procedural and fact-retrieval skills. In keeping with developmental models (Ashcraft, 1982; Siegler, 1986), the academically normal children (i.e., the normal and MD-improved children) showed an increased reliance on fact retrieval and a decreased reliance on counting to solve addition problems, from first to second grade. These children also made fewer counting and fact-retrieval errors and were faster at executing both types of strategies at the end of second grade. The MD children (i.e., the MD-no change children), on the other hand, showed no change in the mix of strategies from first to second grade. These children showed no change in the number of facts that they could remember (about 25%) and when they did remember an answer, they had a very high rate of errors. For instance, at the end of second grade, the MD children made eight times as many fact-retrieval errors as did the normal children (16% vs. 2%, respectively). The solution times for fact retrieval were, again, highly unsystematic. This result is important, because Ashcraft, Yamashita, and Aram (1992) found exactly the same pattern for children who had suffered from some form of early (before age 8 years) injury to the left hemisphere or associated subcortical regions, such as the thalamus (see the Neuropsychological Component section).

Despite these apparent fact-retrieval problems, the MD children did show considerable improvement in the use of counting procedures from the first to second grade. At the end of second grade, the MD children had nearly abandoned the use of the counting-all procedure, in favor of counting on, and did not differ from the normal children in terms of counting accuracy. In all, this pattern of results and related studies (e.g., Garnett & Fleischner, 1983; Geary & Brown, 1991; Gray, 1991; Jordan, Levine, & Huttenlocher, in press), indicate that many young MD children show both procedural and fact-retrieval problems in arithmetic but that the developmental trends for these two skills differ for many MD children. The early procedural skills of MD children are characterized by frequent errors and the use of developmentally immature algorithms, but these differences seem to largely disappear for many MD children by the end of second grade. The fact-retrieval deficit is reflected in a difficulty in retrieving facts from long-term memory. When facts are retrieved from memory, the associated retrieval speeds are highly unsystematic, and there is a high frequency of retrieval errors. These retrieval problems appear to be more persistent than the procedural deficit for most MD children (Garnett & Fleischner, 1983; Geary et al., 1987).

In all, these studies suggest that the procedural skills of many MD children are developmentally delayed. That is, early in the elementary school years, MD children tend to use arithmetic procedures that are more commonly used by younger, academically normal children. However, the procedural skills of many MD children appear to gradually improve, if at a somewhat slower rate than those of normal children. The fact-retrieval problems of many MD children, on the other hand, appear to represent a developmental difference (Goldman et al., 1988) or a more fundamental deficit that many of these children might not completely outgrow. In the next sections, cognitive factors that might contribute to these procedural and fact-retrieval deficits are explored. As noted earlier, these factors include conceptual knowledge, working memory, and speed of processing.

Conceptual Knowledge

In a more recent study, Geary, Bow-Thomas, and Yao (1992) sought to determine if the tendency of many young MD children to use the immature counting-all procedure, and to make a lot of counting errors, is related to an immature understanding of counting concepts. We reasoned, following Ohlsson and Rees (1991), that if the execution of a counting procedure was inconsistent with the children's counting knowledge, then the procedure would be changed so that subsequent executions of the procedure would conform to conceptual rules. In this view, counting ability reflects skill at executing procedures and conceptual knowledge, with counting knowledge providing the standard against which procedural performance is evaluated. For instance, if a child believes that counting always has to start at 1, then this child is not likely to use the counting-on procedure. This is because for counting on, counting starts at some value greater than 1 (except for 1 + 1). From this perspective, the development of counting procedures will be dependent on both counting knowledge and skill at detecting violations of this knowledge. Presumably, even with adequate counting knowledge, if the child does not notice that the counting procedure is inconsistent with counting principles, then the procedure would not be changed so as to conform to principled knowledge.

In this study (Geary, Bow-Thomas, & Yao, 1992), first-grade MD children were again selected on the basis of referral to the Chapter 1 program at the end of kindergarten. As with our other studies (Geary, 1990; Geary et al., 1991), children who had tested out of Chapter 1 at the end of first grade were not included in the MD group. The MD children and a group of normal children were administered a series of counting tasks that were designed to assess their understanding of the three how-to-count principles (Gelman & Gallistel, 1978). The tasks also allowed us to assess the child's understanding of the essential and some of the unessential features of counting (Briars & Siegler, 1984). The tasks involved having the child monitor a puppet who was learning how to count (Gelman & Meck, 1983). Some of the counts were correct, whereas other counts violated one of the how-to-count principles. If the child detected the violation, then it was assumed that the child had at least an implicit understanding of the associated concept. As with the earlier-described studies (e.g., Geary, 1990), all children were also administered the cognitive addition task.

On the addition task, the MD children committed more than twice as many counting-procedure errors as did the normal children and were more likely to use the counting-all, instead of the counting-on, procedure. The MD children also retrieved fewer facts from memory, and when they did remember an answer, it was more likely to be wrong than not (66% error rate). The performance of the MD children on the counting tasks suggested that they had an immature understanding of the essential and unessential features of counting (Briars & Siegler, 1984). For example, many of these children believed that adjacency was an essential feature of correct counting, whereas almost none of the normal children believed that adjacency was essential. The MD children were also poor at detecting certain forms of counting errors. More important, the MD children's immature counting knowledge and poor skills at detecting counting errors appeared to contribute to their use of the counting-all procedure on the addition task, as well as to contribute to their frequent counting errors. Counting knowledge, however, was unrelated to their high rate of fact-retrieval errors.

The results of this study suggest that the delayed procedural skills of many MD children might be due to an immature understanding of the associated concepts. To be sure, this line of research has only addressed the issue of counting knowledge and the use of counting procedures. Other studies, however, suggest that many MD children are not especially skilled at executing other types of arithmetical procedures, such as trading (e.g., Ginsburg, 1989; Russell & Ginsburg, 1984). From this perspective, trading errors might be associated with a poor understanding of place value. Future studies are needed to determine if MD children show a general delay in the understanding of arithmetical concepts or whether different children show selective conceptual and procedural delays.

Working Memory

As noted earlier, working memory is a rather complex topic, especially when applied to the assessment of individual differences in children (e.g., Kail, 1992). Thus, the issue of working memory and arithmetical development in MD children is considered only in very general terms. A more detailed discussion is presented elsewhere (Geary, 1993). Basically, MD children do not perform as well as their academically normal peers on working-memory tasks (Geary et al., 1991; Hitch & McAuley, 1991; D. J. Johnson, 1988; M. D. Levine, 1987; Siegel & Ryan, 1989). This result suggests that MD children are not as skilled as other children in retaining information in working memory while performing other operations. For instance, consider the working-memory requirements for mentally solving a computer-presented, multistep arithmetic problem (see Geary & Widaman, 1992). On the first screen, subjects are shown $3 + 2 = B$. Next, this information disappears (forever), and $B + 1 = C$ is presented. Finally, $D = C/2$ is presented. The subjects are then asked to provide an answer for either B, C, or D. For example, the final screen might present $B = ?$ The ability to remember the value of B while performing the next two steps suggests good working-memory skills.

The punch line is that the relatively poor working-memory skills of many MD children, in addition to conceptual knowledge, appear to contribute to their poor procedural skills (e.g., Geary et al., 1991; Hitch & McAuley, 1991). Moreover, working memory might also be a factor that contributes to their problems in remembering arithmetic facts. This is possible because, as noted earlier, if the child counts to solve a problem and forgets all or part of the problem before getting an answer, then the child might not associate the answer and the problem. So, to solve 2 + 5, if the child counts, "1, 2, 3, 4, 5, 6, 7" but forgets that the augend was 2, then 2 + 5 might not become associated with the answer, 7. This view is probably simplistic, in view of the complexity of working memory. Nevertheless, it seems to be a promising approach to better understanding the cognitive deficits of MD children. At the very least, working memory needs to be considered as a potential factor contributing to both the procedural and fact-retrieval deficits of MD children.

Speed of Processing

On average, MD children take longer to solve arithmetic problems than their normal peers (e.g., Garnett & Fleischner, 1983). To explain this finding, Kirby and Becker (1988) argued that MD children were slower at executing all basic numerical processes. An alternative view is that MD children use a different mix of problem-solving strategies, which in turn leads to differences in overall solution times. Indeed, MD children do tend to use slower counting strategies (e.g., counting all) more frequently, and rely on relatively fast fact retrieval less frequently, than their normal peers (Geary, 1990; Svenson & Broquist, 1975). In this circumstance, MD children will on average be slower than normal children at solving arithmetic problems.

The question then becomes, when MD and normal children use the same strategy, do they differ in the speed of processing? Counting speed is of particular interest, because it might influence the ease with which arithmetic facts are represented in long-term memory. For the verbal counting strategy, Geary (1990) found no difference in the average counting speed of normal and MD first-grade children, although the MD children did appear to be slower at other processes, such as reading and speaking numbers. However, the follow-up study of these children, when they were in second grade, revealed a slower average counting speed for the MD children (see also Geary et al., 1987). Hitch and McAuley (1991) also found slower counting speeds for 8- and 9-year-old MD children, in relation to normal peers. Geary and Brown (1991), on the other hand, found no average-counting-speed differences between groups of normal and MD fourth-grade children. On the basis of these results, it seems that many MD children probably do not count more slowly than normal children. Thus, although it is possible that slow counting speeds contribute to the retrieval problems of some MD children, it seems unlikely that counting speed is a contributing factor to the retrieval deficits of all MD children.

The assessment of retrieval speed for MD children is difficult, because their solution-time patterns for retrieval tend to be highly unsystematic. Any comparisons of MD and normal children on this dimension need to be considered with caution. Nevertheless, the research that has been conducted suggests that there are no retrieval-speed differences between first- and second-grade MD and normal children (Geary, 1990; Geary et al., 1991). However, older MD children do appear to take longer than normal children to retrieve facts from memory (Geary & Brown, 1991). This developmental pattern appears to result from the normal children becoming faster at remembering facts, from grade to grade, with little grade-to-grade change in the retrieval speeds of many MD children.

In all, the results of these studies are mixed, with regard to the question of potential speed-of-processing differences between MD and normal children. The mixed pattern of results quite likely reflects, in part, the heterogeneity of MD groups. It is likely that some MD children are slower at using counting procedures and at retrieving facts. Nevertheless, it does not appear to be the case, contrary to Kirby and Becker's (1988) position, that MD children are slower than normal children at executing all arithmetical processes. In fact, even if speed-of-processing differences exist for some operations, the "mental slowness" of many MD children appears to be related, to a large extent, to their use of more time-consuming problem-solving strategies in relation to their normal peers.

Summary

The cognitive studies indicate that many MD children show two arithmetic-related deficits: procedural and fact retrieval. The procedural deficit (primarily counting procedures in these studies) is manifested by the use of developmentally immature algorithms and by many procedural errors. The procedural deficit appears to be related to an immature understanding of the associated concepts and to poor working-memory resources. The procedural skills of many MD children, however, appear to approach those of their normal peers by the middle of the elementary school years (Geary & Brown, 1991; Geary et al., 1991). On the basis of this pattern, the procedural deficit of many MD children appears to follow a developmental delay model; that is, MD children initially use procedures that are more commonly used by younger, academically normal children but eventually catch up, so to speak, with their same-age peers (Goldman et al., 1988). Larger scale longitudinal studies of MD children that assess many different types of procedural skills, such as counting and trading, are needed before it can be determined if these early procedural deficits encompass all arithmetical procedures or are more selective. Such studies are also needed to determine if most, all, or only some MD children grow out of this type of problem.

The fact-retrieval deficit appears to be more persistent than the procedural deficit for most MD children (Garnett & Fleischner, 1983; Geary, Brown, & Samaranayake, 1991; Geary, Widaman, Little, & Cormier, 1987; Howell et al., 1987). This result suggests that the fact-retrieval and procedural deficits are distinct, although some children will, of course, show both deficits. The retrieval deficit is manifested as problems in remembering arithmetic facts, even after extensive drilling (Howell et al., 1987). Even when answers are retrieved from long-term memory, they are often wrong. Finally, the speed of accessing arithmetic facts, right or wrong, from long-term memory is very variable for many MD children. Sometimes facts are accessed quickly; at other times, fact retrieval is very slow. As noted earlier, this pattern is very similar to the pattern of retrieval speeds that is associated with early lesions of the left hemisphere (Ashcraft et al., 1992).

In all, it would appear that the fact-retrieval deficits of some MD children follow a developmental difference model (Goldman et al., 1988); that is, the pattern of performance is qualitatively different from that of same-age peers and younger, academically normal children. For these children, it seems likely that the retrieval deficits are fundamental and are not likely to improve (see the Genetic Component section), at least not without specialized remedial training (see the Mathematical Disabilities section of chapter 8). For other MD children, the retrieval deficit might simply be related to poor working memory and counting skills. In the latter case, as counting skills and working memory improve, improvements in retrieval skills should follow. For these children, the retrieval deficit might simply be a developmental delay. Future studies are obviously needed to determine if two separate forms of retrieval deficit do indeed exist (Hitch & McAuley, 1991).

Finally, the issue of whether MD children make poor strategy choices when solving arithmetic problems needs to be addressed (Siegler, 1988a). In the Cognitive Perspective section of chapter 4, it was argued that individual differences in children's arithmetic skills were related to the ability to make skilled strategy choices. A skilled strategy choice is reflected in choosing the strategy that is most efficient, in terms of speed and accuracy, for solving each problem. For MD children, the use of the counting-all procedure and the high frequency of retrieval errors suggest that these children often make poor strategy choices (Geary, 1990). For instance, the frequency of retrieval errors might be reduced if these children resorted to backup strategies more frequently. Resorting to a backup strategy when any retrieved answer is likely to be wrong reflects a skilled strategy choice.

At this point, however, it is not clear whether the apparently poor strategy choices of MD children are simply due to their procedural and retrieval deficits or whether they do not have a good intuitive, or metacognitive, understanding of numbers and arithmetic. Future studies, perhaps using measures of nonverbal calculation abilities or subitizing tasks (which appear to assess a child's intuitive sense of arithmetic) as well as metacognitive tasks, should help to clarify this issue (Borkowski, 1992; Gallistel & Gelman, 1992; S. C. Levine et al., 1992).

Neuropsychological Component

There is considerable evidence of a neurological involvement underlying many learning disorders (Hammill, Leigh, McNutt, & Larsen, 1987; Hynd & Semrud-Clikeman, 1989). Thus, neuropsychological studies of the acquired and developmental dyscalculias need to be considered in conjunction with the cognitive studies of MD. In fact, with regard to basic-arithmetic-related deficits, the neuropsychological studies converge on many of the same conclusions as the cognitive studies.

Before this research is presented, note that the results from many of the neuropsychological studies are not clear-cut; that is, it is rare to find case studies in which a focal brain injury produces a distinct mathematical deficit (but see McCloskey, Aliminosa, & Sokol, 1991; Warrington, 1982). This is because even if a brain injury is fairly localized, it is often diffuse enough to affect a variety of cognitive skills and because the measures used in many of these studies are relatively insensitive paper-and-pencil tests (see Ashcraft et al., 1992, for a discussion).

For instance, the assessment of computational skills in dyscalculia has often been based on a summary score for accuracy at solving simple (e.g., 9 + 6) and complex (e.g., 244 + 129) arithmetic problems (e.g., Jackson & Warrington, 1986). Arithmetic computations in the neuropsychological literature are often interpreted as the ability to retrieve basic arithmetic facts from long-term memory. However, as noted in the Cognitive Perspective section of chapter 4, the computations that underlie the solution of simple and complex arithmetic problems involve several component skills, such as fact retrieval and carrying or trading (Geary & Widaman, 1987). In this circumstance, summary scores actually reflect both retrieval and procedural skills, which make these assessments rather imprecise.

Despite these difficulties, there is some consistency in the apparent impact of brain injury on arithmetic skills, which makes neuropsychological research an important complement to the cognitive studies of MD. In particular, double-dissociation studies can be very informative (e.g., Temple, 1991). In these studies, it is usually demonstrated that brain injury in one individual disrupts one arithmetical skill, such as procedural skills, but leaves another skill, such as fact retrieval, unaffected. For another individual, brain injury might result in the opposite pattern: intact procedural skills and disrupted fact retrieval. Badian (1983) and others (e.g., Hécaen, 1962) have classified acquired and developmental dyscalculias into three general categories: alexia and agraphia for numbers, spatial acalculia, and anarithmetria. Temple (1991) recently argued that in terms of basic mathematical deficits, "developmental dyscalculias are analogous to acquired dyscalculias" (p. 155). Thus, the developmental and acquired dyscalculias are discussed together in the next sections.

Alexia and Agraphia for Numbers

Alexia and agraphia for numbers involve difficulties in the reading and writing of numbers, with intact skills in other areas of arithmetic (McCloskey, Caramazza, & Basili, 1985). Hécaen (1962) reported that if this type of problem existed, then it was almost always associated with lesions of the left hemisphere. Alexia and agraphia for numbers are sometimes, but not always, associated with reading and other language-related disorders (Badian, 1983; Boller & Grafman, 1983; Temple, 1989). In young children, difficulties in the reading and writing of numbers are fairly common, because they are still in the process of acquiring these skills. For instance, young children commonly reverse numbers when writing them. Number reversals in and of themselves should probably not be considered an indication of a neuropsychological problem or MD, unless this problem persists well into the elementary school years.

Although more persistent difficulties with the reading and writing of numbers sometimes do occur in children (Kosc, 1974), these appear to be relatively rare, in comparison with spatial acalculia and anarithmetria. Badian (1983), for instance, examined the performance of 50 MD children on a variety of achievement and ability measures. Although some of the children occasionally misread numbers, or operation signs, these errors appeared to be due to inattention rather than to a basic inability to read and write numbers. Thus, alexia and agraphia for numbers are not given further consideration.

Spatial Acalculia

Spatial acalculia is characterized by difficulties in the spatial representation of numerical information and is often associated with damage to the posterior regions of the right hemisphere (Benson & Weir, 1972; Dahmen, Hartje, Büssing, & Sturm, 1982; Luria, 1980). Specific problems associated with spatial acalculia include number omissions, number rotations, misreading arithmetical operation signs, difficulties with place value and decimals, and the misalignment of numbers in multicolumn arithmetic problems, such as 35 + 74 (Grafman, Passafiume, Faglioni, & Boller, 1982; Hartje, 1987; Strang & Rourke, 1985). In the latter case, the 4, for 35 + 74, might be placed under the 3 rather than under the 5. Number reading and writing skills are usually intact, as are basic arithmetic computations (i.e., fact retrieval).

In a set of developmental studies, Rourke and his colleagues examined the pattern of performance on neuropsychological measures for children with an arithmetic-and-reading disability and for children with only an arithmetic disability (Rourke & Finlayson, 1978; Rourke & Strang, 1978). The performance of the arithmetic-and-reading-disability group suggested a left-hemispheric dysfunction, with a common verbal deficit underlying problems in both reading and arithmetic. The arithmetic-disability group, in contrast, showed a pattern of visuospatial deficits that suggested a right-hemispheric dysfunction. Share, Moffitt, and Silva (1988) recently replicated this pattern, but only for groups of low-achieving boys. Arithmetic-and-reading-disabled girls showed the same pattern of relatively better nonverbal than verbal skills, although they still showed nonverbal deficits in relation to an academically normal control group. These studies suggest that at least for boys, visuospatial deficits can disrupt performance in arithmetic but tend to leave reading skills unaffected. Additional research is needed to clarify the relationship between visuo-spatial abilities and arithmetic skills in girls.

Finally, these studies need to be considered in terms of the development of visuospatial skills as they are related to arithmetic. Hartje (1987), for instance, argued that the importance of visuospatial skills for the solution of arithmetic problems might decline as children commit basic facts to memory and as arithmetic procedures become automatized (i.e., automatically executed when used). When first learning to count and with early arithmetic, children often use manipulatives, or fingers, to represent the sets to be counted or added/subtracted (see, e.g., Table 2.1). These visuospatial representations of number sets help children to understand the task and to regulate their own performance (e.g., to keep track of the counting). It is therefore reasonable to expect that visuospatial deficits in preschool children might have a more severe impact on the development of basic number and arithmetic skills than visuospatial deficits associated with brain injury after basic numerical skills have already developed.

In all, these studies suggest that at least for boys, a disturbance of visuospatial functions, perhaps associated with a right-hemispheric dysfunction or injury (Rourke & Strang, 1978), will affect the ability to spatially represent numerical information and skill at interpreting the meaning of these representations (e.g., place value in muticolumn arithmetic problems) while leaving reading and other language-related skills relatively intact. The results of Share et al. (1988), however, suggest that this pattern might not represent the deficits of girls with only an arithmetic disability and normal reading skills. Further studies of gender and developmental differences in visuospatial deficits as related to MD and arithmetic development are clearly needed.

Anarithmetria

The primary deficit associated with acquired anarithmetria in adulthood is a difficulty in the retrieval of basic arithmetic facts from long-term memory (Benson & Weir, 1972; Jackson & Warrington, 1986; Warrington, 1982). This deficit is most commonly associated with damage to the posterior regions of the left hemisphere (e.g., Dehaene & Cohen, 1991; Warrington, 1982). Number reading and writing and the spatial representation of numerical information are typically intact. The understanding of arithmetical concepts is usually intact (Benson & Weir, 1972), although these individuals sometimes have difficulties with operations involving number sequencing, as in executing arithmetic procedures (Spiers, 1987). Nevertheless, Temple (1991) reported a double dissociation between arithmetic-fact retrieval and the ability to execute arithmetic procedures, such as carrying or trading. In all, acquired anarithmetria appears to encompass two distinct deficits, fact-retrieval and procedural, which sometimes are associated with reading or other language-related disorders and sometimes are not (Jackson & Warrington, 1986; Sokol, McCloskey, Cohen, & Aliminosa, 1991).

Children with anarithmetria sometimes confuse arithmetical operation signs, but the most common difficulty involves arithmetic-fact retrieval (Badian, 1983). The children in Rourke and Finlayson's (1978) arithmetic-and-reading-disability group, for instance, showed fact-retrieval deficits that Rourke and Finlayson argued reflected their verbal impairments. Ashcraft et al. (1992), as mentioned earlier, found that about half of the children who had suffered some form of left-hemispheric, or subcortical, injury before 8 years of age appeared to have difficulties retrieving arithmetic facts from long-term memory. In fact, as described in the Cognitive Component section, their unsystematic solution-time patterns for retrieval were very similar to those found in the cognitive studies of MD children. In contrast, none of the older children with similar left-hemispheric or subcortical lesions and only one of nine children in a right-lesion group showed a pattern of unsystematic solution times for retrieval. The young left-lesion group also had problems with arithmetical procedures, such as borrowing in complex subtraction.

In all, the studies with children indicate that the primary deficit associated with anarithmetria is a difficulty in arithmetic-fact retrieval, although difficulties in the execution of arithmetical procedures are sometimes but not always associated with the retrieval deficit (Ashcraft et al., 1992; Temple, 1991). The finding of an occasional dissociation between retrieval and procedural deficits suggests that they are distinct skills (Temple, 1991). Finally, fact-retrieval problems are commonly associated with language-related deficits (Richman, 1983; Rourke & Strang, 1978). At this point, the relationship between procedural deficits and language-related skills is less clear, although the McCloskey, Aliminosa, and Sokol (1991) study suggests that they might not be related.

Summary

Except for occasional studies of the basic ability to read and write numbers (i.e., alexia and agraphia for numbers), the neuropsychological studies have focused on two general but distinct mathematical deficits associated with brain injury or more subtle neuropsychological problems, spatial acalculia and anarithmetria. Spatial acalculia typically occurs with right-hemispheric injury and results in difficulties in the spatial representation of numerical information and in some conceptual problems, such as understanding place value (Benson & Weir, 1972). The most notable feature of anarithmetria is a difficulty in retrieving arithmetic facts from long-term memory (Badian, 1983), although many of these individuals also show deficits in the use of arithmetical procedures. The occasional dissociation between fact retrieval and arithmetical procedures, however, suggests that they are distinct deficits. In all, the deficits subsumed under anarithmetria appear to be the same as the procedural and memory-retrieval deficits discovered with the cognitive research.

Thus, to conclude, the neuropsychological studies suggest three relatively distinct types of arithmetic-related deficits associated with brain injury or dysfunction: fact-retrieval, procedural, and spatial-representation. The fact-retrieval deficits often coexist with certain forms of reading or other language-related disorders, whereas the spatial-representation deficits typically do not (Rourke & Finlayson, 1978; Rourke & Strang, 1978). The relationship between procedural skills and language-related abilities is less clear. Language skills might be important in the use of some procedures, such as the counting-on procedure, and less important for other procedures, such as carrying or trading, at least after the procedures are well learned. If so, then language-related problems might affect procedural skills in children more so than with adults.

Genetic Component

Unfortunately, there have been no studies of the heritability of MD per se. Nevertheless, a consideration of the potential heritability of MD is important, because, as noted in the Behavioral Genetic Perspective section of chapter 4, it appears that roughly half of the variability in basic arithmetical ability is due to genetic factors (Vandenberg, 1966). On the basis of this finding, the possibility that certain forms of MD are also heritable needs to be seriously considered. In particular, if many MD children simply represent the lower end of the distribution of normal variability in arithmetical skills, then the poor performance of these children might have a moderate to strong genetic influence (Gilger et al., 1993; Shaywitz et al., 1992).

Moreover, a review of potential genetic contributions to MD will help to provide a link between MD and RD, because common cognitive and genetic factors appear to underlie individual differences in both reading and basic mathematical skills (Stevenson, Parker, Wilkinson, Hegion, & Fish, 1976; Thompson, Detterman, & Plomin, 1991), because MD and RD coexist in many children, and because many forms of RD appear to be heritable (Olson et al., 1989; Pennington et al., 1991). For instance, phonetic memory, which supports word retrieval and other semantic-memory skills, appears to be an important heritable component of reading skills (Olson et al., 1989). Difficulties in phonetic memory contribute to RD. Geary (1993; Geary et al., 1991) argued that this same memory system supports arithmetic-fact retrieval. If so, then these apparently heritable deficits in semantic-memory skills might also contribute to the fact-retrieval deficits of MD children.

In a related study, Gillis and DeFries (1991), as part of the Colorado Reading Project, examined the relationship between reading and mathematics achievement for a sample of 264 RD twin pairs and 182 matched control twin pairs. Heritability estimates for performance on the mathematics measures were .51 for the RD group and .60 for the control group. For both groups, performance on the reading and mathematics tests were reliably correlated; the better the performance on the reading test, the better the performance on the mathematics test. More important, this relationship between reading and mathematics scores appeared to be due, in part, to common genetic influences. Thompson et al. (1991) also found evidence for common genetic influences for individual differences in reading and mathematics achievement for nondisabled monozygotic and dizygotic twins. This is *not* to say that all genetic sources contributing to reading and mathematics achievement are the same but rather that some of the similarity between performance on reading and mathematics tests might be due to overlapping genetic influences, above and beyond genetic influences that are unique to reading and mathematical ability. The finding of common genetic sources to individual differences in reading and mathematics achievement is important, because it leaves open the possibility that the same genetic sources are responsible for the co-existence, or comorbidity, of MD and RD in many children (see the Relationship Between RD and MD section).

In summary, although it appears that there is a genetic contribution to individual differences in arithmetical ability, the associated studies did not directly address the issue of the heritability of MD. If it were to be assumed that MD represented the lower end of the distribution of basic arithmetic skills, then the individual-difference studies would suggest that there was likely to be a genetic component to MD. If a qualitatively distinct deficit was underlying MD, then the individual-difference studies would be less conclusive, because a distinct skill deficit could be due to genetic influences or to unique environmental experiences, such as brain injury (Gilger et al., 1993). Nevertheless, given the relationship between MD and RD, the heritability of some forms of RD, and the common genetic influences for individual differences in reading and mathematics achievement, it seems likely that certain forms of MD are also heritable, especially those that involve the fact-retrieval deficit (see the Relationship Between RD and MD section).

This is not to say that the genetic sources influencing reading skills are identical to those influencing mathematical skills but rather that there appear to be overlapping genetic influences for reading and mathematical ability and that some of these overlapping genetic influences might contribute to both RD and certain forms of MD. Finally, the possibility that the delayed acquisition of procedural concepts and skills in many MD children reflects genetic, as well as environmental, influences needs to be considered (R. S. Wilson, 1978). Any such genetic contribution to the tempo of arithmetical development might differ from the genetic contribution to the retrieval deficit. Behavioral genetic studies specifically of MD are obviously needed. Any such studies would be especially useful if the focus was on the specific forms of deficit identified with the cognitive and neuropsychological research rather than simply on performance on paper-and-pencil tests.

Relationship Between Reading Disabilities and Mathematical Disabilities

Although some RD children show visuospatial or orthographic deficits, the primary difficulty of most RD children appears to involve auditory memory. The difficulties with auditory memory appear to be secondary to phonological processing deficits (Lindgren, Richman, & Eliason, 1986; Olson, Gillis, Rack, DeFries, & Fulker, 1991; Richman, 1983). These auditory memory problems appear to be related to difficulties in word retrieval, phonological decoding, and a relatively underdeveloped memory network of semantic associations (Bruck, 1992; Olson et al., 1989). Difficulties in phonological decoding are related to difficulties in language segmentation or phonological awareness (i.e., breaking words into component sounds), which, as noted earlier, appear to have a substantial heritable component (Olson et al., 1989). Richman (1983) noted that RD children with auditory memory deficits often have problems with "arithmetic memory functions" (p. 103; i.e., fact retrieval). Rourke (1989), on the other hand, argued that for RD children with an orthographic deficit underlying the reading disorder, "arithmetic and mathematics performance may rise to average or above-average levels when the words used in problems are minimized or learned 'by sight'" (p. 218). In other words, children with orthographic coding difficulties apparently have intact basic arithmetic skills when reading is not required for problem solving (Muth, 1984), whereas RD children with phonological processing and auditory memory deficits often have a specific deficit in arithmetic-fact retrieval.

Indeed, as implicated earlier, the retrieval of arithmetic facts from long-term memory and the underlying memory representations show many of the same characteristics as the representation and retrieval of verbal information in semantic memory (Ashcraft, 1992; Ashcraft & Battaglia, 1978; Campbell, 1987). For instance, arithmetic facts appear to be represented in an associative memory network, with activation spreading from the encoded integers to related numbers and facts (Ashcraft & Battaglia, 1978; McCloskey, Harley, & Sokol, 1991). Geary et al. (1991) found that the facility of retrieving addition facts from long-term memory was correlated with reading skills, but skill at using counting procedures for solving addition problems was not. Studies of acquired dyscalculia suggest that fact-retrieval deficits are most likely to occur with damage to the posterior portions of the left hemisphere, although this is not always the case (e.g., McCloskey, Aliminosa, & Sokol, 1991). Similarly, postmortem studies of dyslexics suggest neuroanatomical abnormalities in these same regions (Hynd & Semrud-Clikeman, 1989).

In all, a tentative conclusion might be drawn from these data: For many children, MD and RD co-occur because of a common underlying neuropsychological deficit, perhaps involving the posterior regions of the left hemisphere. At the cognitive level, this deficit manifests itself as difficulties in the representation and retrieval of semantic information from long-term memory. This would include fact-retrieval problems in simple arithmetic and, for instance, word recognition and phonological awareness difficulties in reading. Moreover, given that these word recognition and phonological awareness difficulties are, at least in part, heritable (Olson et al., 1989) and are often associated with fact-retrieval deficits, it might be the case that the entire constellation of fact-retrieval and verbal deficits is heritable and reflects underlying semantic-memory problems.

These semantic-memory problems appear to be secondary to a poorly developed phonological processing system. Finally, note that the reading skills of adults with childhood RD often improve but that the phonological awareness of these adults does not (Bruck, 1992). Thus, it is possible that long-term arithmetic-fact-retrieval deficits could be associated with relatively good reading skills. Such a finding would not disconfirm this hypothesis, because even with good reading skills, one would still anticipate a relationship between arithmetic-fact-retrieval deficits and poor phonological awareness.

Mathematical Disabilities: Arithmetic and Mathematical Problem Solving

There are two general issues that need to be addressed in this section. First, to what extent do the arithmetic-related disorders, such as fact retrieval, influence the mathematical problem-solving skills of MD children (Zentall & Ferkis, 1993)? For instance, if poor performance on mathematical problem-solving tests was primarily due to an arithmetic-related deficit, then the difficulties with mathematical problem solving would be secondary to this more basic, lower level deficit. Second, are there forms of MD that specifically interfere with some of the more complex features of mathematical problem solving described in chapter 3, such as problem representation, above and beyond any arithmetic-related deficits? In this case, difficulties with problem representation would indicate a primary deficit in mathematical problem solving. Children with this type of disorder may or may not show an arithmetic-related deficit.

With regard to the first issue, it appears that poor basic arithmetic skills contribute to the poor mathematical problem-solving skills of many MD children (Muth, 1984; Zentall & Ferkis, 1993). In particular, the above-described fact-retrieval deficit seems to make the solving of arithmetical word problems more difficult. Zentall and Ferkis argued that the inability to automatically retrieve basic facts from long-term memory makes the solving of word problems more demanding, because the children have to resort to more cumbersome and time-consuming counting strategies to complete the computational aspects of the word problems. The use of counting procedures also increases the working-memory demands of the problem. If the child has to count or use some other form of backup strategy during problem solving, then it is more difficult to keep other important aspects of the problem in mind. As a result, it will be more difficult to simultaneously consider all relevant information presented in the problem, and it is more likely that the child will forget important aspects of the problem, after the problem has been read (Pellegrino & Goldman, 1987). Thus, it is likely that the arithmetic-related forms of MD will also lead to difficulties in solving arithmetical and algebraic word problems.

Other factors also appear to contribute to the poor mathematical problem-solving skills of some MD children, above and beyond poor arithmetic skills. Muth (1984), for instance, argued that basic reading skills will influence skill at solving word problems. In addition to generally poorer reading comprehension, many MD children have difficulties in understanding quantitative words, such as *distance* and *time* (Zentall & Meyer, 1987). Many MD children also fail to understand that letters in algebraic word problems represent unknown variables. Many of these adolescents ignore this information and try to solve the problems only on the basis of the quantitative information presented in the problem (Montague & Applegate, 1993). In all, it appears that difficulties with reading and language comprehension might also be a source of mathematical problem-solving errors for some MD children.

Moreover, it appears that many MD children have particular difficulties in representing the meaning of arithmetical and algebraic word problems (Hutchinson, 1993; Zawaiza & Gerber, 1993). Recall that problem representation involves selecting relevant pieces of information from the stated problem and then integrating these in such a way that they can be translated into an appropriate equation (see the Problem-Solving Processes section of chapter 3). Relative to their nondisabled peers, these adolescents are not as skilled at diagramming the relationships among important components of the problems and have particular difficulty on inconsistent-language problems (e.g., A. B. Lewis, 1989). Zawaiza and Gerber, for example, found that MD children had particular difficulties with subtraction word problems that contained the phrase *more than*. This result suggests that there might be a specific type of learning disorder associated with mathematical problem solving. Future studies of mathematical problem solving in MD children should consider the potential influence of arithmetic-related deficits and reading skills on poor performance, as well as the possibility of a specific representational deficit, which might reflect poor spatial skills.

Conclusion

The research reviewed in this chapter suggests that many children have difficulties in learning basic arithmetic because of a serious cognitive or neuropsychological disorder, certain forms of which might be heritable. The cognitive research suggests that MD children differ from their normal peers on at least two dimensions: skill at using counting procedures to solve arithmetic problems and difficulties in the representation or retrieval of basic arithmetic facts from long-term memory. The development of procedural skills in many MD children appears to be delayed in relation to that in normal children and is manifested in the use of immature procedures and low accuracy (e.g., Garnett & Fleischner, 1983; Geary, 1990). Delayed procedural-skill development, in turn, might be related to a delay in the understanding of the associated concepts and poor skill at detecting when a procedural error has been committed (Geary, Bow-Thomas, & Yao, 1992). Working memory also needs to be considered as a potential contributing factor to the poor procedural skills of many MD children (Geary et al., 1991). The issue of whether the poor counting skills of MD children are related to difficulties in the use of other arithmetical procedures, such as carrying or borrowing, needs to be addressed (Ashcraft et al., 1992). The possibility that these procedural deficits are preceded by even earlier (i.e., before kindergarten) deficits in preverbal calculation skills also needs to be explored (Dehaene, 1992; Gallistel & Gelman, 1992; S. C. Levine et al., 1992).

The second deficit of MD children, at least when solving simple arithmetic problems, involves the representation or retrieval of arithmetic facts from long-term memory (e.g., Geary & Brown, 1991; Geary et al., 1991). Arithmetic-fact retrieval in MD children is manifested by a low overall number of facts (correct or incorrect) that are retrieved from memory. When a fact is retrieved, there is a high error rate, and retrieval speeds are highly unsystematic (Garnett & Fleischner, 1983; Geary & Brown, 1991; Geary et al., 1991). As noted earlier, the working-memory deficits of MD children and the tendency to commit a lot of counting-procedure errors during initial arithmetic learning might contribute to the low frequency of fact retrieval and high rate of retrieval errors, respectively, when facts are retrieved. These factors, however, do not adequately explain the finding of parallel solution-time patterns for retrieval in MD children and children with early left-hemispheric lesions (Ashcraft et al., 1992). This result, combined with the finding that retrieval does not substantially improve for many MD children even with extensive drilling (e.g., Howell et al., 1987), suggests that for many of these children the observed memory-retrieval problems might be indicative of a more fundamental deficit that is not simply a developmental delay.

The studies of acquired and developmental dyscalculia converge on the same conclusions as the cognitive research, strengthening the argument for distinct retrieval and procedural deficits in MD children (McCloskey, Aliminosa, & Sokol, 1991; Temple, 1991). In fact, it seems reasonable to conclude that the children assessed in the developmental dyscalculia and MD studies are to a large extent from the same population. The neuropsychological studies also show a relationship between visuospatial skills and basic arithmetical ability (e.g., Badian, 1983). In particular, visuospatial deficits that affect the ability to spatially represent numerical information disrupt not only functional skills in elementary mathematics (e.g., columnar alignment of numbers to be added) but also the conceptual understanding of the associated representations (e.g., place value). This type of visuospatial deficit has not been identified in the cognitive research, most likely because cognitive researchers have not typically included visuospatial measures in their studies of MD children.

As noted earlier, there have been no behavioral genetic studies specifically of MD. Nevertheless, evidence for the heritability of individual differences in basic arithmetic skills, combined with the finding that certain heritable components of RD are associated with MD, would suggest that certain forms of MD are also heritable. In particular, the memory-retrieval deficit of many MD children would appear to be a good candidate for specific behavioral genetic studies of MD. This argument is based on the finding that auditory memory problems, such as poor phonological awareness, that appear to contribute to certain forms of RD are heritable (Bruck, 1992; Olson et al., 1989) and are often accompanied by problems in arithmetic-fact retrieval (e.g., Richman, 1983).

Indeed, the coexistence of MD and RD in many children (Badian, 1983) might reflect a more general deficit in the representation or retrieval of information from semantic memory. Semantic-memory representation or retrieval problems could influence both reading skills and arithmetic-fact retrieval (e.g., Ashcraft & Battaglia, 1978). This is not to say that for some children, poor mathematics achievement is not secondary to poor reading skills. Some children perform poorly on more complex mathematics tasks, such as word problems, not because of an underlying arithmetical deficit, but rather because they are poor readers (Muth, 1984). Nevertheless, it is argued that for some children, the co-occurrence of MD and RD reflects a common underlying neuropsychological deficit that might be heritable (Gillis & DeFries, 1991).

Note that even if mathematical ability and disability prove to be, to some extent, heritable, genetic influences do not occur in a vacuum but rather are expressed in an environmental context (Detterman, Thompson, & Plomin, 1990; Plomin et al., 1990). Even if certain forms of MD were found to be highly heritable, poor performance on mathematical achievement and ability measures could represent substantial environmental, as well as genetic, influences. A similar argument can be made with regard to the comparison of the cognitive and neuropsychological studies of MD. Even though the neuropsychological and cognitive studies converge on similar conclusions regarding the arithmetic deficits of MD children, this does not necessarily mean that the MD children that have been assessed in the cognitive studies have suffered some type of brain injury.

The deficits of MD children might represent a combination of environmental and early neurodevelopmental problems. The acquisition of any complex skill involves a type of synchrony between the maturation of the supporting neural systems and environmentally appropriate experiences (Greenough, Black, & Wallace, 1987). If the early experiences of the individual do not provide appropriate feedback to the maturing neural systems, then those systems might not develop normally. Thus, even if the same neural systems were involved, for example, in the retrieval deficits of brain-injured individuals and the MD children assessed in the cognitive studies, it does not necessarily follow that the MD children have suffered some form of overt brain injury. Rather, it is possible that any neurological dysfunction represents the additive impact of abnormalities (e.g., in cell migration) in the maturing neural structures supporting the emerging number system and inadequate early experiences.

Regardless of why these deficits exist, the pattern of cognitive and neuropsychological results described in this chapter suggest that the subtyping of arithmetic-related forms of MD is possible, as has been done in the RD area (e.g., Richman, 1983). A tentative framework, expanding on the work of Strang and Rourke (1985), would include three general subtypes. The first subtype involves difficulties in arithmetic-fact retrieval and problems in the memorization of arithmetic tables even with extensive drilling (Geary, 1990; Howell et al., 1987); a summary of the characteristics of these retrieval deficits is presented in Exhibit 5.1.

The cognitive and neuropsychological studies indicate that these retrieval deficits are not an all-or-none phenomenon. As noted in Exhibit 5.1, individuals with this type of deficit are typically able to retrieve some facts from long-term memory but show other performance characteristics, such as unsystematic solution times, that distinguish them from their academically normal peers (Geary et al., 1991; McCloskey, 1992). Children falling within this category might also show difficulties in reading

---

Exhibit 5.1

Mathematical Disabilities, Subtype 1: Semantic Memory

Cognitive/Performance Features

Low frequency of arithmetic-fact retrieval.
When facts are retrieved, there is a high error rate.
Solution times for correct retrieval are unsystematic.

Developmental Features

Appears to follow a developmental difference model. Performance is qualitatively distinct from that of normal children and shows minimal grade-to-grade improvement.

Neuropsychological Features

Appears to be associated with left-hemispheric dysfunction, in particular posterior regions of the left hemisphere.
Possible subcortical involvement, such as the thalamus.

Genetic Features

Unclear, but the relationship with certain forms of RD suggests that this deficit might be heritable.

Relationship to RD

Often covaries with RD, especially if the RD is associated with phonetic deficits.

Note. RD = reading disabilities.

achievement and perhaps be RD (Rourke & Finlayson, 1978; Rourke & Strang, 1978). It is this pattern of deficits that, at this point, appears to be the best candidate for heritability studies. Finally, the neuropsychological research suggests that this type of deficit might reflect some form of dysfunction or injury involving the posterior regions of the left hemisphere and perhaps subcortical involvement (Hynd & Semrud-Clikeman, 1989; McCloskey, Aliminosa, & Sokol, 1991; Warrington, 1982).

The second subtype involves difficulties in the use of arithmetical procedures, such as using counting procedures to solve addition problems (Geary, 1990) or borrowing to solve subtraction problems (Ashcraft et al., 1992). It is not clear at this point whether these procedural deficits reflect a developmental delay in the acquisition of the underlying concepts (Geary, Bow-Thomas, & Yao, 1992) or a more general problem with number sequencing (Spiers, 1987). It seems likely that for some MD children, procedural deficits simply reflect a delay in the acquisition of the underlying concepts (Geary, Bow-Thomas, & Yao, 1992) and that these children might not show a mathematical disability per se but rather are at the

Exhibit 5.2

Mathematical Disabilities, Subtype 2: Procedural

Cognitive/Performance Features

Relatively frequent use of developmentally immature procedures.
Frequent errors in the execution of procedures.
Potential developmental delay in the understanding of the concepts underlying procedural use.

Developmental Features

Appears to follow a developmental delay model. Performance is similar to that of younger, academically normal children and improves from grade to grade.

Neuropsychological Features

Unclear, although some data suggest an association with left-hemispheric dysfunction.

Genetic Features

Unclear

Relationship to RD

Unclear

Note. RD = reading disabilities.

lower end of the distribution for the acquisition of basic number skills. Other children will show persistent difficulties in procedural learning that reflect a real disability (Goldman et al., 1988; Temple, 1991). The work of Ashcraft et al. (1992) suggests that certain procedural deficits might reflect a dysfunction or injury to the left hemisphere. Future prospective studies that use cognitive measures for assessing procedural skills (e.g., VanLehn, 1990) and knowledge of the underlying concepts (e.g., Briars & Siegler, 1984) are needed for the clarification of this issue. An overview of the characteristics of the procedural deficits associated with this form of MD is presented in Exhibit 5.2.

The final subtype—which seems to be associated with a dysfunction of, or injury to, the posterior regions of the right hemisphere—involves a disruption in the use of visuospatial skills for the representation and interpretation of arithmetical information (e.g., Benson & Weir, 1972; Dahmen et al., 1982; Rourke & Finlayson, 1978). It is possible that some forms of procedural deficit noted under Subtype 2, such as borrowing or carrying, might in fact be related to visuospatial deficits. It seems unlikely, however, that all of the procedural problems of MD children, counting errors for instance, are related to visuospatial deficits (Badian,

Exhibit 5.3

Mathematical Disabilities, Subtype 3: Visuospatial

Cognitive/Performance Features

Difficulties in spatially representing numerical information, such as misaligning numbers in multicolumn arithmetic problems or rotating numbers.
Misinterpretation of spatially represented numerical information, such as place-value errors.

Developmental Features

Unclear

Neuropsychological Features

Appears to be associated with right-hemispheric dysfunction, in particular, posterior regions of the right hemisphere.

Genetic Features

Unclear

Relationship to RD

Does not appear to be associated with RD, at least not with the forms of RD that are associated with phonetic deficits.

Note. The described features appear to be more typical of boys than girls. The specifics regarding a potential gender difference in the manifestation of this type of deficit are unclear. RD = reading disabilities.

1983). Nevertheless, cognitive studies of procedural-skill development would be strengthened by the inclusion of measures of visuospatial functions. Finally, the possibility that MD children's difficulties in problem representation, which are found with mathematical problem solving, might be associated with these visuospatial deficits needs to be considered. An overview of the visuospatial deficits associated with this form of MD is presented in Exhibit 5.3.

The presentation of the three potential subtypes of arithmetic-related forms of MD should be interpreted as a personal "best guess," based on the current state of knowledge in this area, rather than as a final conclusion. There are many issues that remain to be addressed in the MD area. In fact, in comparison with studies of RD, research in the MD area is rather primitive. A fundamental issue in RD research, for instance, is how to classify an individual as RD or non-RD. One technique might involve a discrepancy between IQ scores and scores on standard reading achievement measures. With this technique, an individual who scored in the normal range on an IQ test but significantly below normal on a measure of reading achievement would be classified as RD. Another technique might involve classifying individuals as RD if they showed significantly poorer reading scores in relation to same-age peers, without a consideration of IQ (Gilger, 1992). The issue of diagnostic criteria is far from resolved in the RD area (Fletcher, 1992; Pennington, Gilger, Olson, & DeFries, 1992) and has not been systematically addressed in the MD area. Moreover, future research will need to address the validity of the three proposed subtypes as potentially distinct syndromes; whether they co-occur with more general deficits or with other more specific deficits, such as RD; and to what extent they are related to IQ.

In conclusion, in comparison with reading disabilities, relatively little is known about mathematical learning disorders (Sutaria, 1985). Clearly, many factors, such as anxiety or poor reading skills, can lead to poor mathematical achievement (Ashcraft & Faust, 1994; Muth, 1984). The review of cognitive, neuropsychological, and genetic correlates of mathematical achievement and mathematical disorder, however, suggests that there exist real deficits that affect mathematical learning and performance. In this chapter, these factors are tentatively classified into three general types of arithmetic-related cognitive or neuropsychological deficits. Research in this area needs to be expanded to include deficits that might be specifically associated with difficulties in mathematical problem solving, such as difficulties in problem representation (A. B. Lewis & Mayer, 1987).

Gender Differences and the Mathematically Gifted

The primary focus of this chapter is on gender differences in numerical and mathematical abilities. The goal is to integrate research on the pattern of gender differences in mathematical performance and to examine these patterns across cultures. For those areas in which consistent cross-national gender differences in mathematical performance are found, potential cognitive, psychosocial, and biological influences on the gender difference are evaluated and integrated into a dynamic model that accommodates each of these influences. As noted below, the gender difference in mathematical performance is most noteworthy in samples of highly gifted individuals (Benbow, 1988). Thus, the chapter concludes with a discussion of the cognitive and neuropsychological characteristics of the mathematically gifted and how these might be related to the gender difference in certain areas of mathematical performance.

Gender Differences

The discussion of gender differences in numerical and mathematical abilities is presented in two general sections. The first contains an overview of gender differences, or a lack thereof, in the basic domains presented in the first three chapters of this book: Early Numerical Abilities, Arithmetic, and Mathematical Problem Solving. In preview, research in these

---

Summaries are provided at the end of the following subsections on Gender Differences: Early Numerical Abilities (p. 192); Arithmetic (p. 195); and Mathematical Problem Solving (p. 201). In the section on Gender Differences in Mathematical Problem Solving: Potential Causes, a summary is provided of the subsection on Psychosocial Factors (p. 215). Finally, a summary is provided of the Mathematically Gifted section (p. 227).

domains suggests little or no gender difference in early numerical abilities or arithmetic but consistent differences, favoring males, in some areas of mathematical problem solving. Thus, the second general section contains an analysis of potential cognitive, psychosocial, and biological causes of the pattern of gender differences in mathematical problem solving. In each section, an attempt is made to consider relevant cross-cultural research, because much of the better known research on gender differences in mathematical abilities has focused primarily on children and adolescents in the United States (e.g., Hyde, Fennema, & Lamon, 1990). A consideration of cross-cultural patterns is useful for determining the extent to which the magnitude of the gender difference is influenced by cultural and experiential factors (Harnisch, Steinkamp, Tsai, & Walberg, 1986).

Early Numerical Abilities

Although there have been many reviews of gender differences in mathematical abilities and achievement, these reviews have been largely confined to kindergarten and older children (e.g., Hyde, Fennema, & Lamon, 1990; Kimball, 1989). A consideration of potential gender differences in the numerical skills of infants and preschool children is important because many of these skills appear to be fundamental or biologically primary (i.e., inborn; Geary, in press; Gelman, 1990). Evidence for a gender difference in basic numerical abilities would provide important support for the argument that any such differences are strongly influenced by biological factors (e.g., Benbow, 1988). A finding of no gender difference in this area, on the other hand, would argue against the position that boys have a fundamental advantage over girls in elementary mathematics. To determine whether gender differences exist in the number and counting skills that are evident in infancy and during the preschool years, I reviewed much of the research that was described in chapter 1. The results of this review are presented below.

Infancy

Of particular interest with the infancy studies is whether boys and girls differ in their sensitivity to numerosity and in their basic understanding of the effects of addition and subtraction on quantity (e.g., Starkey et al., 1983; Wynn, 1992a). Recall that numerosity studies assess the infant's ability to discriminate small quantities, such as sets of two versus three items. Many of the studies in this area did not examine gender differences in infants' sensitivity to numerosity (e.g., Starkey & Cooper, 1980). Those studies that did examine gender differences are very consistent in their findings: Boy and girl infants do not differ in their ability to discriminate small numerosities (Antell & Keating, 1983; Starkey et al., 1990; Strauss & Curtis, 1981). Starkey et al., for instance, found no gender difference, across five experiments, in the ability of 6- to 9-month-old infants to discriminate small numerosities.

Strauss and Curtis (1981) found no gender difference in the ability of infants to discriminate sets of two versus three items, although they did find that girls were better at discriminating sets of three versus four homogeneous items, whereas boys were better at discriminating sets of three versus four heterogeneous items. This result, however, appeared to be due to a gender difference in the preference for looking at homogeneous versus heterogeneous arrays rather than a gender difference in the sensitivity to numerosity. Although Antell and Keating (1983) found an overall gender difference in looking-time patterns, they found that boy and girl neonates (in the 1st week of life) were equally skilled at discriminating sets of two versus three items. Wynn (1992a), in her assessment of 5-month-old infants' understanding of the effects of addition and subtraction on quantity, did not examine gender differences. In a personal communication, however, she stated that there was no gender difference at all on this task (Wynn, personal communication, August 20, 1993). In a related study, with 18- to 42-month-olds, Starkey (1992) also found no gender difference in the basic understanding of the effects of addition and subtraction on quantity.

Numerical Skills in Preschool Children

As described in the Number and Counting section of chapter 1, an array of important basic skills emerges during the preschool years. These include the ability to count and enumerate (i.e., count objects), as well as a variety of conceptual skills (Fuson, 1988). Important conceptual skills at this age include an understanding of counting concepts, such as cardinality, and number concepts, such as *greater than/less than* (e.g., which is more—4 or 6?). Ginsburg and Russell (1981), in an extensive study of social class and racial (Anglo- vs. African-American) differences in the basic numerical skills of 4- to 5-year-olds, found only a single gender difference: Girls were slightly better than boys on a basic addition and subtraction task. There were no gender differences across race or social class in the ability to count and enumerate or on any task that assessed basic counting and number knowledge, such as cardinality, ordinality, or equivalence tasks. In a similar study, Song and Ginsburg (1987) examined the basic numerical abilities of 4- to 8-year-old children from Korea and the United States. Again, no gender differences for either Korean or American 4- to 5-year-old children were found for basic numerical abilities. Finally, Lummis and Stevenson (1990) examined the pattern of gender differences in reading and mathematical skills from three cross-cultural studies of kindergarten, first-grade, and fifth-grade children from the United States, Taiwan, and Japan. The results showed no gender differences in the counting skills, conceptual knowledge, or the simple arithmetic skills of kindergarten children in any of the three cultures. Results for the first- and fifth-grade children are presented later in the chapter.

Summary

The studies discussed in this section suggest that there are no gender differences in the infants' sensitivity to numerical information or fundamental understanding of arithmetic and no gender differences in preschool children's counting skills and conceptual understanding of counting, number, and arithmetic. This conclusion seems to be especially sound for preschool and kindergarten children, because the results are robust across studies and across cultures. For the infancy research, however, this conclusion must be considered tentative, because the measures used in these studies are probably not sensitive enough to detect any potentially more subtle differences. Nevertheless, given the results for preschool children, there appears to be little reason to suspect that more subtle gender differences exist in these basic skills. The overall pattern of findings for the infancy and preschool studies suggest that boys are not biologically primed to outperform girls in basic mathematics. In other words, the later gender difference in mathematical problem solving, which favors boys (Hyde, Fennema, & Lamon, 1990), does not appear to have its antecedents in fundamental numerical abilities. Nevertheless, as noted in the Biological Factors section, these findings do not preclude secondary biological influences, that is, biologically based skills that have evolved for one reason but are used for another, on the gender difference in mathematical problem solving (Benbow, 1988).

Arithmetic

The overview of research on gender differences in arithmetic skills is presented in three sections: performance on paper-and-pencil ability and

oices, and conceptual knowledge. In all
the solving of arithmetic equations, such
ler differences in the solving of arithmetic
the Mathematical Problem Solving section.

e

girls, is often found on arithmetic achieve-
elementary school and junior high school
mon, 1990; Marshall & Smith, 1987; Ste-
ennema, and Lamon, in a meta-analysis of
natical abilities, found that this advantage is
ard deviation) although robust, at least for
ited States. Cross-cultural studies that have
for solving simple (e.g., 5 + 3) and complex
roblems suggest that the advantage of girls
ted to the solving of complex problems and
U.S.) phenomenon (see Husén, 1967; Mar-

nined gender differences in skill at solving
such as 5 + 7, have produced mixed results.
s, for example, found that Chinese girls out-
a timed simple addition test (e.g., 3 + 5 =
ot in the fifth grade (Stevenson, Lee, Chen,
ontrast, first-grade boys in the United States
t-grade girls on this same test, but there was
he fifth grade. In our studies, we have found
erformance on paper-and-pencil simple addi-
merican kindergarten and first-grade children
l., 1993; Geary, Fan, & Bow-Thomas, 1992).
evenson (1990) found no gender difference in
metic problems for first- and fifth-grade chil-
Taiwan, or Japan. In all, it appears that there
differences in skill at solving simple arithmetic

(1987), in a large-scale (>7,000 children) lon-
thematical development of American children,
rls over boys on basic arithmetic tests from the
advantage appeared to be primarily related to
arithmetic problems, such as 37 − 26 or 67 +
rror patterns provided some intriguing clues as

to why this gender difference emerged. Basically, the advantage of elementary school girls in complex arithmetic was related to a relatively high number of procedural errors in boys' performance. Across grades, the three largest gender differences occurred for complex addition and complex subtraction. For subtraction, boys were much more likely than girls to subtract the smaller number from the larger number regardless of position (e.g., 23 − 7 = 24), and boys committed many more trading errors than girls when solving complex addition and complex subtraction problems. These procedural errors are described in more detail in chapter 2 (see, for example, Figure 2.4) and can occur if the child does not understand the underlying concept, such as place value, or because of procedural bugs. Recall that bugs involve the misapplication of a procedure that might be correct for some problems but is incorrectly used to solve other problems (VanLehn, 1990). The finding of no gender difference in American children's understanding of arithmetic concepts, discussed below, suggests that the common finding (in U.S. studies) of an advantage of girls in basic arithmetic might be related to more procedural bugs in boys' performance.

Strategy Choices

The different types of strategies that children use to solve arithmetic problems, such as finger counting or verbal counting, are described in chapter 2. The question addressed here is whether boys or girls favor one type of strategy over another. Siegler (1988a) found no gender difference in the overall distribution of strategy choices for solving simple addition or simple subtraction problems. Similarly, we found no gender difference in mix of strategies used to solve simple addition problems for Chinese or American kindergarten children (Geary, Bow-Thomas, et al., 1993). However, we did find that at the end of first grade, American girls were more likely than American boys to count on their fingers to solve simple addition problems, although there was no difference in the strategy mix for the Chinese first graders (Geary, Fan, & Bow-Thomas, 1992). In a related study, we found no gender difference in the strategy mix for either Chinese or American children at the end of first grade (Geary, Bow-Thomas, Fan, et al., 1992). Thus, at least for solving simple arithmetic problems, elementary school boys and girls do not appear to differ substantively in strategy choices.

Concepts

Hyde, Fennema, and Lamon (1990) found no gender difference in the understanding of arithmetical concepts in elementary school through high school. Even though most of the studies used in the Hyde, Fennema, and Lamon meta-analysis were conducted in the United States, cross-cultural studies support the same conclusion. Ginsburg et al. (1981a, 1981b) compared the numerical and arithmetical knowledge of elementary school children from the United States with children from two African cultures. There were no substantive gender differences in arithmetical skills or conceptual knowledge in any of the cultures. Song and Ginsburg (1987) and Fuson and Kwon (1992b) found no gender differences in Korean children's understanding of numerical and arithmetic concepts. Similarly, Stevenson and his colleagues found no gender differences in the understanding of arithmetic concepts for elementary school children in the United States, Taiwan, Japan, or mainland China (Lummis & Stevenson, 1990; Stevenson, Lee, Chen, Lummis, et al., 1990).

Summary

Girls often show an advantage over boys on basic arithmetic tests, at least through junior high school (Hyde, Fennema, & Lamon, 1990). Although it cannot be stated definitively, it appears that this advantage is related to the tendency of boys to commit procedural errors when solving complex arithmetic problems. These errors, in turn, appear to reflect bugs, or the use of procedures that are correct for some problems but are inappropriately applied to solve other problems. At this point, it appears that the gender difference in the accuracy for solving complex arithmetic equations is found primarily in the United States and in some other nations, such as Sweden, with low-achieving children (Husén, 1967). Carefully conducted cross-cultural studies in nations with higher achieving children typically do not find a gender difference in this area or sometimes find a male advantage (Husén, 1967; Lummis & Stevenson, 1990). Thus, overall, aside from the advantage of American girls over American boys in the solving of complex arithmetic equations, there appears to be little or no gender difference in arithmetical abilities, including conceptual knowledge and the mix of problem-solving strategies.

Mathematical Problem Solving

The issue of a gender difference in mathematical problem solving is complex, because there are many different types of problem-solving skills associated with mathematics (see chapter 3), some of which show a gender difference, favoring males, and some of which do not (Maccoby & Jacklin, 1974). Moreover, the magnitude and the practical importance of any such gender difference appear to vary with the level of ability of the sample (Hyde, Fennema, & Lamon, 1990). For these reasons, the issue of a gender difference in mathematical problem solving is addressed in two sections. The first section focuses on gender differences in general samples, that is, samples that have not explicitly focused on gifted individuals, and separately examines gender differences in performance on word problems and spatially based domains, such as geometry. The second section focuses on the work of Benbow and her colleagues and addresses the issue of gender differences in gifted samples (Benbow, 1988; Benbow & Stanley, 1983; Stanley, Huang, & Zu, 1986).

In both sections, the assessment of gender differences is based on performance on ability and achievement measures, because the gender difference in mathematical problem solving often emerges on these measures and because performance on these measures is predictive of later academic and job-related performance (Benbow, 1992; Boissiere, Knight, & Sabot, 1985; McGee, 1979; Rivera-Batiz, 1992). However, note that through precalculus courses, girls typically show higher mean grades in mathematics than boys, at least in the United States (Kimball, 1989).

General Samples

The results described in this section are from large-scale studies of what appear to be representative samples of boys and girls from many different nations. These studies represent differences across a broad range of abilities, including giftedness. Hyde, Fennema, and Lamon (1990), and others before them (e.g., Dye & Very, 1968), argued that the gender difference in mathematical problem solving is not evident until adolescence. As described in the Psychometric Perspective section of chapter 4, factor-analytic studies suggest that a mathematical reasoning factor does not emerge until high school, and it is the emergence of this factor that is associated with the emergence of a gender difference in mathematical abilities (Canisia, 1962; Very, 1967).

Nevertheless, a number of cross-cultural studies of elementary school children have found that boys have a performance advantage over girls in several areas of mathematical problem solving, such as the solving of arithmetic word problems. Lummis and Stevenson (1990) found a reliable gender difference, favoring boys, for the solving of word problems for first- and fifth-grade children from the United States, Taiwan, and Japan. Stevenson and his colleagues also found a reliable gender difference, again favoring boys, for performance on arithmetic word problems for fifth-grade children from mainland China and the United States (Stevenson, Lee, Chen, Lummis, et al., 1990). In all of the comparisons, the Asian children outperformed their American peers, but the magnitude of the gender difference was the same across nations. Though robust, the magnitude of the advantage of the elementary school boys over girls for solving arithmetic word problems was small (about ⅕ of a standard deviation).

Harnisch et al. (1986), in a reanalysis of data from a large-scale international study of the mathematics achievement of adolescents (Husén, 1967), found a male advantage on a mathematics achievement measure in each of the 10 nations included in this reassessment, although the magnitude of the difference varied greatly across nations. The smallest overall gender difference was found in the United States and Sweden; the largest was found in Great Britain and Belgium. For 13-year-olds, gender differences were the largest for solving word problems and for geometry in all 10 nations (Steinkamp, Harnisch, Walberg, & Tsai, 1985); for 17-year-olds, this same pattern was found in 8 of the 10 countries. Similarly, Marshall and Smith (1987) found that sixth-grade boys in the United States committed fewer errors than sixth-grade girls when solving arithmetic word problems, although no difference was found in the third grade. Error patterns suggested that girls found the translation of arithmetic word problems into appropriate equations more difficult than boys. From the example provided in their article (Marshall & Smith, 1987), it appears that the translation of compare problems is particularly difficult for elementary school girls (see the Problem-Solving Processes section of chapter 3).

In a nicely conducted series of nine experiments, E. S. Johnson (1984) found that male college students consistently (i.e., in every study) outperformed their female peers for solving algebraic word problems. The problem set used in this study was largely the same as that used in a series of studies conducted in the 1950s. Across studies, the overall magnitude of the gender difference was unchanged from the 1950s to the 1980s (the overall male advantage was about ½ of a standard deviation). However, the magnitude of the male advantage was reduced by about ½ when important features of the word problems were diagrammed. Finally, the male advantage in solving word problems did not generalize to nonmathematical problem-solving tasks (E. S. Johnson, 1984). There were no gender differences for performance on an array of nonmathematical paper-and-pencil and computer-administered problem-solving tasks (E. S. Johnson, 1984; Maccoby & Jacklin, 1974). In all, these studies indicate that beginning in the elementary school years and continuing into adulthood, boys often show an advantage over girls in the solving of arithmetical and algebraic word problems.

Except for the solving of algebraic word problems, there appear to be no other gender differences in algebraic skills (Hyde, Fennema, & Lamon, 1990), but modest gender differences are found in geometry and calculus. As noted earlier, Harnisch et al. (1986) found a consistent male advantage in geometry for both 13- and 17-year-olds across 10 nations. Across nations, male adolescents had about a ½-standard-deviation advantage in geometry over their female peers (Harnisch et al., 1986). Similarly, Lummis, Stevenson, and their colleagues consistently found gender differences on many mathematical tasks that, like geometry, could be solved through the use of visuospatial skills (Lummis & Stevenson, 1990; Stevenson, Lee, Chen, Lummis, et al., 1990). The associated tasks involved measurement, estimation, and the visualization of geometric figures. Differences on these tasks, which always favored boys, emerged as early as the first grade and were found in mainland China, Taiwan, Japan, and the United States. Moreover, despite the finding that large national differences in mathematical performance were found in these studies, the magnitude of the gender difference, which was small to modest, did not vary across nations. Wood (1976) reported this same pattern of gender differences for children in Great Britain. In all, these studies show a consistent advantage of boys over girls in geometry and other spatially related mathematical areas, such as measurement, beginning as early as the first grade.

Gifted Samples

Perhaps the most dramatic gender difference in mathematical problem solving emerges in studies of gifted adolescents (Benbow, 1988; Benbow & Stanley, 1980, 1983). In the early 1970s, J. C. Stanley began a project designed to identify mathematically precocious adolescents (Study of Mathematically Precocious Youth, or SMPY). To achieve this end, children who scored in the top 2% to 5% on standard mathematics achievement tests in the seventh grade were invited to take the Scholastic Aptitude Test (SAT). The Mathematics section of the SAT, SAT–M, assesses the individual's knowledge of some arithmetic concepts, such as fractions, as well as basic algebraic and geometric skills (Stanley et al., 1986). Twelve- and thirteen-year-olds who scored above the mean for high school girls on the SAT–M were considered to be mathematically gifted, that is, in the top 1% of mathematical ability (Benbow, 1988). The SAT–M was chosen as a measure of precocious mathematical skills because, at least in the United States, most 12- to 13-year-olds have not covered much of the content on the SAT–M in their course work. The SAT–M was therefore considered, for adolescents of this age, to be a good index of mathematical problem solving, or mathematical reasoning, and not an index of prior schooling.

The rationale for studying this select group of individuals was that they quite likely represent many of this nation's future mathematicians and physical scientists, that is, they represent an important national resource (Benbow & Stanley, 1982; Gallagher, 1993; McGuinness, 1993). Indeed, a longitudinal study of an early cohort of SMPY adolescents, when they were 23 years old, showed that an unusually high number (84%) of these individuals had graduated from college. Moreover, 72% of the highest ability men and 40% of the highest ability women had a mathematics or science major in college. About half of these individuals were attending graduate school, and 71% and 39%, respectively, of these men and women were working toward advanced degrees in mathematics or science (Benbow, 1992). In short, the adolescents identified through SMPY, and their unidentified high-ability peers, will almost certainly be overrepresented in the United States' pool of future mathematicians and physical scientists. Thus, any gender differences within this highly select group might portend later gender differences in the number of men and women in mathematics and the physical sciences.

The consideration of the gender difference in SAT–M performance for SMPY adolescents has been on two levels: mean differences and the ratio of boys to girls at different levels of performance (Benbow, 1988; Benbow & Stanley, 1983). Across cohorts in the United States, boys, on average, have been found to consistently outperform girls on the SAT–M by about 30 points (about ½ of a standard deviation). This gender difference has also been found in West Germany and in mainland China (Benbow, 1988; Stanley et al., 1986). Stanley et al., for instance, administered the SAT–M to a group of seventh- and eighth-grade Chinese students and found a 15- to 20-point male advantage (about ⅓ of a standard deviation). Note that the mean performance of the Chinese girls ($M = 619$) was between 50 and nearly 200 points higher, depending on the cohort, than the mean of the American boys identified through SMPY. Stanley et al. argued that the advantage of gifted Chinese children over gifted American children on the SAT–M was probably due to the assignment, and completion, of more homework in China and the fact that some of the material covered on the SAT–M is introduced in the seventh

Figure 6.1

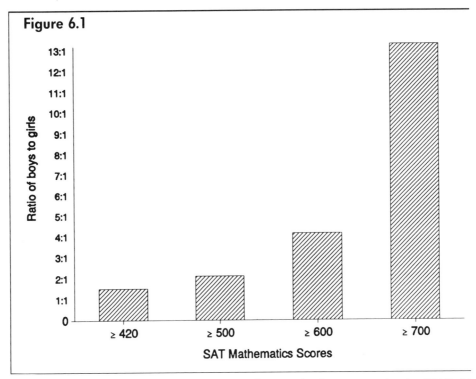

Ratio of mathematically gifted U.S. adolescent boys and girls at varying levels of SAT–M performance.

grade in China, but not until high school in the United States (see the Schooling section of chapter 7).

The gender difference in the ratio of boys to girls at different levels of mathematical reasoning ability, as measured by the SAT–M, is even more dramatic. The ratio of boys to girls in the United States at different levels of mathematical performance (Benbow & Stanley, 1983) is presented in Figure 6.1. As can be seen in Figure 6.1, the ratio of boys to girls at the lower end of SAT–M scores is a rather modest 1.5:1 but increases to 13:1 for those scoring ≥700. Stanley et al. (1986) also found a gender difference in the ratio of high-scoring (≥700) boys and girls in mainland China, but the ratio was only 2.5:1. There were smaller numbers of students assessed in China and the sampling techniques were not as systematic as in the U.S. studies, so the ratio difference across nations needs to be interpreted with caution. Finally, the overrepresentation of boys at the high end of SAT–M performance is not limited to SMPY samples. Dorans and Livingston (1987) reported that across two administrations of the SAT to high school seniors, 96% of the perfect scores (i.e., 800) on the SAT−M were obtained by boys. No matter what the absolute size of the ratio, the finding of a male advantage at the upper end of SAT−M performance probably has long-term implications for the numbers of boys and girls that are likely to enter mathematical and physical science areas (Benbow, 1992). I take a closer look at the cognitive skills of these precocious adolescents at the end of this chapter.

Summary

A consistent gender difference in mathematical problem solving is found in some domains, such as geometry and word problems, but not in other domains, such as algebra (Hyde, Fennema, & Lamon, 1990). It has generally been argued that when a gender difference in mathematical abilities is found, it is typically not found until adolescence (e.g., Benbow, 1988; Dye & Very, 1968; Hyde, Fennema, & Lamon, 1990). This conclusion has been based, for the most part, on comparisons of American children. Multinational studies, in contrast, show that a male advantage in the solving of arithmetic word problems and on tasks that are solvable through the use of visuospatial skills, such as visualizing geometric shapes, is often evident in elementary school (Lummis & Stevenson, 1990; Stevenson, Lee, Chen, Lummis, et al., 1990). Differences across studies primarily conducted in the United States and those conducted in other countries might be related to the overall level of achievement of the associated samples. Harnisch et al. (1986) found that gender differences tended to be smallest in nations with the lowest achieving children, which includes the United States. Thus, the poor mathematical abilities of American children might mask differences that might otherwise be found (see the International Studies section of chapter 7). Studies of gender differences in mathematical abilities within the United States need to take this potential "floor effect" into consideration (e.g., Schratz, 1978; see *Historical trends* section, this chapter).

Even in areas where a gender difference in performance exists across cultures, the difference tends to be modest and selective. For instance, the results of Marshall and Smith (1987) suggest that the advantage of elementary school boys over girls for solving arithmetical word problems might be more pronounced for problems that require the translation of important relationships rather than being a general male advantage in solving word problems. Similarly, Senk and Usiskin (1983), in a large-scale national (U.S.) study, found no gender difference in high school students' ability to write geometric proofs after taking a standard high school geometry course, even though adolescent boys typically perform better than their female peers on geometric ability tests (Hyde, Fennema, & Lamon, 1990). Thus, the male advantage in geometry might also be selective, that is, associated with certain features of geometry rather than the entire domain.

Even if the overall gender difference in mathematical problem solving tends to be modest in the general population, it is often rather dramatic for high-ability groups (Benbow, 1988). The findings of Benbow and her colleagues for mathematically precocious youths show very large differences in the number of mathematically gifted boys and girls, the ratios of which sometimes exceed 13:1 (Benbow & Stanley, 1983). Nevertheless, the results of Stanley et al. (1986), in which gifted Chinese girls significantly outperformed gifted American boys on the SAT−M, suggest that even precocious adolescents in the United States are not performing to their full potential. In other words, the SMPY findings should not be interpreted as the ceiling for the development of mathematical skills, nor should they be interpreted as meaning that only a small minority of girls can develop the mathematical skills necessary to be mathematicians or physical scientists. Stated differently, the performance of SMPY boys should not be considered a threshold for entry into math-intensive careers. Finally, it should be stated again that the male advantage in these studies appears to be limited to *certain* mathematical domains and does not represent a male advantage in general problem-solving skills (E. S. Johnson, 1984; Maccoby & Jacklin, 1974).

Gender Differences in Mathematical Problem Solving: Potential Causes

In this section, factors that appear to contribute to the just-described gender differences in mathematical problem solving are considered. These influences include a gender difference in cognitive style, as well as psychosocial and biological factors (Benbow, 1988; Eccles [Parsons], Adler, & Meece, 1984; Fennema & Sherman, 1977). Research in each of these areas is considered in turn and is followed by an integrative summary.

Cognitive Style

It has been frequently argued that the male advantage in certain areas of mathematical problem solving is related to a male advantage in visuospatial skills (Benbow, 1988; Maccoby & Jacklin, 1974; McGee, 1979; McGuinness, 1993). In this section, a brief overview of gender differences in spatial skills is presented and is followed by an analysis of the relationship between spatial skills and the gender difference in mathematical problem solving. In regard to the former, a gender difference, favoring males, has been found on many different types of psychometric tests of spatial ability, such as those that involve the three-dimensional rotation of images (Linn & Petersen, 1985; Masters & Sanders, 1993). A male performance advantage has also been found on more dynamic measures of spatial skills (Gilger & Ho, 1989; Law, Pellegrino, & Hunt, 1993). Law et al., for example, found that male college students were better at judging relative distance and the relative velocity of moving objects than were female college students. Gender differences on spatial tasks are most pronounced during adolescence but have also been found early in the elementary school years (E. S. Johnson & Meade, 1987; Lummis & Stevenson, 1990).

The development of spatial skills is influenced by both biological and experiential factors. For instance, performance on spatial tasks improves with practice and experience for both males and females, but the magnitude of the male advantage on these tasks does not diminish with practice (Baenninger & Newcombe, 1989; Ben-Chaim, Lappan, & Houang, 1988; Burnett & Lane, 1980; Law et al., 1993; Okagaki & Frensch, 1994). Evidence for biological influences on spatial skill development comes from behavioral genetic studies and from studies of the influence of sex hormones on cognitive abilities. Individual differences on spatial tests tend to be moderately to highly heritable (Vandenberg, 1966). Prenatal exposure to androgens, or male hormones, appears to increase the spatial skills of girls with congenital adrenal hyperplasia, in relation to their unaffected sisters (S. M. Resnick, Berenbaum, Gottesman, & Bouchard, 1986).

Any such biological influences on individual and gender differences in spatial skills probably include both direct and indirect effects (Geary, 1989). Direct influences would include the influence of sex hormones on the development of the neural substrate that supports spatial cognition (e.g., Diamond, Johnson, & Ehlert, 1979). Indirect influences would involve engagement in behaviors that provide spatial-related experiences to the individual, such as building with construction toys (Serbin & Connor, 1979). Engagement in activities that are likely to facilitate the development of spatial skills also appears to be influenced, at least to some extent, by sex hormones (Berenbaum & Hines, 1992). There are also sociocultural and experiential influences on the gender difference in the engagement in spatial-related activities (Linn & Hyde, 1989). Parents appear to reinforce sex-typed behavior, which would increase any biologically based gender difference in the engagement in spatial-related activities (Eibl-Eibesfeldt, 1989; Lytton & Romney, 1991). Participation in drafting, engineering, and certain mathematics courses (e.g., geometry), which typically favors males, also appears to improve spatial skills (Burnett & Lane, 1980). The net result of these biological and experiential factors is that males have an advantage over females in certain forms of spatial cognition (Halpern, 1992).

The relationship between spatial skills and mathematical performance is complex (McGee, 1979; Pattison & Grieve, 1984). Spatial skills might be useful for some types of mathematical tasks, such as visualizing geometric shapes, but are probably relatively unimportant for seemingly similar tasks, such as providing geometric proofs. Also, many mathematical problems that could be solved with the use of spatial representations can probably also be solved with the use of nonspatial strategies (Fennema & Tartre, 1985; McGuinness, 1993), whereas performance in other areas of mathematics, such as solving algebraic equations, might not be strongly facilitated by spatial skills at all (Halpern, 1992). In other words, the relationship between spatial and mathematical skills is probably selective, even within mathematical areas that appear to have important spatial components. To illustrate, Ferrini-Mundy (1987) found that spatial skills were important for solving solids of revolution problems but were less important for solving other types of calculus problems. Thus, it is not surprising that correlational studies sometimes find a relationship between spatial skills and mathematical performance (see McGee, 1979) and sometimes do not (Armstrong, 1981; Pattison & Grieve, 1984). It is likely that the content of mathematical and spatial tests, as well as the skill level of the sample, will strongly influence the degree of correlation between tests of mathematical and spatial abilities.

Nevertheless, there appears to be a moderate overall relationship between spatial skills and certain mathematical skills (e.g., Burnett, Lane, & Dratt, 1979; Fennema & Sherman, 1977; Sherman, 1980). E. S. Johnson (1984), for instance, found spatial skills and word-problem-solving skills to be correlated .52 and .63, respectively, for male and female college students. Many, though not all (e.g., Armstrong, 1981), studies have found a relationship between the male advantage on spatial tasks and the gender difference in mathematical problem solving (McGee, 1979). Burnett et al. statistically eliminated an advantage of male college students over their female peers on the SAT–M by partialing a gender difference on a spatial visualization test (see also Fennema & Sherman, 1977). Cross-cultural studies also show that gender differences in mathematical problem solving primarily reside in tasks that appear to be solvable through the use of spatial representations (Harnisch et al., 1986; Lummis & Stevenson, 1990; Stevenson, Lee, Chen, Lummis, et al., 1990). A relationship between spatial skills and mathematics achievement is sometimes found even within groups of girls. Fennema and Tartre (1985), in a longitudinal study of spatial skills and mathematics achievement from 6th to 8th grade, found that girls with low spatial skills but good verbal skills showed a decline in mathematics achievement as compared with girls with high spatial skills and low verbal ability. Sherman (1981) found that for high school girls, spatial ability in the 9th grade was a stronger predictor of mathematics course enrollment in the 12th grade than were attitudes toward mathematics.

However, one consistent cross-national finding does not, at first glance, appear to fit this pattern. Recall that Stevenson and his colleagues found that elementary school boys have an advantage over girls in the solving of arithmetic word problems in the United States, Taiwan, Japan, and mainland China (Lummis & Stevenson, 1990; Stevenson, Lee, Chen, Lummis, et al., 1990). Harnisch et al. (1986) found that across 10 nations, gender differences in mathematical performance for 13- and 17-year-olds were most consistent for geometry and word problems. It was noted earlier that elementary school girls appear to have particular difficulty in solving word problems that involve relational comparisons and that include confusing key words (Marshall & Smith, 1987). The errors that are associated with solving these types of problems can often be avoided if important relationships described within the problem are diagrammed as part of the problem-solving process, as was shown in Figure 3.3 (A. B. Lewis, 1989; A. B. Lewis & Mayer, 1987). Thus, the gender difference in skill at solving word problems might very well be related to the male advantage in spatial skills.

Indeed, McGuinness (1993) has recently reported that males, from the age of 4 years, are much more likely to resort to spatial-related strategies in problem-solving situations than are females. In particular, males spontaneously use dynamic three-dimensional representations of problem situations much more frequently than do females. This gender difference in strategic approaches to problem solving easily accommodates the cross-national male advantage in certain areas of geometry, as well as performance on estimation and measurement tasks (Harnisch et al., 1986; Lummis & Stevenson, 1990). Moreover, the tendency of boys to resort to the use of spatial strategies in problem-solving situations more frequently than girls might explain why boys make fewer mistakes when solving word problems. E. S. Johnson's (1984) finding that the presentation of diagrams facilitated the solving of algebraic word problems by female, but not male, college students is consistent with this view. Several other studies have also suggested that teaching girls to use spatial-related strategies might improve their performance for solving word problems and for solving certain types of calculus problems (Fennema & Tartre, 1985; Ferrini-Mundy, 1987; Tartre, 1990).

In all, it has been argued that the gender difference in mathematical problem solving is primarily related to a male advantage in visuospatial skills (e.g., McGee, 1979). The tendency of males, from preschool onward, to use spatial strategies in problem-solving situations more frequently than females is consistent with this view, as is the finding that the gender difference in mathematical problem solving is largely confined to areas for which spatial strategies will facilitate performance. In other words, a gender difference in mathematical problem solving is expected only on certain types of tasks, such as dealing with novel geometric problems, but not on other types of tasks that can be solved without the use of spatial strategies, such as providing geometric proofs (Senk & Usiskin, 1983). The pattern of results also suggests that at the cognitive level, the gender difference in mathematical problem solving might be better thought of as a difference in strategy rather than as a difference in mathematical problem-solving skills per se (Byrnes & Takahira, 1993; E. S. Johnson, 1984). This is not to say that the use of nonspatial strategies is always ineffective for solving mathematics problems (Fennema & Tartre, 1985). Rather, the tendency of males to spontaneously use spatial strategies in problem-solving situations more frequently than females appears to contribute to their small to moderate advantage over females in certain mathematical problem-solving areas.

Psychosocial Factors

In this section, psychosocial influences on the magnitude of the gender difference in mathematical problem solving are considered. Any such influence should be considered as complementary to the influence that the gender difference in spatial skills has on mathematical problem solving rather than an alternative explanation. The finding that individual differences in spatial skills are moderately to highly heritable (Vandenberg, 1966) and are influenced by earlier exposure to sex hormones (S. M. Resnick et al., 1986) indicates that it is very unlikely that social experiences create a gender difference in spatial skills. Social influences can, however, influence the magnitude of the gender difference in spatial skills. More important, psychosocial factors do appear to influence the level of participation in mathematics and mathematics-related activities. A gender difference in the level of participation in mathematical activities might increase the male advantage in mathematical and spatial performance but is not likely to create a gender difference in, for instance, the spontaneous use of spatial strategies in problem-solving situations (especially in 4-year-olds; McGuinness, 1993).

Moreover, a child's basic cognitive skills will also influence the types of activities that he or she seeks out (Scarr & McCartney, 1983). Relatively poor spatial skills might make mathematical problem solving, or related activities, relatively difficult and therefore not a preferred activity. Avoidance of spatial-related activities, in turn, will quite likely impede the further development of spatial skills. Thus, simple arguments that the gender difference in mathematical problem solving is solely due to one cause or another are almost certainly incomplete. In this section, three general psychosocial influences on the gender difference in mathematical problem solving are considered in turn: historical trends in the magnitude of the gender difference in mathematical performance, perceived competence and perceived usefulness of mathematics, and classroom experiences.

Historical trends. Several recent meta-analyses have suggested that the magnitude of the gender difference in mathematical performance has declined over the last several decades (Feingold, 1988; Hyde, Fennema, & Lamon, 1990). These trends support the conclusion that the gender difference in certain mathematical skills is responsive to social changes, such as increased participation of girls in mathematics courses and mathematics-related activities (Linn & Hyde, 1989). The historical decline in the gender difference in mathematical performance is selective, however. Hyde, Fennema, and Lamon reported that the overall male advantage in mathematics, averaged across all areas, had been reduced by about one half, comparing studies published before 1973 with studies published in 1974 or later. The authors did not provide analyses for separate mathematical areas, such as arithmetic or geometry, so it is unclear whether this represents a general or selective phenomenon. Also, two data points are certainly not enough to argue for any type of trend.

Feingold (1988), on the other hand, did report separate historical trends, across four time periods, for the Preliminary SAT (PSAT), taken by high school juniors, and for the SAT. For the PSAT, there was a 65% decline in the relative advantage of boys over girls from 1960 to 1983, although boys still showed a slight advantage in 1983. For the SAT, in contrast, the male advantage remained relatively constant from 1960 to

1983. Benbow and her colleagues have found no evidence for a declining male advantage for SMPY adolescents assessed in different years (Benbow, 1988). Similarly, as noted earlier, E. S. Johnson (1984) found that the overall magnitude, across nine studies, of the male advantage in solving algebraic word problems remained unchanged from the 1950s to the 1980s.

The pattern of results in this area is obviously conflicting. It seems likely that in the United States, the magnitude of the gender difference in mathematical problem solving is declining in some areas, especially for the general population (Hyde, Fennema, & Lamon, 1990). This general trend probably reflects a variety of influences, including potentially greater participation of girls in mathematics-related activities and courses (Linn & Hyde, 1989; Travers & Westbury, 1989) and potential "floor effects" (Harnisch et al., 1986; Steinkamp et al., 1985). Recall that Harnisch et al. found that the advantage of 17-year-old adolescent boys over same-age girls was smaller in the United States and in Sweden than in eight other nations. The mathematical achievement of boys from these two countries was, in fact, dramatically lower than the performance of boys in the eight other nations. The mean difference in the mathematical performance of American boys and the performance of boys in the top eight countries was 1.7 standard deviations, whereas the same comparison for girls produced a difference of 1.2 standard deviations.

In other words, the lack of emphasis on mathematics education in American culture (Lapointe, Mead, & Askew, 1992; Stevenson, Lee, Chen, Stigler, et al., 1990) appears to have a larger impact on the mathematical achievement of boys than girls. E. G. Moore and Smith (1987) found a similar gender difference even within the United States: Poorly educated women showed better mathematical skills than their male peers, but better educated men had higher mathematics achievement scores than their female peers. Thus, the historical trend for the magnitude of the gender difference in mathematical performance, which has been generated based primarily on data collected within the United States, needs to be interpreted cautiously. It is very possible that the "disappearance" of some gender differences in cognitive skills actually reflects, at least to some extent, a more general cultural trend in education, which is affecting the achievement of boys more than girls. From this perspective, the gender difference in mathematical problem solving should increase as the level of mathematical performance increases. This is exactly the pattern that has been found with gifted and general samples within the United States

(Benbow, 1988; Dorans & Livingston, 1987; E. G. Moore & Smith, 1987), as well as in cross-national comparisons (Steinkamp et al., 1985).

Nevertheless, these same cross-cultural comparisons show that 17-year-old girls from eight nations had higher mean mathematics achievement scores than the mean of boys in the United States (Harnisch et al., 1986). Sociocultural factors clearly impact the *level* of mathematical development, for both boys and girls, but are probably not primary determinants of the gender difference in mathematical problem solving. The pattern of results across cultures also suggests that motivated and able girls should be able to develop the mathematical skills necessary to succeed in the mathematical and physical sciences. The focus on the gender difference in mathematical problem solving often obscures this fact. Thus, the mathematical achievement of American boys should not, in any way, be considered a standard to be achieved by American girls. The cross-national pattern of results suggests that with appropriate educational experiences, the average American girl can achieve a level of mathematical skill that far exceeds the current skill level of her male peers. This appears to be the case for average as well as gifted children (Stanley et al., 1986). These data also suggest that with any such improved educational experiences, the magnitude of the male advantage in *certain* areas of mathematics will likely "reappear" or increase.

Perceived usefulness and competence of mathematics. In addition to general ability and spatial skills, attitudes about mathematics are related to the development of mathematical skills and to the magnitude of the gender difference in mathematical problem solving (Aiken, 1976). Of particular importance is the influence that attitudes have on participation in mathematics courses. In this section, the gender difference in mathematics course taking is first reviewed and then followed by a consideration of psychosocial factors that influence this gender difference. In preview, the two most important psychosocial influences on later participation in mathematics courses appear to be perceived usefulness of mathematics and perceived mathematical competence (Eccles [Parsons] et al., 1984; Meece, Parsons, Kaczala, Goff, & Futterman, 1982; Meece, Wigfield, & Eccles, 1990).

Participation in mathematics courses is important because the number of mathematics courses taken in high school and beyond will obviously influence the development of mathematical skills and will influence career options in adulthood (Jones, 1987; Sells, 1980; Wise, 1985). Indeed, the probability of attaining long-term goals in science-related areas is, among other things, related to the adequacy of mathematics preparation in high school (Wise, 1985). Many studies, across many different nations, have shown that girls take fewer advanced mathematics courses in high school than do boys (Armstrong, 1981; Husén, 1967; Nevin, 1973; Sherman, 1981; Travers & Westbury, 1989). Armstrong, for instance, found that at the beginning of high school, girls intended to take as many mathematics courses as boys but by the end of high school, they had actually enrolled in fewer advanced courses (e.g., trigonometry and calculus) than boys. This trend lessened somewhat from the 1960s to the 1980s, although in the 1980s male high school students still outnumbered female high school students in advanced courses by about 2 to 1 in the United States and in many other nations (Chipman & Thomas, 1985; Travers & Westbury, 1989).

The relationship between differential course taking and the gender difference in mathematical problem solving is complicated. Overall, it appears that the greater participation of males in mathematics and mathematics-related courses (e.g., physics) contributes to the gender difference in mathematical problem solving, at least in high school, but does not account for all of the difference (Armstrong, 1981; Fennema & Sherman, 1977; Husén, 1967). Wise (1985), for instance, reported that for a group of high school students, mathematics achievement in 12th grade was related to basic mathematical ability, measured in the 9th grade, and the number of mathematics courses taken during high school. Male students had a mathematics achievement advantage over their female peers in the 12th grade, which was related to more mathematics courses taken by the male students than by the female students. Nevertheless, a small gender difference, favoring males, in mathematics achievement was still found even for students who had the same 9th-grade mathematics ability scores and who had taken the same number of mathematics courses during high school.

Even though the gender difference in the number of mathematics courses taken in high school does not entirely explain the gender difference in mathematical problem solving, it is important to understand why women participate in mathematical courses less frequently than do men. As noted earlier, two important influences include a gender difference in perceived mathematical competence and perceived usefulness of mathematics. Eccles, Wigfield, Harold, and Blumenfeld (1993) recently reported that as early as the 1st grade, children differentiate between their perceived competence in various academic areas, including mathematics, and the value or usefulness of the area. Eccles et al. found no gender difference in the perceived value or usefulness of mathematics for 1st-,

2nd-, or 4th-grade children. In a nicely conducted set of studies, Eccles (Parsons) et al. (1984), on the other hand, showed that during the high school years female students begin to value English courses more highly than mathematics courses, whereas male students show the opposite pattern, valuing mathematics courses more than English courses. The gender difference in the relative valuation of English and mathematics mediated a gender difference, favoring males, in the number of upper-level mathematics courses taken in 12th grade. Husén (1967) found the same pattern for male and female adolescents across 12 nations. Thus, during the high school years, male students consistently perceive mathematics as being a more useful skill than do female students.

The perceived usefulness of mathematics appears to be related to long-term career goals (Chipman & Thomas, 1985; Wise, 1985). Not surprisingly, those students who aspire to professions that are mathematics intensive, such as engineering or the physical sciences, take more mathematics courses in high school and college than those students who have aspirations for less mathematics-intensive occupations. Chipman and Thomas showed that women were much less likely to enter mathematics-intensive professions than were equal-ability men. For example, in the United States, by the 1970s, women received just over 3% of the undergraduate engineering degrees and less than 20% of the degrees in computer science and the physical sciences (i.e., chemistry and physics). The proportion of women earning graduate degrees in these areas is even lower (Chipman & Thomas, 1985).

The small number of women in these mathematics-intensive areas is not simply due to the fact that they are male-dominated professions, although this might make some women hesitant to enter these areas. For the most part, it appears that many girls and women do not believe that work in these areas will be especially interesting, even women with very high SAT-M scores (Chipman & Thomas, 1985). Sherman (1982), for instance, assessed the attitudes of ninth-grade girls toward mathematics and science-related careers. One interview question asked the girls to imagine working as a scientist for a day. "A clear majority of girls (53%) disliked that day somewhat or very much" (Sherman, 1982, p. 435). Even those girls who found working as a scientist for a day acceptable did not consider it to be a preferred activity. It was felt by many of the girls that work as a scientist would interfere with family life or be boring. This same question was not posed to ninth-grade boys, so it is not known how many boys of this age would have also imagined that working as a scientist would be unrewarding. Either way, by mid-adolescence there is a clear gender difference, favoring boys, in intentions to pursue scientific or other mathematics-intensive careers (Wise, 1985).

The relative lack of interest in mathematics-intensive careers on the part of girls, in turn, appears to be related to a gender difference in the relative orientation toward people and human relationships (Maccoby, 1988, 1990; McGuinness, 1993). In general, girls show a greater preference for and interest in social relationships than boys, whereas boys are much more likely to be object oriented rather than people oriented (Eibl-Eibesfeldt, 1989; Thorndike, 1911). Object-oriented preferences are related to interests in mathematics-intensive careers, such as engineering or the physical sciences (Chipman & Thomas, 1985; Roe, 1953). Chipman, Krantz, and Silver (1992) found that among college women, an orientation toward objects, rather than people, was related to an interest in a career in the physical sciences. This is not the whole story, however, because female college students are less likely to major in physical science areas than are their male peers, even after controlling for gender differences in object versus people preferences and the number of mathematics courses taken in high school (Chipman & Thomas, 1985). The fact that most mathematics-intensive areas are male dominated may dissuade some women, who might otherwise be interested, from pursuing careers in these areas (Halpern, 1992). Nevertheless, it appears that many girls perceive mathematics-intensive careers (e.g., engineering) to be incompatible with their interests in human relationships, which, in turn, appears to contribute to the gender difference in the relative valuation of mathematical skills and the resulting gender difference in the number of mathematics courses taken in high school.

As noted earlier, the second important psychosocial factor that appears to influence participation in mathematics is perceived competence (Meece et al., 1982). At a general level, perceived competence in academic, and probably nonacademic, areas appears to be related to two factors (Marsh, Smith, & Barnes, 1985). The first is one's performance in relation to peers: The better the relative performance, the higher the perceived competence. The second factor is intraindividuality. For instance, children who are relatively better at reading than mathematics tend to have a higher perceived competence in reading than in mathematics, independent of their skill level in relation to other children (Marsh et al., 1985). In the United States, adolescent boys are typically more confident of their mathematical abilities than are their female peers (Eccles [Parsons] et al., 1984). Harnisch et al. (1986) found that male 17-year-olds had generally more positive attitudes toward mathematics than female 17-year-olds in 8 of the 10 countries included in their assessment. Eccles et al. (1993) and Marsh et al. (1985) found that elementary school boys feel better about their mathematical competence than do elementary school girls, despite the finding that the girls sometimes have higher achievement scores in mathematics.

Lummis and Stevenson (1990) also found that elementary school boys in the United States, Taiwan, and Japan tended to believe that they were better at mathematics than reading, whereas girls showed the opposite pattern. Nevertheless, in Taiwan and the United States, most of the children thought that boys and girls were equally skilled in mathematics. In Japan, roughly one third of the boys and girls thought that there was no gender difference in mathematical ability, one third of the children thought that boys were better, and the remaining one third thought that girls were better. Thus, the finding that elementary school boys have better perceptions about their mathematical skills than do elementary school girls might be related to intraindividual comparisons, that is, because girls tend to prefer reading over mathematics whereas boys tend to prefer mathematics over reading (Lummis & Stevenson, 1990). It is likely that for adolescents, performance differences in high school reinforce these perceptions, as do sex role stereotypes that portray mathematics as a male domain (Linn & Hyde, 1989).

Regardless of why a gender difference in perceived mathematical competence emerges, it has been argued that perceived competence and the associated expectancies for success might influence task persistence after failure and might influence the likelihood that the individual will aspire to a mathematics-intensive career (Chipman et al., 1992). Thus, low confidence in one's mathematical abilities might result in lower task persistence with difficult mathematics problems and perhaps an avoidance of more difficult mathematics courses. For example, Kloosterman (1990) assessed high school students' persistence in algebraic problem solving after a failure to successfully solve several problems. In this study, he found that high school boys tended to increase their efforts after failure whereas high school girls tended to show less effort after failure. The same general pattern was found in a study of seventh-grade children, although the magnitude of the gender difference was smaller. Eccles (Parsons) et al. (1984), however, found no gender difference in the tendency to persist on a mathematical problem-solving task after failure. In a later study, they did find that basic abilities, as indexed by earlier grades, and expectancies both contributed to mathematical performance (Meece et al., 1990). Thus, it is likely that perceived competence and expectancies in mathematics influence, at least to some extent, mathematical performance. The relationship between the gender difference in perceived competence and the gender difference in mathematical problem solving is probably bidirectional, each influencing the other. In all, it seems likely that children who perceive themselves as being competent in mathematics will be more likely than less confident peers to aspire to mathematics-intensive careers and therefore take more difficult mathematics courses in high school (Chipman et al., 1992).

Classroom experiences. It has been suggested that the classroom experiences of boys and girls might impact the development of reading and mathematical skills (Kimball, 1989). Overall, boys tend to receive more contact from teachers, both positive and negative, than girls. Boys also tend to be more assertive than girls in initiating contact with their teachers (Irvine, 1986). In an interesting study, Leinhardt, Seewald, and Engel (1979) found that second-grade teachers spent more individual time in reading instruction with girls and more mathematics instruction time with boys. They estimated that across the academic year, these differences could amount to a 6-hr mathematics instruction advantage for boys. Even so, there was not a gender difference in mathematics achievement by the end of the academic year. Other studies of elementary school children, however, have found no gender difference in the amount of time spent in classroom mathematics activities (Peterson & Fennema, 1985). In high school mathematics courses, a number of studies have found that boys tend to receive more personal contact with teachers than girls, whereas other studies have found no difference (see Kimball, 1989). When differences occur, they appear to be associated with the greater assertiveness of boys in these classrooms and because boys are called on by their teachers more often than girls. Thus, it is likely that the mathematics classroom experiences differ for some boys and girls, especially in high school, and may contribute to the gender difference in perceived mathematical competence. Nevertheless, these experiences are likely to have rather subtle effects overall (of course, the impact might be substantial for some girls), because girls typically receive higher grades than boys in these classes (Kimball, 1989).

Indeed, an intriguing study conducted by Peterson and Fennema (1985) suggests that more general teaching styles might have stronger influences on the mathematical development of boys and girls than individual interactions between students and teachers. In this study, the activities of teachers and students in 36 fourth-grade classrooms were recorded during mathematics instruction for at least 15 days. Student–teacher interactions were observed, such as the amount of time the teacher spent helping individual students, as well as the amount of time each student spent in mathematical and nonmathematical (e.g., socializing) activities. There was no gender difference in the amount of time boys and girls spent learning mathematics and no overall gender difference in the change in mathematics achievement from the beginning to the end of the school year. However, there were important across-class differences in the relative gains of boys and girls in mathematics achievement. These gains appeared to be related to the frequency with which the children engaged in cooperative or competitive classroom activities. Engagement in competitive activities had a considerable negative impact on the mathematics achievement of girls but slightly improved the performance of boys. Engagement in cooperative mathematical activities (e.g., problem solving in small groups), in contrast, was associated with improvements, across the academic year, in the mathematical performance of girls but poorer performance for boys.

Summary. There are a number of psychosocial and cultural factors that influence the mathematical development of children (see chapter 7), as well as the magnitude of the gender difference in mathematical problem solving. The first has received much attention and involves the apparent shrinkage or disappearance, in some cases, of the advantage of male students over female students in mathematical performance (Linn & Hyde, 1989). The shrinkage of the gender difference in mathematical performance, however, appears to be associated with the relatively low valuation of mathematical skills in American culture (Stevenson & Stigler, 1992)—which, when comparisons are made cross-nationally, appears to adversely impact the skill development of boys more so than girls. This perspective implies that for children with poorly developed skills, little or no gender difference in mathematical performance should be found. But as the level of skill development improves, the gap between the performance of males and females in certain areas of mathematics should widen. This is exactly the pattern that is found cross-culturally, as well as for representative and gifted samples within the United States (Benbow, 1988; E. G. Moore & Smith, 1987; Steinkamp et al., 1985).

Aside from historical trends, the two most salient influences on mathematical development and a long-term investment in mathematics are the perceived utility of mathematics and the individual's perceived mathematical competence (Eccles [Parsons] et al., 1984). Adolescents and young adults who aspire to a mathematics-intensive career, such as engineering, are much more likely to take higher level mathematics courses in high school than are their peers who aspire to less mathematics-intensive careers, such as journalism (Wise, 1985). Thus, the pursuit of a mathematics-related education is a rational choice for most adolescents in the United States. Many fewer female than male high school students aspire to mathematics-intensive careers, and therefore, female high school students take fewer higher level mathematics courses than their male peers (Meece et al., 1990).

The gender difference in aspirations for mathematics-intensive careers, in turn, appears to be related to a gender difference in social preferences and perhaps to the fact that these fields are male dominated (Halpern, 1992). In general, females show a greater preference for human relationships and the associated careers than do males, whereas males tend to be more object oriented (e.g., Eibl-Eibesfeldt, 1989; McGuinness, 1993; Thorndike, 1911; Wise, 1985). Object, rather than person, orientation is associated with an interest in mathematics-intensive careers, even among groups of high-ability women (Chipman et al., 1992; Roe, 1953). This argument should not be taken to mean that institutionalized sexism has not restricted the career options of many women. Rather, with equal access to education, it is predicted that ambitious and high-ability women who are interested in science-related careers will be more likely to choose to pursue an MD degree, for instance, instead of a PhD in physics (Chipman et al., 1992).

Perceived competence in mathematics also appears to influence the likelihood that one will choose to pursue a mathematics-intensive career (Chipman et al., 1992) and might influence persistence on difficult mathematics-related tasks (Kloosterman, 1990). As early as the first grade, boys tend to have a higher perceived competence in mathematics than do girls (Eccles et al., 1993). The gender difference in perceived mathematical competence appears to be related to intra- and interindividual comparisons. The fact that boys tend to prefer mathematics over reading and girls tend to prefer reading over mathematics contributes to the finding that boys perceive themselves as being skilled in mathematics, even when they believe that boys and girls have equal mathematical ability (Lummis & Stevenson, 1990). For adolescents, the gender difference in mathematical problem solving probably also contributes to this difference in perceived competence, as does the general stereotype of a "male superiority" in mathematics (Hyde, Fennema, Ryan, Frost, & Hopp, 1990). Either way, perceived competence does appear to have a small influence on mathematical performance (Meece et al., 1990). Finally, classroom experiences can influence the relative mathematical development of boys and girls. Any such influence, however, is probably related to the extent to which mathematics instruction is socially competitive or cooperative within the classroom, rather than to a difference in how teachers treat boys and girls within the classroom (Peterson & Fennema, 1985).

Biological Factors

A consideration of potential biological influences on the gender difference in mathematical problem solving is important, because it is often argued that the gender difference in this area has biological origins (e.g., Benbow, 1988; McGee, 1979). In my view, any such consideration of the biological correlates of the gender difference in mathematical problem solving, or any other area, needs to be considered within the wider perspective of human evolution, not simply in terms of biological correlates. According to Gould and Vrba (1982), there are two general ways in which cognitive, social, or physical attributes might be adaptive, with respect to evolution.

The first way involves the traditional view of natural selection; that is, individuals who have skills that are well suited for a particular ecology will survive and reproduce in greater numbers than those individuals with less developed skills. The second way involves the co-optation of a skill or attribute. Here, a particular attribute that evolved for one reason now provides the individual with an adaptive advantage in an unrelated area. Cognitive or social skills that follow the first route would be considered biologically primary; that is, they provided an advantage over other individuals, in terms of human evolution, and were therefore directly selected for in the traditional sense. Co-opted skills, on the other hand, would be considered biologically secondary for the co-opted task (Geary, in press). For instance, language acquisition is biologically primary. Reading involves many of the same biological systems that support language (Luria, 1980), but reading is biologically secondary with respect to these systems. In other words, the basic language system enables us to learn how to read, but the system did not evolve for this reason.

In terms of mathematics, it appears that a sensitivity to numerosity, an understanding of basic number concepts, and a basic understanding of the effects of addition and subtraction on quantity are biologically primary skills (Gallistel & Gelman, 1992; Geary, in press; Starkey et al., 1991). The consistent finding of no gender difference on any of these basic numerical dimensions strongly suggests that the gender difference in mathematical problem solving is *not* biologically primary. In other words, the evolution of cognitive skills has not provided males with a primary advantage in mathematics: It is very unlikely that there is anything like a "male math gene."

Thus, any biological correlates of the gender difference in mathematical problem solving are probably biologically secondary (McGee, 1979). With this in mind, we need to consider the conditions under which selection pressures might have differed for males and females, that is, the conditions that might have resulted in cognitive, social, and physical differences between males and females. Evolutionary pressures were probably more similar than different for males and females, with the exception of reproductive strategies and the associated division of labor (Buss & Schmitt, 1993; Eibl-Eibesfeldt, 1989; Frayer & Wolpoff, 1985; Ghiglieri, 1987; Ruff, 1987; Trivers, 1972). In fact, it is likely that many gender differences in psychological and physical attributes are related to the different reproductive pressures that males and females were confronted with during our evolutionary history (i.e., sexual selection). The development of gender differences in those attributes that are related to reproduction will quite likely be controlled by sex hormones (Arnold & Gorski, 1984), as well as by sex-typed child rearing practices (MacDonald, 1988). Any biological bias in social and cognitive skills would presumably be modifiable, by means of early socialization practices, to allow for the associated systems to be "fine-tuned" to better meet the demands of the larger society (MacDonald, 1992).

Briefly, in terms of our evolutionary history, it appears that reproduction has been less certain for males than for females. As a result, males compete with one another for status and resources, the acquisition of which is a prerequisite for marriage in many cultures (Buss, 1989). Females, in contrast, are thought to be especially sensitive to a potential father's likelihood of financially and emotionally investing in the marriage and future children. Thus, from this perspective, it is predicted that females will be more sensitive than males to the social cues that are likely to be predictive of long-term relationships (e.g., MacDonald, 1988) and will develop social styles that facilitate the maintenance of social relationships. Males, in contrast, will tend to be more dominance oriented in their social interactions. It is therefore expected that males will be more competitive (especially with other males) than females, whereas females will be more sensitive to social relationships and more cooperative within those relationships (e.g., MacDonald, 1988; M. Wilson & Daly, 1985). Males, of course, cooperate with one another, but this is often in the context of competing with other groups of males (Ghiglieri, 1987). Likewise, females are competitive with one another, but this competitiveness typically does not involve the same level of intensity as is found among males (M. Wilson & Daly, 1985).

As noted earlier, there also appears to be a gender difference in the relative interest in objects versus people. This gender difference is found in many different cultures (Eibl-Eibesfeldt, 1989). For instance, one study, described by Eibl-Eibesfeldt, of !Ko bushmen found a large gender difference in the subject of children's drawings. Boys depicted technical objects (e.g., wagons and airplanes) in about 20% of their drawings, as compared with about 2% of the girls' drawings. This finding is important because these technical objects were not part of the !Ko children's natural environment. The children could not have been drawing objects that they had seen their same-sex parents using. Rather, in their interactions with Western culture, these objects captured the attention of boys much more readily than that of girls. From an evolutionary perspective, the relative lack of interest in human relationships by males is probably related to a relatively greater concern for achieving status and dominance within all-male groups, as opposed to a concern for developing relatively more egalitarian social structures, which are more commonly seen in all-female groups. The male interest in objects might also be related to a stronger concern by males, as compared with females, for the acquisition of material possessions (Eibl-Eibesfeldt, 1989).

Developmentally, a gender difference on these social dimensions begins to emerge during the preschool years (Maccoby, 1988) and appears to be influenced by early exposure to sex hormones (Ehrhardt & Meyer-Bahlburg, 1981), as well as by early child-rearing practices (MacDonald, 1988). Gender differences in play activities and social styles probably reflect, at least to some extent, gender differences in evolutionary pressures (Geary, 1992). This is not the whole story, of course. The point is, gender differences in physical development, certain social behaviors, and certain cognitive skills are probably biologically primary and are probably related to a gender difference in reproductive pressures (i.e., sexual selection). It might, therefore, be fruitful to consider whether differences in reproductive strategies might somehow be related to the gender difference in mathematical problem solving and interest in mathematics.

In fact, several of the earlier-described psychosocial influences on the gender difference in the relative valuation of mathematics and mathematical performance might be secondary to this fundamental gender difference in social style. The finding that the mathematical development of elementary school boys and girls is influenced by the relative degree to which the classroom environment is competitive or cooperative is clearly consistent with the evolutionary perspective (Peterson & Fennema, 1985). From this perspective, socially competitive environments are more similar to the behavioral styles that emerge within boys' groups, whereas socially cooperative environments are more similar to the behavioral styles that emerge within girls' groups (Maccoby, 1988). These early gender differences in social behaviors probably reflect, at least to some extent, gender differences associated with the evolution of reproductive strategies (Geary, 1992). Thus, it is not surprising that creating relatively competitive or cooperative classroom environments will impact the achievement of boys and girls differently (Peterson & Fennema, 1985). The second psychosocial influence that might be understandable within the evolutionary perspective concerns the gender difference in the preference for objects versus human relationships. In short, the gender difference in preference for mathematics-intensive careers, such as engineering, and the associated perceived utility of mathematics might be secondary to more fundamental differences, described above, between men and women in the relative degree of object versus people interests.

The gender difference in cognitive style, as related to mathematical problem solving, might also be understandable from this perspective. In particular, the findings that spatial skills are moderately to highly heritable (Vandenberg, 1966), are influenced by exposure to sex hormones (Diamond et al., 1979; S. M. Resnick et al., 1986), and show a consistent gender difference favoring males strongly suggest that gender differences in certain forms of spatial cognition are related to the gender difference in reproductive strategies or the associated division of labor. The fact that social factors, such as sex-typed parenting, might also influence engagement in spatial-related activities does not mean that any gender difference in this area is of a cultural origin. Rather, the level of spatial skill development, as well as the magnitude of the gender difference, is responsive to ecological pressures. It has been argued that good spatial skills provided males with an advantage over other, less skilled males, for example, by facilitating resource acquisition (e.g., hunting) or by providing an advantage during more direct or ritualized competition with other males (Eibl-Eibesfeldt, 1989). Whatever the reason, the male advantage in certain spatial skills is probably biologically primary, but these same skills are probably biologically secondary with regard to mathematical problem solving. In other words, the gender difference in mathematical problem solving is related to males co-opting spatial strategies to solve mathematical problems, not because males have somehow evolved to be better at mathematical problem solving than females.

Summary and Integrative Model

In this section, I integrate research on the cognitive, psychosocial, and biological influences on the gender difference in mathematical problem solving, in an attempt to provide a more dynamic understanding of this phenomenon than is possible with the consideration of only a single influence. A model that acknowledges all of these influences is probably stating the obvious. I hope not to state the obvious, but to propose a pattern of relationships that better organizes our understanding of this area. The proposed pattern of relationships is shown in Figure 6.2. The basic goal is to understand the emergence of the gender difference in mathematical problem solving, represented by the rightmost circle in Figure 6.2, but not the level of mathematical development. In other words, the goal is to understand why boys and girls differ from one another in certain areas of mathematical problem solving, not why children in some cultures, for instance, develop mathematical skills that far exceed those of children in other cultures (see the International Studies section of chapter 7).

With this in mind, it appears that the two most salient direct influences on the gender difference in mathematical problem solving are gender differences in spatial skills (McGee, 1979) and in the number of mathematics-related activities (Fennema & Sherman, 1977). Both appear to act in concert to increase the magnitude of the gender difference in mathematical problem solving, because males often have better developed spatial skills and participate more frequently in mathematics-related activities than do females (Burnett et al., 1979; Wise, 1985). At the same time, spatial skills and mathematics-related experiences should be considered to have bidirectional influences on one another. Individuals with well-developed spatial skills are likely to take more mathematics-related courses (e.g., drafting and engineering) on average than those with relatively less developed spatial skills (Scarr & McCartney, 1983). Participation in such courses will further improve their spatial skills. In terms of gender differences, the spatial skill advantage of males over females

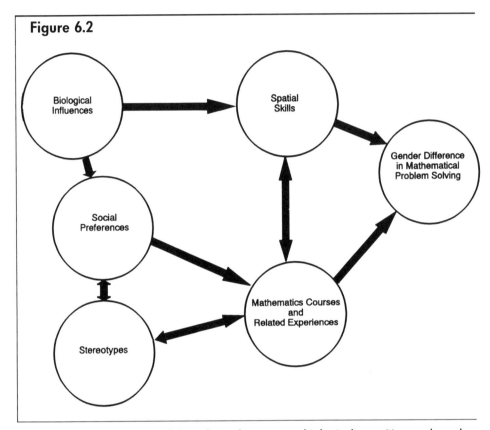

Figure 6.2

Schematic representation of the relationships among biological, cognitive, and psycho-social influences on the gender difference in mathematical problem solving.

not only will influence the gender difference in mathematical problem solving but also might lead males, more so than females, to seek experiences that will lead to the further development of their mathematical and spatial skills. These experiences, in turn, will increase the magnitude of the male advantage in mathematical problem solving.

Participation in mathematics-related activities also appears to be influenced by social preferences and sex role stereotypes (Chipman et al., 1992; Linn & Hyde, 1989). The gender difference in the relative preference for object-oriented versus people-oriented activities appears to contribute to the tendency of males to aspire to mathematics-intensive careers more frequently than do females (Chipman & Thomas, 1985; McGuinness, 1993). A relatively greater people orientation than object orientation appears to contribute to the perception of many girls that mathematics- and science-related careers would be relatively unrewarding

(Sherman, 1981). The gender difference in the aspiration to mathematics-intensive careers, in turn, also appears to contribute to the tendency of male high school students to take more mathematics courses than their female peers (Travers & Westbury, 1989). It has been argued that sex role stereotypes might also impact differential course taking, perhaps by influencing perceived competence in mathematics, although it appears that perceived usefulness of mathematics has a more important influence than stereotypes on mathematics course taking (Fennema, Wolleat, Pedro, & Becker, 1981).

Regardless, stereotypes might dissuade some girls from aspiring to mathematics-intensive careers and might influence some boys to consider mathematics-intensive careers (e.g., engineering) as a more viable alternative than other careers (e.g., nursing). In relation to its effect on girls, the influence of stereotypes on boys' attitudes and course taking in mathematics has received very little attention. Again, social preferences and stereotypes probably have mutual influences on one another, as do stereotypes and mathematics course taking. Mathematics is stereotyped as a male domain, because it is largely a male domain. The stereotype, in turn, might serve to keep mathematics more of a male domain than it otherwise would be. Overall, as compared with social preferences, stereotypes are probably a minor influence on the magnitude of the gender difference in mathematical problem solving, because stereotypes are the smallest in those groups (i.e., high-ability groups) where the gender difference in mathematical problem solving is the largest (Lubinski & Humphreys, 1990; Raymond & Benbow, 1986).

The finding that there were essentially no gender differences in the most fundamental numerical skills found in humans, that is, for a basic understanding of number, counting, and arithmetic, strongly suggests that any biological correlates of the gender difference in mathematical problem solving are secondary. That is, males have not evolved to be better at mathematical problem solving than females. Rather, biological influences on the gender difference in mathematical problem solving are indirect. In the proposed model, biological influences (probably sex hormones) are seen as directly influencing the gender difference in spatial skills, as well as the gender difference in social preferences. The evolution of good spatial skills in males, as compared with females, was not to aid males in mathematical problem solving. Nevertheless, these skills can be co-opted in problem-solving situations (McGuinness, 1993). Similarly, the relative object versus people orientation of the average male and the people versus object orientation of the average female probably evolved for reasons associated with a gender difference in reproductive strategies and, in this sense, is completely unrelated to mathematics. Nevertheless, this bias probably does influence the gender difference in the relative attractiveness of mathematics-intensive careers and therefore indirectly contributes to the gender difference in participation in mathematics-related activities.

The Mathematically Gifted

As noted in the introduction, this chapter concludes with a discussion of mathematically gifted children and adolescents, because the gender difference in mathematical problem solving is most pronounced for these individuals. In this section, a series of rather interesting studies of the cognitive and neuropsychological characteristics of SMPY adolescents will be highlighted (Benbow, 1986; Dark & Benbow, 1990, 1991). However, before this work is discussed, a brief overview of the cognitive skills of gifted children in general is in order.

From a cognitive perspective, gifted children differ from their average-ability peers on several dimensions (Sternberg & Davidson, 1986). First, gifted individuals generally process information more quickly than their average-ability peers. For instance, gifted individuals can retrieve words more quickly from long-term memory than can nongifted individuals. In fact, by the end of the elementary school years, most gifted children appear to process information as quickly as average-ability adults (Keating & Bobbitt, 1978).

Second, gifted children begin to use adultlike problem-solving strategies many years before their average-ability peers, although there tends to be no qualitative difference between the types of problem-solving strategies used by gifted and nongifted children, just a developmental difference (Siegler & Kotovsky, 1986). For instance, a gifted 8-year-old with an IQ of 150 will use problem-solving strategies that are more commonly used by average-ability 12-year-olds. This developmental difference is, in part, due to the fact that gifted children learn more quickly than nongifted children and because they are able to "pull together" seemingly disparate pieces of information to construct new strategies at a much younger age than nongifted children (Sternberg & Davidson, 1986). Finally, gifted individuals usually have a better conceptual understanding of basic domains, such as mathematics, than nongifted individuals. From a cognitive perspective, then, many mathematically gifted children are not "qualitatively" different from their nongifted peers but rather develop the same basic mathematical skills at an accelerated rate (e.g., Geary & Brown, 1991; Lubinski & Humphreys, 1992). Nevertheless, there is some evidence that extremely mathematically gifted adolescents, and eminent adults in many areas, might differ from the typical gifted child (Benbow, 1986; Siegler & Kotovsky, 1986).

For instance, some—but not most—gifted children will grow up to make long-term contributions to their chosen field (Terman, 1954), whereas other individuals who have made eminent contributions to their fields might not have been considered gifted as children (Siegler & Kotovsky, 1986). The gifted adult, in addition to being rather intelligent, also tends to be creative, confident, introverted, and *very* motivated to achieve in his or her field (Albert & Runco, 1986). Similarly, the SMPY youths represent the upper range of gifted children and might therefore differ in some important respects from other gifted adolescents. In the sections to follow, the patterns of cognitive and neuropsychological skills of very mathematically gifted adolescents will be contrasted with the cognitive and neuropsychological profiles of their average-ability and verbally gifted peers. These comparisons were undertaken to determine if some gifted individuals have a "qualitatively distinct" pattern of skills that facilitates mathematical development and might distinguish them from the typical high-IQ child.

Cognitive Characteristics

In a series of well-done studies, Dark and Benbow (1990, 1991) sought to determine whether mathematically gifted adolescents identified through SMPY differed from their verbally gifted peers on the component skills needed to solve algebraic word problems, such as problem translation, and in terms of the associated cognitive skills, such as spatial memory (see the Algebra section of chapter 3). In the first set of studies, Dark and Benbow (1990) assessed the problem-solving and cognitive skills of groups of mathematically and verbally gifted junior high school SMPY students, a group of average-ability peers, and a group of college students. The mathematically gifted adolescents had a mean SAT−M score of 651 and a mean of 452 on the Verbal section of the SAT (SAT−V). The verbally gifted group had mean scores of 499 and 553 on the SAT−M and SAT−V, respectively. The tasks used in these studies assessed the ability to translate algebraic word problems into equations; memory and understanding of important features of the word problems; and some of the associated cognitive skills, such as the ability to retain and manipulate numerical and spatial information in working memory.

For the algebra problems, the mathematically gifted adolescents were more skilled than individuals in all other groups in the ability to translate problem features into appropriate equations but did not differ from the verbally gifted adolescents or the college students in the understanding of, or memory for, important features of the word problems (Dark & Benbow, 1990). A second experiment showed that the mathematically gifted youths were better than their verbally gifted and average-ability peers in the ability to retain and manipulate numerical and spatial information in working memory. For instance, the mathematically gifted adolescents were especially skilled at remembering the spatial location of briefly presented information and in combining information that was presented across different spatial locations.

In a second set of experiments, Dark and Benbow (1991) contrasted the basic verbal, numerical, and spatial skills of mathematically gifted and verbally gifted SMPY junior high school students. Again, these adolescents were placed in their respective groups based on SAT−M and SAT−V scores. The mathematically gifted adolescents scored, on average, about 200 points higher on the SAT−M than the verbally gifted adolescents (the respective means were 656 and 453), whereas the verbally gifted adolescents scored, on average, 128 points higher on the SAT−V than the mathematically gifted group (the respective means were 536 and 408). Again, the mathematically gifted youths were better than their verbally gifted peers at remembering numerical and spatial information and at manipulating this information in working memory. The verbally gifted youths, in contrast, were very skilled at retaining words in memory and were faster than the mathematically gifted adolescents in retrieving word meanings from long-term memory. Across all experiments, there were more boys than girls in the mathematically gifted groups, and where gender differences could be assessed, the boys outperformed the girls on many of these important dimensions. For example, Dark and Benbow (1991) found that mathematically gifted boys had a better spatial memory than their female peers.

Neuropsychological Characteristics

In a series of related studies, Benbow and her colleagues also examined the physiological and neuropsychological correlates of mathematical giftedness (Benbow, 1986; O'Boyle et al., 1991; O'Boyle & Benbow, 1990).

Of particular interest was the finding that the mathematically precocious adolescents assessed in these studies showed a greater utilization of right-hemispheric resources during the processing of verbal and visual information than average-ability peers, as indicated by both behavioral and electroencephalographic measures. O'Boyle and Benbow also found a moderate correlation between an index of right-hemispheric involvement during a visual task and overall performance on the SAT. The greater the right-hemispheric involvement in this task, the higher the overall SAT score.

Summary

Though preliminary, the just-described studies suggest that the mathematically gifted adolescents identified through SMPY might differ in several important respects from their average-ability and verbally gifted peers. In terms of performance, these mathematically gifted adolescents are very skilled at translating word problems into appropriate equations but are not exceptional, in relation to other gifted adolescents, in their understanding or memory of important features of word problems (Dark & Benbow, 1990). At a cognitive level, these mathematically gifted youths show an enhanced memory for numbers and spatial information and an exceptional skill at manipulating numerical and spatial information in working memory, but their memory for verbal information is not exceptional (Dark & Benbow, 1990, 1991).

At a neuropsychological level, mathematically gifted adolescents appear either to have an unusual pattern of hemispheric specialization or to be better able to use skills that are primarily localized in the right hemisphere (O'Boyle & Hellige, 1989). Although these preliminary results are intriguing and merit further investigation, at this point it is premature to speculate any more specifically than this. Finally, note again that the adolescents identified through SMPY represent the upper range of the gifted population (Benbow, Stanley, Kirk, & Zonderman, 1983) and therefore the above-described results might not be applicable to the more typical gifted child. In fact, Lubinski and Humphreys (1990, 1992) did not find the same cognitive and physiological patterns as did Benbow and her colleagues with mathematically gifted high school students who were not quite as precocious as the SMPY youth.

The students identified by Lubinski and Humphreys (1990, 1992) appear to fit the pattern described at the beginning of this section; that is, they are intelligent individuals who appear to be developmentally advanced but not qualitatively different, from a cognitive perspective, than the average-ability high school student. Thus, the basic cognitive systems are the same in gifted and nongifted children, but these systems work more efficiently in gifted children. Many of the higher ability SMPY adolescents, in contrast, might have a somewhat different configuration of cognitive systems (O'Boyle & Hellige, 1989). This difference appears to be found much more frequently in gifted males than in gifted females (Benbow, 1986) and might contribute to the large gender difference at the high end of SAT–M performance for SMPY adolescents (Benbow, 1988).

Conclusion

Research on gender differences in early numerical abilities, arithmetic, and mathematical problem solving was reviewed to determine if a consistent cross-national gender difference emerged in any of these areas. For the most part, this research suggests little or no gender difference in basic numerical skills, such as infants' sensitivity to numerosity or preschool children's understanding of counting, number, and arithmetical concepts. Similarly, across cultures, there appears to be no consistent gender difference in elementary school children's basic arithmetical skills (e.g., Song & Ginsburg, 1987). The finding of no consistent cross-cultural gender differences in basic numerical and arithmetical skills is especially important because many of these abilities are probably biologically primary (Gallistel & Gelman, 1992; Geary, in press). It therefore seems reasonable to conclude that males are not biologically primed to outperform females in mathematics and that the gender difference in mathematical problem solving does not emerge from earlier gender differences in fundamental numerical abilities.

Nevertheless, small to moderate cross-national gender differences are often found on mathematical problem-solving measures (Hyde, Fennema, & Lamon, 1990). The gender difference in mathematical problem solving is found as early as the first grade and includes the solving of arithmetic and algebraic word problems, geometry, and other spatially related areas such as measurement and calculus but does not include algebra (except algebraic word problems). Even when gender differences in mathematical problem solving are found, they tend to be modest and selective. For instance, high school boys consistently outperform their female peers on measures that can be solved through the use of spatial strategies, such as plane geometry, but do not show an advantage in seemingly similar areas, such as providing geometric proofs (Harnisch et al., 1986; Senk & Usiskin, 1983). Statements about a broad male advantage in mathematical problem solving are not warranted, although males clearly have an advantage over females in *certain* mathematical problem-solving areas (Hyde, Fennema, & Lamon, 1990).

The reason for the male advantage in mathematical problem solving is complex and appears to involve a confluence of cognitive, psychosocial, and biological factors. The male advantage in certain forms of spatial cognition appears to be an important contributing factor to the gender difference in mathematical problem solving. In particular, the tendency of males to spontaneously resort to the use of spatial representations in problem-solving situations more frequently than females appears to account for a large proportion of the magnitude of the gender difference in mathematical problem solving (McGuinness, 1993). Psychosocial factors appear to increase the magnitude of the male advantage in mathematical problem solving by creating a gender difference in the number of mathematics courses taken by high school students (Wise, 1985). In particular, gender differences in perceived usefulness of mathematics and in perceived mathematical competence operate to increase the participation of boys in upper-level high school mathematics courses, which, in turn, quite likely increases the male advantage in mathematical problem solving.

Biological influences on the gender difference in mathematical problem solving are probably secondary (McGee, 1979). That is, biological correlates of this gender difference are not related to mathematical problem solving per se but might nevertheless influence mathematical performance (Gould & Vrba, 1982). First, the gender difference in the perceived usefulness of mathematics appears to be directly related to long-term career goals (Wise, 1985). Female high school students aspire to mathematics-intensive careers, such as engineering, much less frequently than their male peers, which, in turn, contributes to the gender difference in the number of mathematics courses taken during high school. Career aspirations, in turn, appear to be influenced by one's general orientation toward people or toward objects (Chipman et al., 1992; Roe, 1953). Object orientation is related to an interest in mathematics-intensive careers and is probably influenced by the biological factors that are associated with the gender difference in reproductive strategies (McGuinness, 1993; Thorndike, 1911). Similarly, the gender difference in spatial skills is also biologically primary but is secondary to the gender difference in mathematical problem solving. In other words, males did not evolve good spatial skills to solve mathematics problems, but these skills can be co-opted in problem-solving situations.

Finally, the gender difference in mathematical problem solving is especially pronounced in gifted adolescents and adults (Benbow, 1988, 1992). The ratio of mathematically gifted boys to girls sometimes exceeds 13:1 at the extreme end of mathematical giftedness. Though preliminary, the studies of Dark and Benbow (1990, 1991) suggest that the highest ability mathematically gifted youths differ in some important respects from their verbally gifted peers. In terms of problem-solving performance, the mathematically gifted excel in the translation of word problems into appropriate equations but not at other features of mathematical problem solving. The mathematically gifted appear to have exceptional memories for numerical and spatial information and are especially skilled at combining and manipulating this information in working memory. In contrast, the mathematically gifted are not exceptional in their memory for words or in their understanding of basic problem-solving concepts. It remains to be seen whether the constellation of cognitive skills of these youths simply represents the upper end of spatial and working-memory skills or whether they are somehow "qualitatively" distinct, cognitively and neuropsychologically, from other gifted individuals (Benbow, 1986).

Cultural Influences on Mathematical Development

An understanding of how and why children of different nations vary in the pace and level of mathematical development is essential, because the basic mathematical skills of the subsequent work force will strongly influence the economic health of the respective nations (Bishop, 1989). Basic reading and mathematical skills will influence the economic productivity of the typical worker and will influence each worker's wages and employability (Boissiere et al., 1985; Rivera-Batiz, 1992). Beginning in the 1960s, the reading and mathematics achievement scores of American children, and of the subsequent labor force, began to decline (Bishop, 1989). Bishop argued that the decline in basic skills resulted in a decline in productivity growth and cost the U.S. economy \$86 billion in 1987. Furthermore, Bishop estimated that the relatively poor reading and mathematical skills of much of the work force will cost the U.S. economy nearly \$170 billion each year by the year 2000. Improving the basic academic skills of America's children and subsequent labor force is arguably one of the nation's most pressing needs. In fact, the importance of improving the mathematical skills of American children was highlighted in *America 2000: An Education Strategy*, where one of the six primary goals was the following: "U.S. students will be first in the world in science and mathematics achievement [by the year 2000]" (U.S. Department of Education, 1991, p. 3).

Cross-cultural research was presented in the first two chapters to highlight the universals of children's understanding of number, counting, and arithmetic and to review how these basic mathematical skills are

---

Summaries are provided for the following subsections on Potential Causes of International Achievement Differences: Race, Intelligence, and Mathematical Achievement (p. 243) and Schooling (p. 253).

expressed differently in different cultures. In chapter 6, the pattern of gender differences in numerical, arithmetical, and mathematical problem-solving skills was examined across cultures to determine the extent to which any associated gender differences were influenced by sociocultural factors. In contrast, the focus of the first section of this chapter is on cross-national differences in the *level* of mathematical achievement. The second section contains an analysis of potential causes of any such differences. These potential causes include intelligence, language influences, schooling, family, and general cultural attitudes toward academic achievement. In both sections, the central focus is on the mathematical achievement of American children and how this achievement differs from international norms. The chapter closes with an integrative discussion of the relationships among the influencing factors and how they might affect international differences in mathematical achievement.

International Studies

The International Project for the Evaluation of Educational Achievement (IEA), conducted in 1964, was the first large-scale comparative study of mathematical achievement (Husén, 1967). The goal of this research was to assess a broad range of mathematical skills of 13- and 17-year-old adolescents and to examine potential influences on mathematical achievement, such as the amount of homework and family background. Thirteen-year-olds were assessed because they were at the end of compulsory schooling in many of the nations that participated in the study, whereas 17-year-olds were at the end point of secondary schooling in most of the participating countries (e.g., the end of high school in the United States). In all, participating in the study were 132,775 adolescents from 12 industrialized nations: Australia, Belgium, England, Finland, France, Germany, Holland, Israel, Japan, Scotland, Sweden, and the United States.

The achievement test that was administered to the 13-year-olds primarily assessed basic skills in arithmetic, algebra, and geometry, whereas the test that was administered to the 17-year-olds included additional items on trigonometry, calculus, probability, and logic. Because children started elementary school at different ages across the participating nations, two sets of cross-national comparisons were conducted for the 13-year-olds. The first was for adolescents with an actual chronological age of 13 years, and the second was based on the grade level that contained most of the 13-year-olds. In both comparisons, adolescents from Sweden scored lower than adolescents from all other nations, followed very closely by American adolescents in the first comparison and by Australian and American adolescents in the second comparison. The performance of Japanese adolescents was substantially above that of adolescents from all other nations in the first comparison, whereas the scores of adolescents from Israel, Japan, and Belgium clustered together at the top for the second comparison. For both comparisons, the mean scores of American adolescents were between $\frac{1}{10}$ and $\frac{1}{2}$ of a standard deviation below the international mean on *every* mathematical area that was assessed. In contrast, the mean scores of adolescents from Israel, Japan, and Belgium were considerably above the international mean for all mathematical areas.

Two sets of comparisons were also conducted for the 17-year-olds. Both comparisons were for students who were studying "in schools from which the universities or equivalent institutions of higher learning normally recruit their students" (Husén, 1967, Vol. 1, p. 46). The first comparison included only those adolescents for whom mathematics was an essential part of their curriculum; the second included students for whom mathematics was a secondary feature of their curriculum. For both comparisons, the overall mathematical performance of American students was substantially (about 1 standard deviation) below the international mean and was, in fact, considerably lower than the mean performance of adolescents from the 11th-ranked nation. Moreover, for both comparisons, the mean scores of American students were considerably ($>\frac{1}{2}$ *SD*) below the international mean on most of the individual mathematical areas assessed. At the other end of the scale were adolescents from Israel for the first comparison and adolescents from Germany for the second comparison.

The cross-national comparison of the 17-year-olds was confounded by the fact that the educational system was more comprehensive in some countries than in others. For instance, only about 11% of the 17-year-old adolescents from Germany were enrolled in a preuniversity school setting, whereas 70% and 57% of their peers in the United States and Japan, respectively, were enrolled in comparable settings. The extremely poor performance of American 17-year-olds was due, in part, to the fact that the American sample included a larger proportion of lower ability students than did the samples from most other nations. Nevertheless, American adolescents still performed poorly when comparisons were made of the most elite students in each country—those who were taking a lot of mathematics and science courses in preparation for a mathematics-intensive career (e.g., engineering). Only 18% of America's best prepared 17-

year-olds scored at or above the international mean; 37% of the elite adolescents from the 11th-ranked country, Australia, scored at or above the international mean. In contrast, 10% or more of the elite students from Belgium, England, and Japan scored at or above the 95th percentile; less than 4% of the U.S. students scored in this range.

A second IEA study, which included two dozen countries, was conducted in the early 1980s (Crosswhite, Dossey, Swafford, McKnight, & Cooney, 1985). Germany did not participate in this assessment, and several Third World countries were added, such as Nigeria and Swaziland. As a result, the international norms were probably lower for this group of nations, in relation to the first IEA assessment. Thus, the most important question with this assessment was whether American 13- and 17-year-olds were performing any better in mathematics in 1981, in relation to the groups that were tested in 1964.

The overall mathematical test scores of American 13-year-olds showed a modest decline from 48% correct in 1964 to 45% correct in 1981. There were slight improvements in some areas, such as measurement, and modest declines in other areas, such as geometry. For the 17-year-olds, there was a modest improvement from 32% correct in 1964 to 38% correct in 1981 (Crosswhite et al., 1985). This gain, however, was primarily due to improved performance on the calculus portion of the test, which, in turn, was probably due to a fourfold increase in the number of American high school students taking calculus from 1964 to 1981. With the exception of these calculus students, the overall pattern of results for the 17-year-olds was basically the same as was found for the 13-year-olds. Even with the relatively lower international standards, in comparison with the first IEA study, America's best trained students (the top 5%) performed at about the international average in algebra, geometry, and calculus (M.D. Miller & Linn, 1989). About 60% of the American students for whom mathematics was an integral part of their high school curriculum (the top 13%) scored substantially (≥ 1 SD) below the international mean in algebra, geometry, and calculus.

A more recent (i.e., 1991) comparison of 9- and 13-year-old students from 20 nations yielded similar results (Lapointe et al., 1992). For 13-year-olds, the overall mathematical performance of American children ranked 13th (tied with Spain) out of the 15 nations that included comparable samples. Students from Korea, Taiwan, and Switzerland were ranked 1st, 2nd, and 3rd, respectively, for this group of 15 nations; Japan did not participate in this assessment. Students from mainland China were also assessed but were not included in the primary comparisons because the researchers were not able to assess students from all areas of China. Those Chinese students who did participate in the study substantially outperformed the students from Korea and Taiwan. American 9-year-olds did not fare any better: Their average score ranked 9th out of 10 nations. The overall performance of children from Korea, Hungary, and Taiwan ranked 1st, 2nd, and 3rd, respectively, for this age group; mainland China and Switzerland did not participate in the assessment of 9-year-olds.

Despite relatively poor average scores, the top-scoring (the top 5%) 13-year-old American children performed about as well as the top-scoring 13-year-old children from Korea and Taiwan. However, the comparison of the performance of these top-scoring children appears to have been confounded by a ceiling effect (i.e., the test was too easy and therefore underestimated the mathematical skills of Korean and Chinese children; see Bracey, 1992, for an alternative explanation).

A series of important studies conducted by Stevenson and his colleagues also showed that the mathematical development of American children lagged behind that of children from Japan, Taiwan, and mainland China by the first grade and sometimes during kindergarten (Stevenson, Chen, & Lee, 1993; Stevenson, Lee, Chen, Stigler, et al., 1990; Stevenson, Lee, & Stigler, 1986). In the first of these studies, which was conducted in 1980, groups of kindergarten, first-, and fifth-grade children from Japan, Taiwan, and the United States were administered tests of basic mathematical and reading skills (Stevenson et al., 1986). The tests were carefully constructed so as to include materials that were representative of the mathematics and reading curriculum of each of the respective nations.

In kindergarten, the Japanese children outperformed the American and Chinese children on the mathematics achievement test; there was no overall difference between the performance of the Chinese and American kindergarten children (Stevenson et al., 1986). Toward the end of first grade, the Chinese children had mathematics achievement scores that were slightly better than those of the Japanese children and substantially better (by about ½ of a standard deviation) than those of the American children. By the fifth grade, the gap between the mathematical performance of the American children and their peers from Taiwan and Japan had widened. In other words, the relative performance of American children became increasingly worse with each successive grade. In the first grade, 15 American children scored in the top 100 on the mathematics test, but only 1 American child scored in the top 100 in the fifth grade.

By the fifth grade, children in the best American classroom had, on average, mathematics test scores that were below the mean of the lowest scoring Japanese classroom and better than only 1 of 20 Chinese classrooms. In fact, fifth-grade children in the lowest scoring American classroom had, on average, only slightly better mathematical test scores than children in the best *first-grade* Chinese classroom.

In a follow-up study, many of the first graders that were assessed in the just-described research were retested when they were in 5th grade and again in the 11th grade (Stevenson et al., 1993). The assessment of these students, and of some additional 5th and 11th graders, was conducted in 1990, nearly 10 years after the movement in the United States to reform mathematics education. The 1990 assessment provided an excellent opportunity to determine whether the changes in educational policy that occurred in the United States during the 1980s had actually influenced the mathematical development of American children. From 1980 to 1990, the relative performance of Japanese and American fifth graders was unchanged, but "the difference between the performance of the Chinese and American children was greater in 1990 than in 1980" (Stevenson et al., 1993, p. 54). The average test scores of American children improved slightly from 1980 to 1990, whereas the average test scores of Chinese children improved considerably (by about 20%) during this 10-year period. The average scores of the Japanese fifth graders were about the same in 1980 and 1990.

In the 11th grade, the American students scored substantially lower than their peers from Japan and Taiwan. In fact, "the achievement gap in mathematics increased between the first and eleventh grades" Stevenson et al., 1993, p. 55). For both the 5th and 11th grades, the best American students (i.e., the top 10%) had mathematics test scores that were at about the average of same-age Japanese and Chinese students. A separate study that compared 1st- and 5th-grade children from the United States and mainland China produced the same result: American children were behind their Chinese peers on nearly every dimension of mathematical competence in the 1st grade, and the gap only widened by the 5th grade (Stevenson, Lee, Chen, Lummis, et al., 1990).

In summary, the results of the just-described research and of other studies (e.g., Geary, Fan, & Bow-Thomas, 1992; Song & Ginsburg, 1987) are very clear: In the industrialized world, American children are among the most poorly educated children in mathematics. In terms of international standards, American children have poorly developed mathematical skills before the end of the first grade (Song & Ginsburg, 1987; Stevenson et al., 1986). The gap between the mathematical performance of American children and their future competitors in the world economy appears to widen with each successive grade (Husén, 1967; Stevenson et al., 1993). The increased political attention to this issue during the 1980s has had little, if any, effect on narrowing the gap between the mathematical development of American children and children from other nations. The overall mathematical skills of contemporary American children are no better and, in some cases, are worse, than the skills of American children at the time of the first IEA assessment. With this historical pattern, it seems almost certain that the *America 2000* goal for U.S. children to be "first in the world in science and mathematics achievement [by the year 2000]" (U.S. Department of Education, 1991, p. 3) will not be reached or even approached.

Potential Causes of International Achievement Differences

As noted earlier, intelligence, language, schooling, family, and cultural attitudes toward mathematics will be considered in this section as potential causes of the just-described cross-national differences in mathematics achievement.

Race, Intelligence, and Mathematical Achievement

As noted in the first section, cross-national differences in mathematical achievement are often, though not always, the largest when the performance of American children is compared with the performance of children from Asian countries. It has been suggested that the achievement difference between Asian and American students is due, at least in part, to racial differences in intelligence (e.g., Lynn, 1982; Rushton, 1992). Proffered reasons for these differences in intelligence range from differences in the level of early environmental stimulation (Lynn, 1982) to genetic differences (Rushton, 1992). Either way, if the children of Asian countries are more intelligent than their American peers, then, as compared with American children, Asian children will show accelerated mathematical development—in fact, they will show accelerated development in most cognitive areas (see the section entitled The Mathematically Gifted in chapter 6). The merit of this claim is considered in this section. The section concludes with a discussion of the racial and ethnic differences in mathematical achievement that are found within the United States.

It seems that many Americans believe that performance in mathematics is largely determined by intelligence rather than by hard work (Stevenson, Lee, Chen, Stigler, et al., 1990). It follows that many individuals might therefore conclude that racial or national differences in mathematical achievement stem primarily from racial or national differences in the level of intelligence. However, making inferences about cross-national differences in the level of intelligence solely on the basis of achievement test scores is problematic, despite the finding that IQ scores predict later achievement in school (e.g., Stevenson et al., 1976). In other words, even though intelligence will influence how quickly academic skills are learned, it does not necessarily follow that differences in achievement reflect differences in intelligence. The difficulty in making such inferences can be illustrated by trends even within the United States. Flynn (1987), for example, noted that from the 1960s to the 1980s the SAT scores of American high school students dropped about 90 points, but at the same time IQ scores increased.

Nevertheless, Lynn (1982) argued that the high levels of academic achievement that were found in Japan stemmed primarily from higher IQs in Japanese individuals as compared with Americans and Europeans. In one study, Lynn (1982) concluded that Japanese children had a mean IQ of 111, compared with a mean of 100 for American and European children. Mean IQ differences of this magnitude would result in roughly five times more mentally gifted (i.e., IQs ≥130) Japanese than American and European children. For the "population as a whole, 77% of Japanese [would] have a higher IQ than the average American or European" (Lynn, 1982, p. 223). Lynn's conclusion was based on comparisons of the Japanese and American standardization samples for the Wechsler intelligence scales (e.g., Wechsler, 1974). Basically, raw scores for the Japanese standardization samples were rescored using American norms. Across 27 cohorts, the means for the Japanese samples were between 2 and 15 points higher than the means for the associated American samples. Moreover, the more recent the cohort, the larger the difference in mean IQs between the Japanese and American samples.

Stevenson and Azuma (1983) argued that Lynn (1982) had overestimated the IQ scores of Japanese children because, among other things, higher ability individuals were overrepresented in the standardization samples for the Japanese Wechsler intelligence scales. The Japanese samples included too many urban and higher socioeconomic status children in relation to Japan as a whole. Because children from urban areas score higher on IQ tests than children from rural areas (e.g., Coon, Carey, &

Fulker, 1992), the Japanese standardization samples produced inflated IQ scores. Flynn (1984, 1987) also showed that overall IQ scores tend to increase with successive cohorts. For example, people born in the 1940s score, on average, lower on IQ tests than people born in the 1950s, who in turn score lower, on average, than people born in the 1960s. Because the American standardization samples for the Wechsler scales are earlier cohorts than the associated Japanese standardization samples, the upward drift in IQ scores across cohorts should produce higher scores in Japan than in the United States, even when overall intelligence levels are the same. Lynn (1983) attempted to take these factors into account and reestimated the mean IQ of Japanese children to be 104. Even so, the comparison of Japanese and American IQ scores is rather indirect with this type of study.

Stevenson and his colleagues sought to compare the cognitive abilities of Japanese, Chinese, and American elementary school children more directly (Stevenson et al., 1985). In this study, the performance of a group of urban Anglo-American children on a battery of cognitive ability tests was compared with the performance of groups of urban children from Japan, Taiwan, and Hong Kong. The cognitive battery included tests of auditory memory, verbal memory, memory for numbers, spatial skills, vocabulary, and general information, among others. All children were also administered reading and mathematics achievement tests. If Asian children were more intelligent than American children, then the performance of the American children would have been lower than that of their Asian peers on most or all of the cognitive ability tests (Spearman, 1927). The overall results supported no such pattern. For some of the tests, such as spatial relations, the Japanese children outperformed the Chinese and American children. On other tests, such as memory for numbers, the Chinese children outperformed their Japanese and American peers. On still other tests, such as verbal memory, the American children outperformed the Japanese and Chinese children. From this pattern of results, Stevenson et al. (1985) concluded, "this study offers no support for the argument that there are differences in the general cognitive functioning of Chinese, Japanese, or American children" (p. 733).

Despite no apparent difference in basic cognitive ability, moderate to large cross-national differences were found for reading and mathematics achievement (Stevenson et al., 1985). Across several indexes of reading skills, the Chinese children showed an advantage over their Japanese and American peers. For the most part, the reading skills of Japanese and American children were comparable. The performance of

American children, however, was consistently inferior to the performance of Chinese and Japanese children on the mathematics achievement test. In a more recent study, Stevenson et al. (1993) found that American children scored, on average, slightly better on a general information test than children from Taiwan and Japan, but as a group the American children also showed the lowest mathematical achievement scores. The pattern of poor mathematical achievement scores in combination with somewhat above-average scores on the general information test is important, because performance on general information tests are strongly correlated with overall IQ scores (Sattler, 1988). These studies indicate that cross-national achievement differences are often associated with no differences in basic cognitive ability (i.e., intelligence).

This point is further illustrated by a nicely conducted study by Song and Ginsburg (1987). As you know, in this study, groups of American and Korean 4- to 8-year-olds were compared on tests that assessed their formal and informal mathematical skills. Formal mathematical skills are taught in school, whereas informal skills emerge before formal schooling and include a basic understanding of counting, number, and arithmetic (see chapters 1 and 2). Performance on the test of informal mathematical knowledge was especially interesting, because this measure likely provides an index of biologically primary numerical skills (Geary, in press). If the high mathematics achievement of Asian children is due to higher levels of intelligence, or to a biological advantage in the mathematical area, then differences between Asian and American children should be found for tests of informal mathematical knowledge.

Before entering elementary school, the American children *outperformed* their Korean peers on the test of informal mathematical knowledge, but before the end of first grade, the Korean children outperformed their American counterparts on this test (Song & Ginsburg, 1987). By the second grade, the gap between the Korean and American children in both formal and informal mathematical knowledge had widened. Thus, in comparison with Korean children, American children have a head start in the understanding of basic mathematical concepts, but by the end of first grade, Korean children outperform American children in most mathematical areas (Song & Ginsburg, 1987). It is very unlikely that Korean children experience a rapid growth of intelligence between 6 and 7 years of age. Rather, the national difference in mathematical achievement that was found in this study for the 7-year-olds was almost certainly due to cultural factors, such as schooling (Song & Ginsburg, 1987). In all, these studies suggest that Asian children are not more intelligent than American children. Even though IQ scores are predictive of later academic achievement, cross-national academic achievement differences should not be used to make inferences about cross-national differences in the level of intelligence.

Another area that has been used to make inferences about differences in the intelligence, or in the basic ability to learn mathematics, of Asians and Americans is the academic performance of Asian-American children. Tsang (1988) points out that direct comparisons of Asian Americans and other Americans in mathematics achievement, and in other areas, are complicated. For the most part, immigration to the United States tends to be selective. Since 1968, highly skilled and educated individuals have been favored. These individuals are likely to be of higher ability than those individuals who do not immigrate to the United States, and the immigrants probably have higher than average educational expectations for their children. Earlier in this century, on the other hand, lower ability Asian individuals were more likely to immigrate to the United States than higher ability individuals, because of racial discrimination in the availability of professional-level jobs (Tsang, 1988). These social policy and historical trends make direct comparisons of ethnic groups within the United States very complicated, because these trends will almost certainly influence academic performance as well as performance on IQ tests.

Nevertheless, it is important, for the sake of completeness, to overview the mathematical achievement of Asian Americans and other ethnic and racial groups within the United States. Tsang (1984, 1988) reviewed research on the mathematical achievement of Asian-American elementary school and high school students as compared with their Anglo-American peers. Before changes in the immigration policy, noted above, Asian-American children (primarily of Chinese and Japanese ancestry) had slightly lower mathematical achievement scores than Anglo-American children (Mayeske, Okada, Beaton, Cohen, & Wisler, 1973; cited in Tsang, 1984). More recently, however, Asian-American children scored higher, on average, on mathematics achievement tests than other ethnic groups within the United States (Mullis et al., 1991). For instance, in 1982 and 1983, the mean SAT−M performance of Anglo-American students was 484, whereas that of Asian students was 514. Similarly, Asian Americans obtain the highest mean scores, as compared with other ethnic groups, on the Quantitative section of the Graduate Record Examination (GRE), but they score lower, on average, than Anglo-Americans on the Analytic scale of the GRE.

In another set of comparisons, it was found that the highest scoring high school seniors on a mathematics achievement test were Asians who had lived in the United States between 6 and 10 years (Tsang, 1988). In other words, Asians who had received most or all of their elementary school training in a country other than the United States showed the highest mathematics achievement scores at the end of high school (Tsang, 1988). These students outperformed Asian-American students who had lived in the United States their entire life (by about $\frac{1}{5}$ of a standard deviation), who, in turn, performed slightly better than their Anglo-American peers (by about $\frac{1}{10}$ of a standard deviation). On the basis of the results of the above-described cross-cultural studies (Song & Ginsburg, 1987; Stevenson et al., 1993) and the achievement patterns before and after changes in U.S. immigration policy, it is not likely that these differences arise from racial differences in intelligence. Any differences in IQ or mathematical achievement that are found within the United States might, in part, be attributable to higher ability Asian parents' immigrating to the United States in greater proportions than lower ability parents (Tsang, 1988), not to more general racial differences in intelligence.

For the most part, the strong mathematical achievement of Asian-American children appears to result from a much greater investment in academic activities in relation to other American children (Chen & Stevenson, 1993). The average Asian-American student in the United States spends considerably more time doing homework and extracurricular academic activities than other American children and adolescents (Caplan, Choy, & Whitmore, 1992; Tsang, 1984). The more assimilated the Asian-American student is in American culture, the lower the grade point average in school and the lower the achievement test scores in mathematics and other areas, in relation to Asian students with more traditional values toward academic achievement (Caplan et al., 1992). In Asian culture, the traditional emphasis is on the importance of study and hard work, rather than native ability, for academic achievement (see the Culture section; Stevenson, Lee, Chen, Stigler, et al., 1990).

The same general conclusion can probably be drawn with respect to other ethnic or racial differences in mathematical performance that are found within the United States. The average mathematics achievement test scores of African-American and Hispanic-American children are often found to be lower than the scores of Anglo-American children (e.g., MacCorquodale, 1988; Mullis et al., 1991; Schratz, 1978). However, it is not likely that these differences arise from fundamental differences in basic mathematical abilities. Ginsburg and his colleagues, for instance, showed that with formal schooling, the basic mathematical skills of elementary school children from the United States and Africa are comparable in terms of developmental patterns and mean levels of performance (e.g., Ginsburg et al., 1981b). Moreover, African-American and Anglo-American children do not appear to differ in informal mathematical skills but do differ in formal mathematical skills (Ginsburg & Russell, 1981; Jordan et al., 1992).

Children from Spain appear to outperform their Spanish-speaking peers in the United States in mathematics and perform as well as American children in general (Lapointe et al., 1992). Low socioeconomic status children from Peru perform as well as Anglo-American children and Asian children on cognitive ability tests but have much lower reading and mathematics achievement test scores (Stevenson, Chen, Lee, & Fuligni, 1991). As with Asian-American and Anglo-American differences in mean mathematical performance, differences in the mathematical achievement of Anglo-American children and children of other ethnic groups within the United States mirror differences in the number of mathematics courses taken in high school, time spent on homework, and so on (Mac-Corquodale, 1988).

Summary

If racial differences in intelligence exist, then they are likely to be rather small and not a primary source of mathematical achievement differences between Asian- and Anglo-American children. Although IQ tests provide an excellent measure of individual differences in intelligence, raw score comparisons of different national and racial groups are very problematic and should probably be avoided (Ceci, 1991; Flynn, 1987). This is because performance on IQ tests, and intelligence itself to a lesser extent, is influenced by a variety of cultural and environmental factors, such as the quantity of schooling (Ceci, 1991; Coon et al., 1992).

Moreover, the research of Stevenson and his associates (e.g., Stevenson et al., 1985) suggests that there are no mean IQ differences between Asian and American children, despite the finding of consistent differences between the mathematics achievement of Asian and American children. Song and Ginsburg's (1987) study suggests that American children begin school with an advantage over their Korean peers in informal mathematics but that by the second grade the Korean children have the advantage in informal as well as formal mathematics. At this age, the advantage of the Korean children over their American peers is almost certainly due to schooling and not to intelligence. As detailed in the following sections, Asian children have a greater opportunity to learn mathematics than American children, which, in turn, is the most likely source of mathematical achievement differences between Asian and American children, as well as the source of differences among ethnic groups within the United States.

This is not to say that the ethnic and racial differences in mathematical achievement that are often found within the United States are not real or important. They are. As noted earlier, basic mathematical skills will influence later employability and productivity in the work place (Boissiere et al., 1985; Rivera-Batiz, 1992). Nevertheless, it does not necessarily follow that any such achievement differences are due to more fundamental differences in the ability to learn mathematics.

Language

In chapters 1 and 2, the relationship between the structure of Asian-language and European-language number words and children's number, counting, and arithmetical development was discussed (e.g., Fuson & Kwon, 1991). The associated research is briefly reviewed in this section and then considered in terms of international differences in mathematical achievement. Recall that number words in most Asian languages are structured around the base-10 system, whereas number words in most European-derived languages are more arbitrary, at least until the 100s. For instance, in Chinese, the number word for 34 is *three ten four*. For the Chinese language, it is very obvious that 34 is composed of 3 tens and 4 ones, but this structure is not at all obvious in English or in most other European-based languages. An array of studies demonstrated that the structure of Asian-language number words facilitates Asian children's understanding of counting, number, and arithmetic (Fuson & Kwon, 1992b; K. F. Miller & Stigler, 1987; Miura, 1987; Miura et al., 1993).

More precisely, because of the difference in the structure of Asian-language number words and European-derived number words, Asian children make fewer counting errors; understand counting and number concepts at an earlier age; make fewer problem-solving errors in arithmetic; and understand basic arithmetical concepts, such as place value and trading, at a much younger age than their American and European peers (Fuson & Kwon, 1992b; K. F. Miller & Stigler, 1987; Miura, 1987; Miura et al., 1993; Paredes & Miller, 1993). This difference in the structure of number words most likely gives Asian children an early edge in basic mathematics in relation to American and most European children. In particular, language structure might be an important contributing factor to the very rapid development of Asian children's basic mathematical skills during the first few years of formal schooling, where instruction is focused on basic number and arithmetic skills (Fuson & Kwon, 1992b; Geary, Bow-Thomas, et al., 1993; Song & Ginsburg, 1987; Stevenson et al., 1986).

Asian children also outperform American children in mathematical areas that presumably would not be influenced by the structure of number words, such as interpreting graphs and tables, as early as the first grade. This result indicates that the structure of number words is only one of many influences on international differences in mathematical achievement and development (Stevenson, Lee, Chen, Stigler, et al., 1990). Moreover, language is not likely to be an important factor in the large differences in mathematical achievement comparing adolescents from Japan and the United States. This is because the tests administered to adolescents assess skills that should not be strongly influenced by number words (e.g., calculus). Finally, language cannot be used to explain the finding that by 9 years of age, children from many European countries outperform their American peers in mathematics (Husén, 1967; Lapointe et al., 1992).

Schooling

In this section, factors that contribute to or influence the child's experience in school are considered, because they are potentially related to international differences in mathematical achievement. These factors include the opportunity to learn mathematics, homework, curriculum, classroom experiences, and financial support for education, each of which is considered in turn.

Opportunity to Learn Mathematics

One important finding of the first IEA study was that cross-national differences in the opportunity to learn mathematics, which was defined as the relative emphasis on mathematics instruction in the classroom, was moderately to strongly related to international differences in mathematical achievement (Husén, 1967). "Thus, national differences [in mathematical achievement] can in part be explained by differences in emphasis in curriculum" (Husén, 1967, Vol. 2, p. 300). Except for one aspect of geometry, it was also found that achievement in different mathematical areas varied directly with the relative degree of emphasis in the mathematics curriculum. For example, students with the best scores in basic algebra came from nations where the mathematics curriculum emphasized basic algebra. Walker and Schaffarzick (1974) also concluded that the most important influence on academic achievement was the content of what is taught. Bahrick and Hall (1991) demonstrated that the long-term (over 50 years) retention of algebraic skills was strongly related to the degree to which the material was rehearsed and practiced in high school and college (see also K. R. Johnson & Layng, 1992). Other factors, such as teaching style, also appear to influence learning and achievement and are discussed in the *Classroom Experiences* section.

The most important question to be addressed in this section is whether the quantity of mathematics instruction differs substantially for American children and their international peers. To put these comparisons in perspective, one needs to consider also the overall amount of time children from different nations spend in school. The most recent international comparison of mathematical achievement and the associated school practices indicated a wide range in the average number of days of instruction per year: from 173 days (Ireland) to 251 days (mainland China; Lapointe et al., 1992). For this most recent assessment, it was found that American children attended school for 178 days per year, on average, whereas children from the four top-scoring nations (for 13-year-olds) attended school each year for between 207 days (Switzerland) to 251 days (mainland China). This is not the whole story, however, because there is great variation in the amount of time that children spend in classes each day. Of the nations included in this study, 13-year-old American children had the longest school day. Considering the total amount of potential instruction time, across the academic year, children in the United States had about as much potential instruction time as children from Korea and Switzerland and 175 to 275 hours less potential instruction time than children from Taiwan and mainland China.

More important, however, is the amount of time that the average child spends doing mathematics rather than the average amount of instructional time allotted for mathematics. The time on each task, not simply the time in class, influences actual achievement (e.g., Peterson & Fennema, 1985). Fortunately, Stevenson and his colleagues have conducted extensive observational studies of classroom practices in Japan, Taiwan, and the United States (Stevenson, Lee, Chen, Stigler, et al., 1990; Stevenson et al., 1986). From these observations, they have been able to estimate the amount of time that teachers spend teaching mathematics, as well as the amount of time Japanese, Chinese, and American children spend engaged in academic (mathematics and reading) and nonacademic (e.g., socializing) activities.

Stevenson et al. (1986) found that during potential instructional time, American first-grade children were engaged in academic activities about 70% of the time, compared with 85% and 79% of the time for their Chinese and Japanese peers, respectively. By the fifth grade, American children spent about 65% of their time engaged in academic activities, compared with 92% and 87% for Chinese and Japanese fifth graders, respectively. In the first grade, American teachers spent, on average, between 1½ and 2 hr less time per week on mathematics instruction than did teachers in Taiwan and Japan (Stevenson, Lee, Chen, Stigler, et al., 1990). By the fifth grade, the typical American teacher spent almost 4½ hr less time per week on mathematics instruction than their Japanese peers and about 8 hr per week less time than their Chinese peers. Across the school year, the net result was that the average first grader from Taiwan, for example, received about 63 (97 vs. 160) more hours of mathematics instruction than their peers in the United States. By the fifth grade, the average student in Taiwan would receive about 346 (122 vs. 468) more hours of mathematics instruction than the typical American fifth grader. The overall pattern of results indicates that American children receive considerably less mathematics instruction than do children from nations with the highest mathematical achievement scores.

Homework

Within nations, the first IEA assessment found only a weak, though positive, relationship between the amount of mathematics homework and mathematical achievement scores (Husén, 1967). Across nations, however, there was a moderate relationship between the amount of homework and mathematical achievement. Adolescents in those nations with higher levels of mathematics homework tended to have higher mean mathematical achievement scores than adolescents from nations with lower levels of mathematics homework. The most recent international study also found a positive relationship between the amount of time spent on mathematics homework and mathematical achievement scores for 10 of 15 nations; they did not do cross-national comparisons in this area (Lapointe et al., 1992). Moreover, in most of the nations that participated in this study, a positive relationship was found between time spent in extracurricular mathematics activities (e.g., practice at home) and mathematical achievement (Lapointe et al., 1992). In all, it appears that mathematical activities outside of the classroom, such as homework, contribute to children's mathematical development.

Stevenson and his colleagues reported substantial differences among the amounts of homework that American, Chinese, and Japanese children did each week (Stevenson, Lee, Chen, Stigler, et al., 1990). Across all areas, American first-grade children did just over 1 hr of homework each week, whereas their peers in Taiwan and Japan did about 8 and 4 hr, respectively, of homework each week. In the fifth grade, American children spent about 4 hr per week on homework, as compared with about 13 and 6 hr per week in Taiwan and Japan, respectively. Despite their considerable amount of homework, by American standards, Chinese and Japanese children's attitudes toward homework were generally positive. The attitudes of American children toward homework were generally negative. In all, given the above-described relationship between homework and mathematical achievement, and the relationship between the long-term retention of mathematical material and the degree of initial learning (Bahrick & Hall, 1991), it seems very likely that these national differences in amount of homework contribute to international differences in mathematical achievement.

Curriculum

In this section, the mathematics curriculum of American children is contrasted with that of children from other nations. The focus is on information presented in textbooks, because this information appears to be reflective of curriculum differences in general (Travers & Westbury, 1989). In comparison with international standards, mathematics curriculum in the United States is developmentally delayed (Fuson et al., 1988; Stevenson & Bartsch, 1992; Stigler et al., 1986; Stigler, Lee, Lucker, & Stevenson, 1982). Stevenson and Bartsch in an extensive analysis of Japanese and American elementary and secondary school mathematics textbooks found that for the most part, textbooks in the two nations present the same concepts. However, in many areas, the material is conceptually more difficult and presented in an earlier grade in Japanese, in relation to American, textbooks. Moreover, American textbooks are cumbersome and probably include too much unnecessary material. In Japanese textbooks, the basic concept and material are presented, and it is expected that the teacher will elaborate on the material in class or that the children will figure out how to solve the problem on their own. In contrast, step-by-step procedures, along with ancillary pictures, are typically presented for the child in American textbooks.

These differences are not restricted to Japanese and American textbooks (Travers & Westbury, 1989). Fuson et al. (1988) found that many arithmetic topics that are introduced in the fifth or sixth grade in the

United States are introduced in the third grade, or sooner, in Japan, mainland China, the former Soviet Union, and Taiwan. Stigler et al. (1986) showed that in comparison with textbooks in the former Soviet Union, American textbooks did a very poor job of presenting arithmetic word problems. In American textbooks, the presentation of the word problems (such as those shown in Exhibit 3.1) is haphazard and incomplete, with only the easiest types of problems presented with any frequency. In contrast, textbooks in the former Soviet Union present word problems in a more comprehensive and systematic manner and include more complex problems. Problems that are similar in Soviet and American texts are presented in earlier grades in the former Soviet Union than in the United States. In conclusion, the mathematics curriculum in the United States is poorly conceived and too easy and, as a result, almost certainly contributes to the delayed mathematical development of American children.

Classroom Experiences

In the first part of this section, the organizational and structural features of American classrooms, as well as more qualitative facets of the child's classroom experiences, are contrasted with those of children in other nations. The section concludes with a discussion of American and Asian children's subjective evaluation of their classroom experiences.

By international standards, the physical infrastructure of American schools and classrooms is, on average, among the best in the world (Lapointe et al., 1992). The average classroom in many parts of the world includes between 20 and 30 students, as do classrooms in the United States. The most notable exceptions are classrooms in Asian countries, which often include 40 to 50 students. More important than the number of students and the physical quality of the classroom is the organization of student activities.

Stigler, Perry, and their colleagues have conducted a series of important studies of the classroom experiences of American, Japanese, and Chinese children. These studies were undertaken to determine if classroom experiences contributed to the mathematical achievement differences between American and Asian children (Perry, VanderStoep, & Yu, 1993; Stigler, Lee, & Stevenson, 1987; Stigler & Perry, 1988). In the first of these studies, Stigler et al. conducted extensive observations, using rigorous time-sampling techniques, of mathematics instruction periods in first- and fifth-grade classrooms in Japan, Taiwan, and the United States. It was first noted that the organization of American and Asian classrooms differed considerably. Children in American classrooms usually sit in desks that are arranged in small groups, whereas children in Japanese and Chinese classrooms sit in rows of seats that face the front of the class, where the teacher presents the lessons. As noted earlier, children in American classrooms are off-task more often than children in Japanese and Chinese classrooms. For instance, it was found that during mathematics lessons, American children were out of their seat, on average, more than 20% of the time, compared with less than 4% of the time during mathematics instruction in Japanese and Chinese classrooms (Stigler et al., 1987).

In American classrooms, teachers spend much more time working with individual students or with small groups of students than in Japanese or Chinese classrooms (Stigler et al., 1987); Japanese and Chinese teachers spend most of the mathematics instruction time lecturing or discussing mathematical material. As a result, the average American student spends more than 50% of the mathematics instruction time engaged in seat work or other activities that are not directed by the teacher. In contrast, children in Taiwan and Japan spend 90% and 74%, respectively, of their mathematics instruction time engaged in activities that are directed by the teacher. Stigler et al. (1987) also examined the relationship between classroom features and student achievement within each nation. Some of the same features that distinguished the organization of American classrooms from classrooms in Japan and Taiwan, such as the amount of time the child spent in activities directed by the teacher, were the same features that distinguished effective and ineffective classroom practices within the United States. "Loosely organized classrooms, classrooms in which there was a high frequency of irrelevant activities, and ones in which less time was spent on substantive material were the ones that tended to have children with lower levels of achievement" (Stigler et al., 1987, p. 1283).

More qualitative assessments of the nature of teacher-student interactions also reveal potentially important differences between the classroom experiences of American and Asian students during mathematics instruction (Perry et al., 1993; Stevenson, 1992b; Stigler & Perry, 1988). Perry et al. compared first-grade classrooms in Japan, Taiwan, and the United States to determine if systematic differences existed in the types of questions that teachers in the respective countries asked their students. This analysis was conducted because it had been shown that students' achievement could be improved by teachers asking them conceptually difficult questions (Redfield & Rousseau, 1981). Across nations, six types of questions were asked by the teachers: *computational/rote recall* ("How much is . . ."), *rule recall* ("What is the procedure for . . ."), *computing in context* (word problems), *make-up problems, problem-solving strategies,* and *conceptual questions.*

In all three countries, at least 50% of the lessons included computational and rote recall questions (Perry et al., 1993). There were no cross-national differences in the frequency with which teachers asked their students to make up problems or recall rules. Teachers in Japan and Taiwan, however, asked their students computing-in-context questions nearly twice as often as American teachers. The Asian teachers almost always used contexts that the children were familiar with, such as buying a notebook at the store. American teachers were much less consistent. Oftentimes the context of the problems was unfamiliar or nonsensical. For example, one American teacher asked her class the following question: "She walked to the store with 8¢, earned 4 more, how much does she have now?" (Perry et al., 1993, p. 36). The protagonist in this word problem, "she," apparently got a job on the way to the store. Logical inconsistencies, as is illustrated by this word problem, might lead children to think about how, for instance, "she earned 4 more," rather than thinking about how to solve the problem.

Japanese and Chinese teachers also asked questions about problem-solving strategies (e.g., "What is the best way to solve this problem?") more than twice as often as American teachers. Similarly, American teachers almost never asked their students questions that tapped their conceptual understanding of the material, whereas this type of question was asked in 20% of the Chinese lessons and 37% of the Japanese lessons (Perry et al., 1993). Stigler and Perry (1988) also noted that Japanese and Chinese teachers often spent an entire lesson discussing and solving a few problems, whereas lessons in U.S. classrooms often focus on how many problems can be solved in a given lesson. In other words, mathematics lessons, particularly in Japan, involve discussing and demonstrating procedural and conceptual features of one or two problems. The children then practice solving related problems as part of their homework.

Briefly, note that Japanese and Chinese classrooms differ in many respects. Japanese teachers have a more "reflective" approach to mathematics. For example, errors are discussed extensively during classroom lessons to ensure that the students understand the problem conceptually. Chinese teachers are more performance oriented, focusing on speed and accuracy of problem solving (Stigler & Perry, 1988). No matter what the relative emphasis, the achievement levels of Chinese and Japanese children are comparable but greatly exceed those of American children (Stevenson et al., 1993).

A recent study suggests that despite high expectations, Chinese and Japanese children do not, on average, find their school experiences to be stressful. "Critics of the academic success of Chinese and Japanese students often suggest that their high levels of performance come at great psychological cost" (Stevenson et al., 1993, p. 57). To test this belief, Stevenson et al. asked 11th-grade students from Japan, Taiwan, and the United States to indicate how frequently they experienced feelings of stress, depression, aggression, sleeping problems, and so on. American students reported the highest frequency of these stress-related symptoms, Japanese students reported the lowest. This study suggests that the focus on academic excellence in Asian culture might not be at "great psychological cost." Similar conclusions appear to be warranted for Japanese and Chinese elementary school children (Stigler & Perry, 1988).

Finally, note that the attitudes of Japanese, Chinese, and American students toward mathematics are comparable and generally favorable (Husén, 1967; Stevenson, Lee, Chen, Stigler, et al., 1990). In fact, American children tend to feel more confident of their mathematical abilities than their Japanese or Chinese peers (Stevenson, Lee, Chen, Stigler, et al., 1990). This greater confidence might stem, in part, from American teachers giving more praise, for lower levels of performance, to their students than do teachers in Japan and Taiwan (Perry et al., 1993). The fact that the mathematics curriculum in the United States is easier than the same-grade curriculum in Asian, and other, countries probably also contributes to this phenomenon. In all, compared with American children, children in Asian classrooms spend much more of their mathematics instruction time engaged in activities that are directed by the teacher, are on-task much more frequently, and are asked more difficult questions by their teachers. These differences in the classroom experiences of Asian and American children are very likely to be contributing factors to the cross-national achievement differences in mathematics.

Financial Support

With the first IEA assessment, a negative relationship between spending on education and mathematics achievement was found (Husén, 1967). The greater the expenditures on education, the lower the overall achievement scores. This relationship was due primarily to the fact that although the United States spent more per student on education ($545 per year) than any other nation, American students' performance was very poor in mathematics. In contrast, Japan spent less per student on education

($81 per year) than any other nation, and Japanese adolescents showed very high mathematics achievement scores. In 1964, the United States spent 4.5% of its gross national product (GNP) on education (Husén, 1967). By 1991, the United States was spending 7.5% of its GNP on education (Lapointe et al., 1992). Yet, SAT scores were lower in 1991 than in 1964, and the overall mathematical skills of American children changed little—if they did not worsen—from the 1960s to the 1990s. Of course, the impact that expenditures have on educational achievement is moderated by many factors, such as neighborhood and family (Bronfenbrenner, 1986). Nevertheless, it is not likely that unfocused increases in expenditures on education will substantially improve the mathematical skills of American children (Stevenson & Stigler, 1992).

Summary

In terms of mathematics instruction, the classroom experiences of American children differ on several important dimensions from those of children from many other nations. In comparison with children from Asian countries, American children receive substantially less instruction in mathematics. Over the course of 12 years of schooling, these instructional time differences could amount to several thousand hours. The instructional time differences are compounded by the fact that American children do much less mathematics homework than children in most other industrialized countries. Clearly, in terms of quantity, American children have considerably less opportunity to learn mathematics than do children in many other nations. Moreover, the time that is spent on mathematics instruction in American classrooms is qualitatively different from that in some other countries. The overall result is that American teachers, on average, do not challenge their students to the same extent as teachers in other nations. Finally, the mathematics curriculum in the United States is haphazardly organized and too easy. In combination, it seems very likely that these schooling differences underlie most of the international differences in mathematical achievement. Family and cultural factors that contribute to these schooling differences are discussed in the following sections.

Family

There are a number of important differences between Asian and American families that appear to contribute to the mathematical achievement differences of Asian and American children. These differences include parental expectations for academic excellence, relative emphasis on ability versus effort, the home environment, and the relative frequency of disruptive family influences (e.g., divorce), each of which is briefly considered in turn.

The research of Stevenson and his colleagues indicates quite clearly that in terms of mathematics, Chinese and Japanese parents have much higher expectations for their children's achievement than do American parents (Stevenson et al., 1993; Stevenson, Lee, Chen, Stigler, et al., 1990). Despite the frequent reports in the popular press that American children perform very poorly in mathematics in comparison with children in many other nations, nearly half of American parents report that they are very satisfied with their child's academic skills. In contrast, very few Chinese or Japanese parents report that they are very satisfied with their children's academic skills (Stevenson et al., 1993). About 10% of American parents are not satisfied with their children's academic achievement, as compared with between 30% and 40% of Asian parents. Stevenson argued that these differences, at least in part, might be the result of vague feedback that most elementary school teachers provide to parents about the relative performance of their children (Stevenson, 1992a; Stevenson et al., 1993). Whatever the reason, American parents, for the most part, appear to be satisfied with the low levels of mathematical performance of American children. This satisfaction almost certainly translates into lower implicit and explicit standards for achievement for American, as compared with Asian, children (see also Caplan et al., 1992).

Moreover, American parents tend to believe that innate ability, as well as effort, is an important influence on academic achievement, as contrasted with Asian parents' belief that effort and hard work are of primary importance. "The belief that increased effort pays off in improved performance is an important factor in accounting for the willingness of Chinese and Japanese children, teachers, and parents to spend so much time and effort on children's academic work" (Stevenson, Lee, Chen, Stigler, et al., 1990, p. 67). In other words, if a child is having difficulty with mathematics, then Asian parents will tend to emphasize the need to devote more time to mathematics, whereas American parents appear to be more likely to attribute the difficulty to a lack of "natural talent for mathematics." As a result, once mathematics begins to become difficult, Asian parents encourage increased effort on the part of their children, whereas many American parents appear to accept poor performance as inevitable (Stevenson, Lee, Chen, Stigler, et al., 1990).

These attitudes are also reflected in the overall home environment of Asian and American families. The average family in Japan, Taiwan, and mainland China has considerably less living space than the typical family in the United States. Nevertheless, many more Asian than American families set aside quiet work spaces, with desks, for their children's schoolwork (Stevenson, Lee, Chen, Stigler, et al., 1990). One set of studies also found that two to four times as many parents in Japan and Taiwan than parents in the United States directly assisted their children with mathematics (Crystal & Stevenson, 1991). The everyday lives of Chinese and Japanese children are focused on academic activities (e.g., reading or playing chess), whereas the majority of the after school hours of American children are spent engaged in nonacademic social activities. It is not that Asian children spend all of their out-of-school time engaged in some type of academic pursuit. Rather, Asian parents, for the most part, allow their children to watch television, for example, but only after homework has been completed. In fact, by the fifth grade, the average Japanese child watches slightly more television per week than the average American child.

These differences in family attitudes toward academic achievement almost certainly contribute to children's relative valuation of school versus other activities (e.g., sports or social activities). A focus on academic activities is implicitly and explicitly rewarded much more in Asian than in American families. Moreover, it is possible that some of the earlier-described differences in the classroom behavior of Asian and American children, such as time spent performing academic tasks, reflect these differences in family expectations, although temperamental differences between Asian and American children might also be contributing factors to these cross-national differences in classroom behavior (Freedman & Freedman, 1969).

Finally, note that American children are much more likely to experience disruptive family events, such as divorce, that will adversely impact academic achievement than are children in nearly all other industrialized nations (Amato & Keith, 1991). These events are not likely to be a primary source of mathematical achievement differences between the typical American child and children from other nations, although they are almost certainly contributing factors. The more likely result of problems within the family is to create greater variability in the achievement of American children in comparison with children in other nations (cf. Rutter, 1982). The overall effect is lower achievement scores for American children, on average, and a large number of American children with very poor academic skills. The families of American children, on average, do not foster academic excellence to the same extent as Asian families, and children from those American families that experience a reduction in parental investment in their children's academic development, because of divorce or family conflict, for example, will probably be at an even greater disadvantage than other American children.

Culture

It is clear from the above-noted research that schools in Asian countries, as well as Asian parents, place a much higher value on the acquisition of mathematical skills than do their counterparts in the United States (Stevenson, Lee, Chen, Stigler, et al., 1990). The differences in the relative emphasis on children's mathematical development in school and at home reflect wider cultural values (Stevenson & Stigler, 1992). Within different cultures there are different ways in which social status and self-esteem can be achieved (Hatano, 1990). Children within these different cultures are therefore either implicitly or explicitly rewarded for acquiring different types of skills. A consideration of the types of activities that are valued in different cultures is especially important when considering activities, such as mathematics, that are not likely to be inherently interesting or fun for most children (Geary, in press).

"Asian culture emphasizes and gives priority to mathematical learning; high achievement in mathematics is taken by mature members of the culture to be an important goal for its less mature members" (Hatano, 1990, pp. 110–111). Mathematics is not necessarily inherently more interesting or fun for Asian children than it is for American children. Rather, adults in Asian culture expect and support their children's mathematical development. As a result, doing well in mathematics and in other academic areas is likely to be a source of self-esteem for Asian children and their families. American culture clearly does not value mathematical skills and knowledge to the same extent as Asian culture. In fact, Eccles et al. (1993) found that by the second grade, American children value sports more than mathematics, reading, or music. Achievement in sports is an important goal for many American children because it adds to feelings of self-competence and almost certainly social status, especially for boys. Yet, there are many more jobs in the United States for engineers than for professional basketball players. The greater valuation of sports over academic activities in American culture is not likely to be in the best long-term interest of most American children.

American culture is also much more open than Asian culture in the extent to which individuals are allowed to pursue their own self-interests or engage in activities that are inherently interesting (Hatano, 1990). Except for basic counting and number activities, most mathematical areas are not likely to be inherently interesting for most individuals, Asian or American (see the Educational Philosophy section of chapter 8). Thus, the acquisition of complex mathematical skills will not occur for a large segment of a given population without strong cultural values that reward mathematical development. Given the American cultural milieu, which allows for the pursuit of easier activities and activities that are inherently more enjoyable than mathematics, it is not surprising that very few American children achieve the same level of mathematical competence that is found in Asian children. These cultural differences in the relative valuation of mathematics and other activities probably underlie cross-national differences in children's experiences in school and at home. Moreover, these differences in the relative valuation of academic versus nonacademic activities are probably important reasons as to why it is so expensive to educate American children and so relatively inexpensive to educate Asian children.

Conclusion

The first large-scale comparative study of mathematical achievement showed that by the end of high school, American adolescents were among the most poorly educated students in mathematics in the industrialized world (Husén, 1967). Every study that has been conducted in the roughly 30 years since this assessment has shown that the mathematical skills of American students have changed little, if any, since the 1960s (Crosswhite et al., 1985; Lapointe et al., 1992; Stevenson et al., 1993). The poor mathematical skills of American children are evident in the first grade and become increasingly worse, by international standards, with each successive grade (Stevenson et al., 1993). The consequences of the poor mathematical skills of America's children and subsequent labor force will quite likely be lowered economic productivity and a declining standard of living (Bishop, 1989). Thus, it is not surprising that recent educational reforms in the United States have focused on improving the mathematical skills of American children. In fact, one of the primary goals of *America 2000: An Education Strategy* is to have American children first in the world in mathematics and science achievement by the year 2000 (U.S. Department of Education, 1991). Nevertheless, in consideration of the lack of progress over the last 30 years, the *America 2000* goal for American chil-

Figure 7.1

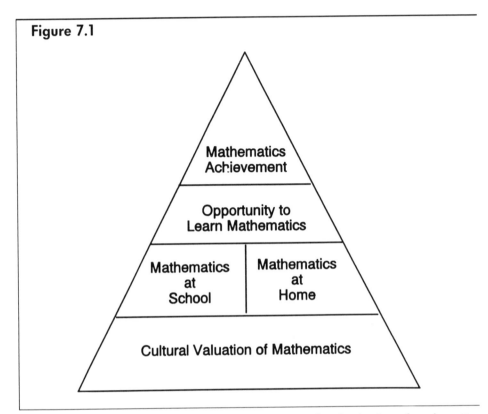

Schematic representation of the relationship among cultural valuation of mathematics, children's mathematical experiences at school and at home, and mathematical opportunities and achievement.

dren to be first in mathematical achievement by the year 2000 almost certainly will not be achieved.

A variety of factors appear to contribute to the poor mathematical development of American children in comparison with that of children in many other nations. As shown in Figure 7.1, the bedrock of these differences appears to be the relatively little value that American culture places on mathematical achievement, as compared with many other cultures, especially Asian culture (Stevenson, Lee, Chen, Stigler, et al., 1990). The relative valuation of mathematics and other academic activities, as opposed to social or nonacademic pursuits such as sports, permeates the child's experiences at home and at school. Cultural valuation and tradition determine the different ways in which children can achieve self-esteem and social status, because these traditions will determine the relative reward value of achieving above-average skills in various areas (Hatano, 1990). Stevenson and his colleagues (e.g., Stevenson & Stigler, 1992) have examined these issues, comparing Asian and American culture, and have shown how they appear to contribute to international differences in mathematical achievement (see the Social Ideology section of chapter 8).

Asian culture supports and rewards the mathematical development of children, whereas American culture does not. In fact, one recent study indicated that by the second grade, American children value sports activities more than academic activities (Eccles et al., 1993). Schools in the United States devote considerably less time to mathematical instruction than do schools in many other nations (Stevenson, Lee, Chen, Stigler, et al., 1990; Travers & Westbury, 1989). The development of mathematics textbooks is driven by market forces in the United States, that is, by what children and teachers prefer, rather than by what is in the best long-term interest of American children. Thus, the material presented in these textbooks is too easy and poorly organized, by international standards (Stigler et al., 1986). Also, Asian teachers appear to challenge their students in mathematics to a much greater extent than do American teachers (Perry et al., 1993), but American parents are generally satisfied with their children's low level of mathematical development (Stevenson et al., 1993).

The relatively low priority of mathematics in American culture influences the child's social, school, and home experiences. The home environment, for instance, is much less supportive of mathematical achievement in the typical American family than in its Asian counterpart (Crystal & Stevenson, 1991). The typical American parent devotes relatively fewer resources and less time to the academic development of the children than does the typical Asian parent (Caplan et al., 1992). The net result of this pattern of school and home activities is that children in the United States have much less opportunity to learn mathematics than do children in Asian, as well as other, nations (Travers & Westbury, 1989). These national differences in the overall opportunity to learn mathematics, that is, the quantity and quality of mathematics instruction and activities, appear to be the primary source of international differences in mathematical achievement (Husén, 1967). (See Stevenson & Stigler, 1992, for an excellent and more comprehensive discussion of these and related issues.)

Cross-national differences in other factors, such as intelligence or inherent mathematical ability, are likely to be much smaller than is often believed (Lynn, 1982; Rushton, 1992) and do not appear to be primary determinants of international differences in mathematical achievement. The one exception is the structure of number words in Asian- and European-derived languages, such as English. The fact that the base-10 system is reflected in Asian-language number words, and is not at all obvious in European-based languages, probably facilitates the early numerical and arithmetical development of Asian children (Fuson & Kwon, 1992b). Nevertheless, in comparison with national differences in the overall opportunity to learn mathematics, the structure of number words is probably a relatively minor influence on international achievement differences in mathematics.

Improving Mathematical Instruction and Remediation

8

The goal of this chapter is to present ideas on how to improve the mathematical education of American children. Most generally, changes in mathematics education will involve instructional changes that affect all children and improvements in remedial education techniques for children and adolescents with specific disabilities. The Instruction section contains a consideration of general philosophical issues that influence mathematics education in the United States, as well as general suggestions as to how to improve mathematical instruction. The Remediation section deals specifically with mathematics anxiety and mathematical disabilities and the different methods that might be used to treat these disorders.

Instruction

We begin with an overview and critique of basic philosophical themes that currently guide educational practice in the United States. The critique is followed by a discussion of sociocultural influences on mathematical development and by a discussion of the instructional implications that follow from the understanding of children's mathematical development.

Educational Philosophy

The relatively poor mathematical skills of American children, described in chapter 7, indicate a need for a concerted effort on the part of scientists and educators to better understand mathematical development and to

---

Summaries are provided at the end of each of the main sections: Instruction (p. 272) and Remediation (p. 284).

develop better ways to teach mathematics. Yet, the "community of people concerned with research in mathematics education is increasingly divided into specialized groups and cliques that are not always tolerant of each other" (Sierpinska et al., 1993, p. 274). Discussions of mathematics education are often philosophically or ideologically based rather than directly focused on how to improve the mathematical skills of American children. In fact, there is no clear agreement, at this point, as to the types of mathematical skills that American children should be acquiring, nor is there agreement on how to best measure those skills that are acquired. A comprehensive discussion of these issues is beyond the purview of this book, but a brief glimpse at current educational philosophy, in particular the constructivist approach, seems to be necessary. This is because constructivist views of mathematics education appear to be growing in influence and therefore will likely impact the future mathematical development of American children.

Researchers from a variety of disciplines make reference to some of the basic themes of constructivism, described below, but also study mathematical development and mathematical cognition by means of methods and assumptions that follow the information-processing tradition (e.g., Greeno, 1989, 1993; Saxe, 1991). The discussion in this section is confined to constructivist researchers and philosophers who have rejected the basic assumptions of the information-processing approach to mathematical development and therefore might be considered rather extreme in their philosophical views of mathematics education (e.g., Cobb, Yackel, & Wood, 1992; von Glaserfeld & Steffe, 1991). Some of the specific features of the constructivist approach, as they are related to arithmetical development, are discussed in chapter 2 and will not be repeated here (see the Schema-Based Model section). Rather, the basic assumptions that appear to underlie this approach will be briefly reviewed and critiqued.

Von Glaserfeld and Steffe (1991) argued that there were two general approaches to educational research and practice: mechanistic and organismic. The mechanistic approach is exemplified by traditional learning theory and, in von Glaserfeld and Steffe's view, by many contemporary information-processing approaches to cognitive development. The basic assumption of the mechanistic approach is that the learner passively receives information from the environment, most notably the teacher. The information results in reflexive changes in the child's overt behavior, such as the number of problems solved correctly, or in the child's mental representations of mathematical information. The latter might be illustrated by the development of memory representations for specific arithmetic facts (Siegler, 1986). It is further assumed that any such changes in overt behavior or mental representations generally occur without the child conceptually understanding the material (Cobb et al., 1992). It follows from this assumption that many features of the information-processing system that are emphasized by cognitive psychologists, such as automaticity (that overlearned processes occur automatically without conscious effort), are viewed as detrimental to the child's academic development. Likewise, the route to automaticity, that is, drill and practice, is interpreted as the bane of children's mathematical growth.

The constructivist perspective offers an alternative, organismic approach to mathematical development. This organismic approach is exemplified by the theories of Piaget and Vygotsky. The basic assumption is that children are active learners and must construct for themselves mathematical knowledge. To completely understand mathematical material, the child must rediscover basic mathematical principles. The teacher provides appropriate materials and a social context within which the material is discussed but does not lecture or guide discussion in the traditional sense (Cobb et al., 1992; Lampert, 1990). "In constructivism, a *zone of potential construction* of a specific mathematical concept is determined by the modifications of the concept children might make in, or as a result of, interactive communication in the mathematical learning environment" (Steffe, 1992, p. 261). Mathematical learning is a social enterprise. Social disagreements about the meaning of mathematical materials or concepts provide the grist for mathematical development, because these disagreements provide the impetus to change or accommodate one's understanding of such concepts. Any such change serves to make the child's understanding of mathematics more consistent with the understanding of the larger social community (Steffe, 1990). With the development of appropriate social mathematical environments, "it is possible for students to construct for themselves the mathematical practices that, historically, took several thousand years to evolve" (Cobb et al., 1992, p. 28).

The constructivist's mechanistic versus organismic conceptualization of educational theory and research follows quite closely the debate in the 1960s about how to best conceptualize human cognitive development (e.g., Reese & Overton, 1970). Of course, cognitive development is very complex and involves mechanistic as well as organismic changes. To be sure, there is much to be gained by understanding social contextual influences on mathematical development (e.g., Saxe, 1991), but to assume that all development follows this route and to reject outright the idea that there are mechanistic changes in children's cognitive growth is naive.

Moreover, in addition to building basic skills, mechanistic approaches to mathematical tasks probably do influence children's conceptual development. The research of Briars and Siegler (1984), for instance, suggests that preschool children induce counting concepts by noting regularities associated with the act of rote counting.

Constructivist philosophers and researchers also fail to distinguish between biologically primary and biologically secondary cognitive skills (Geary, in press; Gould & Vrba, 1982; see the Biological Factors section of chapter 6). To illustrate, language is a biologically primary social cognitive skill (Dunbar, 1993; Steele, 1989). Humans are born with specialized neurobiological systems for the processing of language-related information (Witelson, 1987) and find engaging in activities that facilitate the acquisition of language, that is, social activities, inherently enjoyable. The combination of prewired neurobiological systems to support language and an inherent enjoyment of language-related activities ensures that language is acquired by all normal children without the need for direct instruction. Language acquisition is therefore a relatively effortless endeavor that occurs within a social context.

Reading, on the other hand, is a biologically secondary cognitive skill. The failure to distinguish biologically primary from biologically secondary skills has lead to the development of the *whole-language* approach to reading. Here, it is assumed that children will acquire reading skills in the same way that they acquire language skills. It is assumed that through natural curiosity and exposure to appropriate materials, children will become proficient readers. Even though many of the neurobiological systems that support language also support reading (Luria, 1980), these systems have not evolved to automatically acquire reading skills. Children have to be taught how to read. The belief that reading acquisition will occur in much the same way as language acquisition is almost certainly wrong, and the associated instructional techniques, such as whole reading, are very likely to be a disservice to many children (Rayner, 1993). This argument should *not* be taken to mean that children's inherent interest in social activities, as well as their curiosity, cannot be used in the service of education. They certainly can. Having children read and discuss stories in social groups can be an important aspect of practicing reading skills, once basic skills have been acquired (e.g., phonological decoding; Wagner & Torgesen, 1987). It cannot be assumed, however, that these basic reading skills will be acquired in social discussion groups or through the children's inductions from story context (e.g., matching a word with a picture).

Back to mathematics: One of the implicit assumptions of the constructivist approach is that mathematics is a biologically primary domain. That is, given an appropriate social context and materials, children will be motivated and able to construct mathematical knowledge for themselves (Cobb et al., 1992). In fact, this is probably not an unreasonable assumption for certain mathematical areas—such as number, counting, and some features of arithmetic (see chapter 1). Saxe et al. (1987), for instance, showed that counting and other basic numerical activities are indeed part of the everyday social interactions between parents and children. These social interactions, in turn, facilitate the child's understanding of number and counting. However, many other mathematical skills, such as solving algebraic word problems, are biologically secondary. The conditions under which these skills are acquired are likely to be very different from the conditions under which counting and other biologically primary numerical skills are acquired, just as language acquisition and reading acquisition differ (Geary, in press). Changing classroom practices to try to make complex mathematics resemble a biologically primary skill can probably only be achieved at the expense of content.

Moreover, even if complex mathematical skills can be learned through "conflicts that arise in the course of social interactions and the generally unnoticed mutual appropriations of meanings" (Cobb et al., 1992, p. 18), it is not likely that all children will be equally skilled or motivated in the construction of mathematical knowledge. One basic dimension of human personality is intellectual curiosity (Goldberg, 1993). However, there are large individual differences for this and other personality traits. There might be many children who are curious and motivated enough to pursue mathematical knowledge in the way envisioned by constructivists, but there will quite likely be many more children who are not so curious and motivated.

Finally, the argument that drill and practice and the development of basic cognitive skills, such as fact retrieval, are unnecessary and unwanted in mathematics education fails to appreciate the importance of basic skills for mathematical development. As noted earlier, drill and practice provide an environment in which the child can notice regularities in mathematical operations and glean basic concepts from these regularities (Briars & Siegler, 1984). Much of mathematics involves being able to use procedures, equations, and so on. Except for basic numerical and arithmetical skills, most children are not likely to be able to develop mathematical procedures solely on the basis of their conceptual knowledge.

Instead, some procedures and equations simply need to be taught and practiced (Sweller et al., 1983). Practice to the extent that the equation is automatically remembered will facilitate complex problem solving by reducing the working-memory demands that constructing an equation would otherwise make. Children also need to understand how the equation manipulates numbers or represents mathematical relationships. The point is that mathematical expertise requires both mechanical skills and a deep conceptual understanding of the domain. The processes that facilitate the acquisition of conceptual knowledge appear to differ from those processes that facilitate the acquisition of more mechanical skills (Sweller et al., 1983). The constructivist approach focuses on the former, but at the expense of the latter. Because of this, constructivism is not likely to lead to substantial long-term improvements in the mathematical skills of American children.

Instructional Implications

A comprehensive discussion of the relationship between children's mathematical development and mathematical instruction is beyond the scope of this book, as these issues have been extensively addressed in many other sources (e.g., Brophy, 1986; Grouws & Cooney, 1988). Nevertheless, some general themes and a few specific suggestions as to how our understanding of children's mathematical development might be used in instructional settings are offered. This section begins with a consideration of a number of very general issues, such as a national curriculum, that will influence how children are taught mathematics within the United States. The sections following the first provide instructional themes for each of the major mathematical domains discussed in chapters 1 through 3, that is, early numerical abilities, arithmetic, and mathematical problem solving.

Social Ideology

One of the social policy issues that influence children's mathematical development is the presence or absence of a national curriculum in mathematics (Stevenson & Stigler, 1992). The United States is one of the few industrialized nations without a national curriculum in mathematics (Lapointe et al., 1992). The lack of a national curriculum results in great variability in the quantity and quality of mathematics instruction that different American children receive in school. Stevenson et al. (1986), for instance, found that American children received, on average, much less instruction in mathematics than children from Japan and Taiwan.

Moreover, there was greater variability in the proportion of time spent on mathematics in American classrooms in comparison with classrooms in Japan and Taiwan. Some American teachers devoted the same proportion of time to mathematics as their Chinese and Japanese peers. (Because American children spend less time in school, the overall amount of time devoted to mathematics was still lower in these U.S. classrooms, relative to Asian classrooms.) Other American teachers apparently spent very little time teaching mathematics.

There is also much more variability in American schools, relative to schools in many other nations, in the level of difficulty of the implemented mathematics curriculum (Travers & Westbury, 1989). Students in some parts of the United States receive relatively rigorous mathematics instruction, whereas other students receive very weak mathematics instruction. The only way to ensure uniform improvements in the mathematical skills of American children is to ensure that they all receive upgraded and comparable levels of mathematics instruction, which, in turn, is probably only achievable through a national curriculum in mathematics. These arguments, of course, are contrary to democratic beliefs in autonomy and local control over curriculum (Stevenson & Stigler, 1992). Nevertheless, the national goal of substantially improving the mathematical skills of American children (U.S. Department of Education, 1991) is probably not achievable without a national curriculum or at least national performance standards.

Another social factor that influences American children's mathematical development is the relatively low priority that many Americans place on academic achievement, in comparison with individuals in many other nations (Stevenson & Stigler, 1992). Children's role models in the United States tend to be sports figures or entertainers, whereas the activities of role models in Asia are much more intellectual or prosocial in nature. This was not always the case:

> The United States has had many well-known models. Earlier in this century every American schoolchild was aware of the frugality and inventiveness of Franklin, the compassion of Nightingale, the creativity of Edison. . . [but] for the most part, such cultural models have been displaced in the United States today. (Stevenson & Stigler, 1992, pp. 85–86)

If we want to promote the academic and mathematical development of American children, then discussion of such role models needs to find its way back into American schools. American children need to be socialized so as to view mathematics and other intellectual activities as important and socially valued pursuits, not the domain of "nerds." (For an excellent discussion of these and related issues, see Stevenson & Stigler, 1992.)

Early Numerical Abilities

The development of early numerical abilities is discussed in chapter 1 and includes the child's understanding of quantity, counting, and arithmetic (Fuson, 1988). The development of early numerical skills appears to occur in a manner very similar to that described by constructivist philosophers, that is, in the context of parent-child social interactions (Saxe et al., 1987). In fact, given that many of these early abilities are probably biologically primary, the constructivist perspective on mathematical development makes some sense in understanding the development of these basic abilities. Parents tend to be sensitive to what children know and do not know and gear number-related activities to the child's level of understanding. If a mother and child are walking up some stairs, and the child counts "1, 2, 4," then the mother might recount "1, 2, 3" (Saxe et al., 1987). The child's understanding of number, counting, and basic arithmetic probably emerges from inherent numerical abilities and from knowledge gleaned from these types of social interactions.

If we consider these early abilities with respect to the goal of improving American children's mathematical development, then there is probably not too much that needs to be changed. During the preschool years, American parents tend to emphasize basic numerical skills more than Asian parents (Stevenson & Stigler, 1992). As a result, American preschool children often show a better understanding of basic number concepts than their peers in some Asian nations (Song & Ginsburg, 1987). Nevertheless, two changes that might be implemented during kindergarten include ensuring that American children learn basic word names and are introduced to the base-10 system (Fuson, 1988). Number words in all languages are arbitrary; that is, the sounds associated with a number word provide no clue as to the quantity it represents. Even though children implicitly understand that counting and quantity are related (Gelman & Gallistel, 1978), they still have to memorize number names.

Educators might also want to consider teaching kindergarten children to count using the same structure of number names that is used in Asian languages. These children might be taught, in addition to standard English number words, to count, "ten one, ten two, ten three," and so on. Teaching children that *eleven* also means ten one (1 ten and 1 one) and having them count using this word structure will likely facilitate their understanding of the base-10 system and the acquisition of the associated skills (e.g., trading; Fuson & Kwon, 1991).

Arithmetic and Mathematical Problem Solving

For the most part, many arithmetic skills and probably most mathematical problem-solving skills are biologically secondary. The general conditions under which these skills are learned are probably going to be different from the conditions under which early numerical skills are learned, although knowing the basic goals of the activity seems to facilitate learning in both biologically secondary and biologically primary domains (Siegler & Crowley, in press). In addition to knowing the goal of the activity, across both arithmetic and mathematical problem solving, there are two basic types of competencies: conceptual and procedural. The former might be represented by the child's understanding of the base-10 system. It was shown in the Subtraction section of chapter 2, for instance, that many of the common trading errors that are associated with American children's arithmetic appear to be related, at least in part, to a poor understanding of the base-10 system (see Figure 2.4). Procedural skills, on the other hand, are associated with the use of specific algorithms or equations for solving arithmetic or mathematics problems. Learning the rules for finding derivatives in calculus or using the counting-on method in simple addition are two examples of procedural competencies. Even though procedural and conceptual competencies might influence one another, from an educational perspective they should probably be considered distinct (Silver, 1987). As noted earlier, current educational philosophy focuses on conceptual knowledge and deems procedural skills unnecessary and even a detriment to children's mathematical development (e.g., Cobb et al., 1992).

Regardless of this philosophy, children need to understand mathematical concepts, *and* they need to know how and when to use mathematical procedures. These different competencies probably require different forms of instruction (G. Cooper & Sweller, 1987; Novick, 1992). Procedural learning requires extensive practice on a wide variety of problems on which the procedure might eventually be used. Practice should not, however, involve the use of the same procedure on the same type of problem for an extended period of time. Wenger (1987) argued that this form of practice resulted in the development of procedural bugs, that is, procedures that are correct for some problems but are incorrectly extended to other problems (see Table 3.1). Rather, practice should involve the use of a mixture of procedures that are practiced on a variety of problems. Practice should also occur in small doses (e.g., 20 min/day) and over an extended period of time. A procedure that is practiced on one or two work sheets for a day or two will quite likely be forgotten rather quickly (Bahrick, 1993). Basically, the procedure should be practiced until the child can automatically execute the procedure with the different types of problems that the procedure is normally used to solve.

I am recommending, of course, a modified form of drill and practice. Although the bane of many contemporary educational researchers, it is probably the only way to ensure long-term retention of basic procedures. The bottom line is, "If you want somebody to know something, you teach it to them" (Detterman, 1993, p. 15). If you want somebody to know something and retain it for a long period of time, then you have them practice it (Symonds & Chase, 1929)! The practice of basic procedures, especially when the practice is mixed with other types of procedures, should also provide the child with an opportunity to come to understand how the procedure works. Moreover, once procedures are automatized, then they require little conscious effort to use, which, in turn, frees attentional and working-memory resources for use on other, more important features of the problem (Geary & Widaman, 1992; Silver, 1987).

A deep conceptual understanding of a mathematical problem or domain probably also requires a lot of experience but does not appear to require drill and practice per se (G. Cooper & Sweller, 1987). Conceptual knowledge reflects the child's understanding of the basic principles of the domain and allows the child to see similarities across problems that have different superficial features (e.g., Morales et al., 1985; Perry, 1991). A child might demonstrate a good conceptual understanding of counting, for instance, when he or she knows that counting can occur from left to right, from right to left, or haphazardly and, as long as all of the items are counted, still yield the same answer. One way that appears to be useful for promoting conceptual knowledge is to ask students to come up with as many ways as possible to solve a particular problem or class of problems (Sweller et al., 1983). For this example, instruction might involve having the children count in as many different ways as they can (e.g., left to right and right to left).

In fact, this is a common approach used in elementary school classrooms in Japan (Stevenson & Stigler, 1992). Here, mathematics lessons often begin with a practical problem or a word problem. The teacher guides the student's attention to important aspects of the problem and asks for suggestions as to how the problem might be solved. The merits of different solution strategies, suggested by different children, are considered in turn. Problem-solving errors are also given much attention by the teacher, because it is assumed that such errors provide important insights into the child's conceptual misunderstandings. First, it is determined why the child made the error, and then other children are asked to provide alternative ways of solving the same problem. The lesson ends with the teacher providing an overview and summary of important concepts that were introduced that day.

In some respects, this approach is similar to the constructivist philosophy of teaching, but it differs in other respects. First, even though the teacher guides the children in their exploration of the problem, the goal is for each child to develop the same basic conceptual understanding of the problem. The constructivist approach stresses that each child will develop his or her own unique understanding of mathematics and that even subtle attempts to "impose" the teacher's view of the problem should be avoided. Japanese teachers also present the lesson as part of a problem-solving task and try to develop techniques to make the task interesting and engaging for the children (Stevenson & Stigler, 1992). This is an important goal, but it is very different from the constructivist assumption that children will inherently be able to construct mathematical knowledge for themselves (Cobb et al., 1992). Because much of mathematics is biologically secondary, it is therefore not likely to be inherently interesting for many children. Nevertheless, mathematical topics can be presented in such a way as to make them relatively more interesting for children to learn. It does not follow, however, that mathematics learning will occur in much the same way as the learning of biologically primary skills (e.g., language acquisition).

Other important instructional features that might facilitate children's development of conceptual knowledge include presenting problems in familiar contexts, those that the child can relate to personal experiences (Perry et al., 1993), and teaching mathematics teachers more about how children understand and solve mathematics problems (Carpenter, Fennema, Peterson, Chiang, & Loef, 1989; Fennema, Franke, Carpenter, & Carey, 1993). Carpenter, Fennema, and their colleagues have shown that the mathematics lessons of American teachers become much more similar to the Japanese lessons described above, after the teachers have been taught about children's mathematical development. These American teachers are much more likely to ask questions that require the child to develop a conceptual understanding of the problem and appear to be much more sensitive to children's conceptual misunderstandings. More-

Exhibit 8.1

Suggestions for Improving Mathematics Instruction

Stating Goals

The goal or end point of problem solving should be explicitly stated when the topic is first introduced.

The stated goals should be immediate, "The goal for this type of problem is to find the answer for *X*," as well as long term, "This type of problem solving is used in many different types of jobs, including . . . ."

Teaching Procedures

Mastery of mathematical procedures requires extensive (sometimes boring) practice.

Practice should include the following:

Small doses (e.g., 20 min) over an extended period of time.

Practice on a variety of problem types mixed together. Do not practice on only a single problem type—this leads to procedural bugs.

Practice until the procedure is executed automatically, that is, until the child can use it without having to think about it.

Once automaticity is reached, include some additional practice of the procedure as part of review segments for more complex material. This facilitates the long-term retention of the procedure.

Teaching Concepts

When possible, present the material (e.g., word problems) in contexts that are meaningful to the child.

Make one goal to solve the problem in as many different ways as possible rather than simply teaching a problem-solving algorithm. This goal can be achieved as a feature of class discussion.

Discuss problem-solving errors. Use errors as a diagnostic for conceptual misunderstandings and an opportunity to clarify the misconception.

over, teachers who understand children's mathematical development are better able to present lessons that clear up these conceptual errors.

Summary

Educational practices in the United States are often driven by philosophical beliefs rather than empirical research on children's learning. The basing of educational practices on philosophical approaches often results in an extreme reliance on one form of instruction or another. At times, mathematics education focuses on the basics, that is, the acquisition of mechanical or procedural skills. At other times, the focus is on children's understanding of concepts, with a complete de-emphasis on procedural skills. The current constructivist approach to mathematics education represents one example in this trend and focuses almost exclusively on the acquisition of conceptual knowledge. These swings from emphasizing one form of mathematics instruction at the expense of the other seem to be based, in part, on a failure to realize that mathematics requires the acquisition of *both* procedural skills and conceptual knowledge. Extreme approaches fail, because they facilitate the learning of one class of competency at the expense of the other.

Psychological research suggests that different teaching techniques are needed for children to acquire procedural and conceptual competencies. Thus, emphasizing the learning of, for example, conceptual knowledge will not facilitate the learning of procedural skills. The research on procedural and conceptual learning also has more practical implications. I have translated the implications of this research into instructional suggestions that can be used in classroom settings and for curriculum development (see Exhibit 8.1). Finally, as described in Exhibit 8.1, it is important to note that explicitly stating the goals of the instructional activity seems to facilitate the acquisition of both procedural skills and conceptual knowledge.

Remediation

Aside from the instructional and philosophical issues described earlier, the two most common blocks to mathematical development are mathematics anxiety and mathematical disabilities. Even though both are associated with relatively poor mathematical performance, mathematics anxiety and mathematical disabilities are two distinct phenomena. Potential remedial techniques for both of these disorders are presented in this section.

Mathematics Anxiety

In this section, basic research on the mathematics anxiety construct will be presented. The basic phenomenon is described first, followed by a discussion of how mathematics anxiety appears to affect mathematical development and performance. The section closes with a consideration of the effectiveness of different types of treatments for mathematics anxiety.

Phenomenon

Mathematics anxiety is a state of fear or apprehension that is associated with many mathematical endeavors, including test taking, course taking, homework, and so on (Hembree, 1990; Ramirez & Dockweiler, 1987; Tobias, 1978). Even though mathematics anxiety is correlated with test anxiety and trait anxiety, it appears to be a separate phenomenon (Hembree, 1990). Stated differently, many individuals show acute anxiety about mathematical activities, above and beyond any more general tendency to be anxious. Mathematics anxiety is slightly to moderately more intense, on average, in females than in males (by about ⅙ of a standard deviation), is first evident during the elementary school years, and is most extreme (i.e., mathematics anxiety test scores are highest) in the 9th and 10th grades (Hembree, 1990; Hyde, Fennema, Ryan, et al., 1990). Lower levels of mathematical ability are associated with higher levels of mathematics anxiety, but mathematics anxiety does not appear to be strongly related to general intelligence (Hembree, 1990; Hyde, Fennema, Ryan, et al., 1990; Sewell, Farley, & Sewell, 1983). Finally, Hispanic and Native Americans appear to show the highest levels of mathematics anxiety, on average, in comparison with other ethnic groups within the United States, whereas there appears to be no difference in the tendency to show mathematics anxiety among Anglo-Americans and African Americans (Hembree, 1990; Ramirez & Dockweiler, 1987).

Effects on Performance

Mathematics anxiety appears to influence mathematical development and performance in two ways. First, individuals with high levels of mathematics anxiety do not take many mathematics courses in high school, do not prepare adequately for examinations when they do take a mathematics course, and tend not to aspire to mathematics-intensive careers (Chipman et al., 1992; Hunsley, 1987; Tobias, 1978). The general avoidance of mathematical activities will, of course, result in the poor development of mathematical skills.

Second, mathematics anxiety appears to directly influence mathematical performance. On the basis of an extensive meta-analysis, Hembree (1990) estimated that high levels of mathematics anxiety resulted in about a 7% reduction (about ½ of a standard deviation) in mathematical test scores. Mathematics anxiety, and test anxiety in general, appears to increase levels of emotionality and worry, both of which negatively influence performance (Arkin, Detchon, & Maruyama, 1982; Ellis, Varner, & Becker, 1993; Sarason, 1984). Increases in the level of emotionality are associated with increases in the level of physiological arousal, which, in turn, tends to make it more difficult to attend to the task at hand (Ellis et al., 1993). For instance, high levels of physiological arousal will probably result in frequent automatic shifts in attention due to extraneous noises or other distractions in the immediate environment. Because high levels of arousal will be uncomfortable for many people, this might also motivate the individual to complete the task as quickly as possible, without regard for accuracy. Indeed, Ashcraft and Faust (1994) found that individuals with high levels of mathematics anxiety solved simple and complex arithmetic problems more quickly and made many more errors than individuals with more moderate levels of anxiety. The problem-solving strategies of the highly anxious individuals were also more rigid and algorithmic than the strategies used by individuals with low levels of mathematics anxiety.

Worry involves the cognitive component of mathematics and test anxiety. Arkin et al. (1982) found that individuals with high levels of test anxiety experienced intrusive thoughts, or irrelevant internal dialogue, much more frequently than less anxious individuals. The intrusive thoughts were typically negative attributions about their test performance. For instance, highly test-anxious individuals reported the following types of thoughts during a problem-solving task: "I thought about how poorly I was doing"; "I thought about how often I got confused" (Arkin et al., 1982, p. 1114). These types of attributions might result in the individual's not attempting more difficult problems (Dweck, 1975) and interfere with the solving of problems that are attempted. Intrusive thoughts will pull attentional and working-memory resources away from the task at hand, which will almost certainly result in poorer performance (Ellis et al., 1993). Poor performance, in turn, will quite likely contribute to the individual's concern over her or his future ability to succeed in mathematics, especially if the individual attributes the poor performance to a lack of ability rather than, for example, a lack of adequate preparation (Dweck, 1975; Randhawa, Beamer, & Lundberg, 1993).

Treatment

Hembree (1990) found that reductions in mathematics anxiety could result in significant (about $\frac{1}{2}$ of a standard deviation) improvements in mathematical test scores and in grade point average in mathematics courses.

However, not all treatments are equally effective for reducing mathematics anxiety or for improving mathematical performance.

Traditional individual- or group-counseling techniques appear to be relatively ineffective in reducing mathematics anxiety or improving mathematical performance (Hembree, 1990). Similarly, changes in classroom mathematics curriculum, such as providing calculators or microcomputers to aid in problem solving, appear to be largely ineffective in reducing mathematics anxiety in most individuals (Hembree, 1990). The one exception appears to be curricular changes that increase the student's mathematical competence. Hutton and Levitt (1987) improved feelings of competence, or self-efficacy, by focusing on the relationship between mathematical performance and good study habits and by improving basic skills. These goals were achieved, in the context of an algebra class, through the use of an especially designed textbook. For each algebraic topic, the textbook presented a review of the basic arithmetic skills needed to solve the associated algebra problems. These basic skills were then practiced. Lectures and the text material were also synchronized so that the basic foundation of each lecture was presented as "skeletal notes" in the textbook. This feature was designed to improve the student's note taking and to focus the student on essential features of the lecture. The overall intervention resulted in significant reductions in mathematics anxiety, as well as in improved algebraic skills (Hutton & Levitt, 1987).

Other treatment methods have focused on reducing the emotionality or worry aspect of mathematics anxiety. Treatments that specifically seek to reduce emotionality or arousal, such as relaxation therapy, are associated with moderate (about ½ of a standard deviation) reductions in mathematics anxiety but do not appear to be particularly effective in improving mathematical performance (Ellis et al., 1993; Hembree, 1990). However, relaxation as a component of systematic desensitization does appear to be effective in reducing mathematics anxiety (by about 1 standard deviation) and improving mathematical performance (by about ½ of a standard deviation). Systematic desensitization involves associating a state of relaxation with an anxiety-provoking context, such as taking a mathematics test (Wolpe, 1958). With this procedure, the client first develops an anxiety hierarchy in which fear-producing activities are arranged from least to most anxiety provoking. The client then imagines the least anxiety-provoking situation, such as signing up for a mathematics course, while in a state of deep relaxation. Relaxation is then conditioned while imagining the next activity in the hierarchy. This process continues until the client can imagine the most anxiety-producing situation in the hierarchy, such as taking a mathematics final, while in a state of relaxation.

Cognitive therapies focus on the worry component of mathematics anxiety (Ellis et al., 1993) and are associated with moderate declines (about $\frac{1}{2}$ of a standard deviation) in mathematics anxiety, as well as modest (about $\frac{1}{3}$ of a standard deviation) improvements in mathematical performance (Hembree, 1990). These therapies focus on reducing the frequency of intrusive thoughts during mathematical activities and on changing the individual's attributions about his or her performance. Poor performance that is attributed to a lack of ability will often result in an avoidance of, and a lack of persistence on, difficult mathematical tasks. Changing attributions so that they focus on more controllable factors, such as preparation and hard work, often results in more persistent task-related behaviors and improvements in performance (Dweck, 1975). Treatments that combine cognitive approaches with systematic desensitization appear to be somewhat more effective in reducing mathematics anxiety than the use of systematic desensitization alone (Hembree, 1990).

Finally, Ellis et al. (1993) argued that the treatment of mathematics and test anxiety should include building the student's basic competencies, knowledge, and skills. Increasing the competencies of students appears to reduce both the emotionality and worry components of mathematics anxiety, in addition to being an important goal in and of itself (e.g., Randhawa et al., 1993). From this perspective, to reduce mathematics anxiety, the individual's level of mathematical competence should first be assessed. Next, the individual practices solving problems that are at about this level. General success at this level should improve the individual's feelings of competency in mathematics (Dweck, 1975), reduce the intensity of the individual's mathematics anxiety, and reduce the deficits that are likely to be interfering with performance on more difficult tasks (Hutton & Levitt, 1987). The level of difficulty of the problems should then be increased. Gradually increasing the level of difficulty of the problems, building on recently reinforced skills (K. R. Johnson & Layng, 1992), is in a sense an in vivo type of systematic desensitization. As these component skills improve, success on more difficult problems should become more frequent, and as a result these problems should become less anxiety provoking.

Mathematical Disabilities

In this section, potential remedial education strategies for the various forms of mathematical disability (MD) that were described in chapter 5 are presented. These forms of MD include semantic memory, procedural, and visuospatial, as well as difficulties with mathematical problem solving.

Semantic Memory

The semantic-memory form of MD is associated with, among other things, a low frequency of arithmetic-fact retrieval and a high rate of errors when facts are retrieved from long-term memory (see Exhibit 5.1). This disorder tends to persist from one grade to the next, even with remedial education, and is often associated with reading difficulties. It is very likely that this retrieval problem is heritable and represents a neuropsychological, as well as cognitive, disorder (Geary, 1993). In this section, I overview the normal route through which arithmetic facts are memorized, discuss the effectiveness of remedial education procedures for teaching basic facts to children with this type of disorder, and then suggest an alternative remedial technique.

As described in chapter 2 (see the Strategy-Choice Model section), the learning of basic arithmetic facts appears to occur automatically for most children, as they use counting and other types of procedures to solve arithmetic problems (Siegler, 1986). For example, after solving 3 + 2 by counting "3, 4, 5," most children, after many such counts, will begin to remember that the answer is 5. Many MD children, however, do not automatically develop memory representations for many basic arithmetic facts, even after years of using counting or other types of strategies to solve arithmetic problems (Geary et al., 1987). This pattern suggests that the normal processes that underlie the memorization of arithmetic facts are somehow disrupted in many MD children (see Geary, 1993). It appears that many of these children will have lifelong difficulties in this area without remedial treatment (Dockrell & McShane, 1993).

Howell et al. (1987) examined the usefulness of several types of remedial interventions for improving the arithmetic-fact recall of an MD adolescent. The first type of intervention involved the use of a software program designed to provide computer-based drill and practice of basic multiplication facts. A second, more sophisticated, software program, which was designed specifically to aid the individual in associating multiplication problems with the correct answer, was also used in a follow-up study. Initially, with both software programs, drill and practice were associated with a significant reduction in errors and faster solution times. However, once the drill and practice stopped, error rates and solution times increased to almost preintervention levels. During these interventions, the authors did not systematically record the types of strategies the adolescent was using to solve the multiplication problems. Thus, it is not clear whether the initial improvements in performance were due to the better use of multiplication procedures, such as repeated addition (see Table 2.3), or whether some facts were memorized but then quickly forgotten. Either way, the repeated solving of multiplication problems did not lead to the learning of basic multiplication facts.

The second intervention involved the use of the more sophisticated drill-and-practice software combined with individualized instruction from the teacher (Howell et al., 1987). The teacher instructed the adolescent on how to solve simple multiplication problems with the "rule of nines" procedure. This procedure was used because the student found remembering multiples of 9 particularly difficult. "The Rule of Nines is an algorithmic approach to problem solution in multiplication in which the product of nine with any single digit number (n) is such that the tens digit is one less than (n) and the sum of the digits is nine" (Howell et al., 1987, p. 338). For instance, for 9×6, the tens value is 50, because $6 - 1 = 5$, and the ones value is 4, because $9 - 5 = 4$. Once this procedure was taught and practiced, long-term improvements in the adolescent's multiplication performance were found. Specifically, he committed fewer errors and solved problems with 9 as a multiple much faster.

In all, this set of studies suggests that the use of standard remedial education techniques will not likely facilitate the learning of arithmetic facts for MD children with a semantic-memory deficit (Howell et al., 1987). This research also suggests that remedial approaches with these children should probably focus on teaching the child to use efficient procedures for performing basic calculations. Once the procedure is understood, then the procedure should be practiced to reduce error rates and solution times.

An intriguing technique that was used for the remediation of reading disabilities (RD) suggests another possibility for teaching MD children basic arithmetic facts (Rozin, Poritsky, & Sotsky, 1971). The RD children in this study did not know all of the alphabet-sound correspondences in English and therefore had great difficulty reading. Nevertheless, with about 4 hr of instruction, Rozin et al. taught these children to read 30 words by using Chinese symbols. Chinese symbols map directly onto words (e.g., there is one symbol for *cat*). In contrast, the basic unit for reading in English is the phoneme, that is, the basic language sound, such as "*ka*." Because RD is often associated with difficulties in memory for phonemes (see the section of chapter 5 entitled, Relationship Between RD and MD), the use of Chinese symbols to read might have bypassed the children's difficulty in associating phonetic sounds with letters.

Many MD children appear to have the same basic memory deficit as the RD children in the Rozin et al. (1971) study. Thus, it is very possible that teaching MD children a method for performing basic calculations that bypasses the apparently defective semantic-memory system (which appears to underlie phonetic memory and arithmetic-fact retrieval) might lead to long-term improvements in their basic ability to perform calculations. One such potential method is the abacus. The abacus is a Chinese instrument that consists of columns of beads. Place value is represented by the columns. For instance, ones values might be represented by the leftmost column and tens values by the adjacent column. There are five beads in each column. The bottom four beads represent unit values and are physically separated from the top bead, which represents a fives value. Calculations are performed by physically moving the beads to different positions (for an excellent description, see Stigler, 1984).

In a series of well-done studies, Stigler (1984) demonstrated that the processes underlying the use of a mental abacus to solve simple and complex addition problems differed from the processes typically used by Americans to solve the same problems. During calculations, the use of a mental abacus was analogous in many respects to the physical manipulation of an actual abacus. Individuals who are skilled in the use of the abacus apparently generate a mental image of the abacus and then mentally manipulate the beads to perform simple and complex arithmetical calculations (Hatano, Miyake, & Binks, 1977).

Many children with the semantic-memory form of MD appear to have normal visuospatial skills (Rourke & Finlayson, 1978). In theory then, the MD children with a semantic-memory deficit should be capable of learning to use the physical abacus and, with practice, a mental abacus. Stated differently, the cognitive, and probably neurological, systems that underlie arithmetic-fact retrieval and the use of a mental abacus appear to differ (Stigler, 1984). The systems that probably support the generation and use of a mental abacus appear to be intact for many children with difficulties with arithmetic-fact retrieval. It might therefore be possible to remediate this deficit by bypassing the semantic-memory system and teaching the child to rely on an image-based system, such as the abacus, to perform the same functions. Learning the abacus is a difficult and time-intensive undertaking (Stigler, 1984). However, because the basic procedural skills, such as trading, of many of these children appear to be normal, a tool such as the abacus might only be needed to learn basic facts (e.g., 5 + 8). Therefore, teaching these children to use the abacus for solving more complex problems might not be necessary.

Procedural

As noted in Exhibit 5.2, the procedural deficits of some MD children are characterized by the use of arithmetic procedures that are more commonly used by younger, academically normal children and by a lot of procedural errors. In some cases, the procedural errors appear to be related to a poor understanding of the associated concepts and in other cases to a lack of attention to the task at hand (Geary, 1993; Russell & Ginsburg, 1984). Either way, the procedural deficits of many MD children appear to be short-term (Geary et al., 1991). That is, even though many of these children appear to be delayed in the understanding and use of many arithmetical procedures, most of these children eventually catch up to their academically normal peers (Russell & Ginsburg, 1984). Because there does not appear to be a long-term cognitive or neuropsychological deficit underlying these delays in procedural use, no special remediation techniques are likely to be needed for most of these children.

Rather, any remediation attempts should be directed toward facilitating the development of the child's procedural skills. Although many children will discover efficient procedures on their own, many MD children do not make these discoveries as readily as their academically normal peers. Thus, these MD children should be directly instructed in the use of efficient arithmetic procedures and given practice in their use (Schloss, Smith, & Schloss, 1990). Any such remedial attempts should probably first focus on the child's conceptual understanding of the procedure (Dockrell & McShane, 1993). For instance, MD children who count from 1 to solve 3 + 2, rather than counting "3, 4, 5" appear to have a poor understanding of counting concepts (Geary, Bow-Thomas, & Yao, 1992). These children appear to believe that counting should start at 1. Instruction that focuses on the flexibility of counting, that is, for example, that counting can occur from left to right or from right to left, might lead to a more mature understanding of counting. A better conceptual understanding of counting should, in turn, facilitate the child's use of more sophisticated counting procedures to solve arithmetic problems. Moreover, a conceptual understanding of the procedure appears to reduce the frequency of procedural errors (Geary, Bow-Thomas, & Yao, 1992). Once the concept is understood, then the child should practice using the associated procedure to solve as many different types of problems as possible.

Figure 8.1

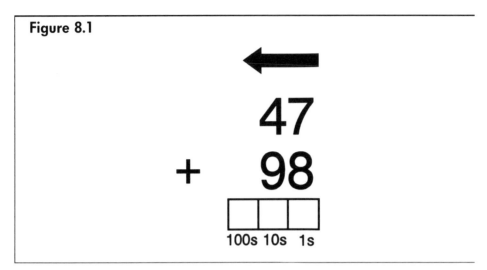

Proposed instructional aid for teaching multicolumn arithmetic to children with visuospatial deficits.

Visuospatial

The visuospatial form of MD involves difficulties in the spatial representation of numerical information and in some conceptual difficulties, such as understanding place value (see Exhibit 5.3). The developmental course of these deficits is not clear at this point. Therefore, it is not clear whether specific remediation techniques need to be developed for children with poor visuospatial skills. Nevertheless, Schloss et al. (1990) suggested that the use of concrete prompts that direct the flow of problem solving might be useful for children with this form of difficulty. For instance, with the solving of multicolumn addition problems, children with a visuospatial deficit often have problems putting the columnar information in the correct position. Perhaps structuring the written form of such problems, as shown in Figure 8.1, might reduce the frequency of columnar errors. The prompt at the top of the problem directs the sequence of columnar addition (Schloss et al., 1990); the labeled boxes explicitly note columnar values. With experience, the prompts might be removable (Schloss et al., 1990).

As with the abacus suggestion noted above, this proposed remedial technique is speculative but certainly testable. Finally, note that difficulties with columnar arithmetic are not the only type of problem that is associated with the visuospatial deficit. The point is that the spatial structuring and sequencing of problem solving might need to be explicitly presented for children with visuospatial deficits, at least for initial learning.

Mathematical Problem Solving

Arithmetical deficits, such as fact-retrieval problems, appear to be one factor that contributes to the poor mathematical problem-solving skills of many MD children and adolescents (Zentall & Ferkis, 1993). This is because arithmetical deficits render the solving of word problems slower and more error-prone. Nevertheless, as noted in chapter 5, many MD children have difficulty with the solving of mathematical word problems above and beyond any problems that are associated with lower level arithmetic deficits or poor reading skills (see the MD: Arithmetic and Mathematical Problem Solving section). More precisely, MD adolescents appear to find the problem-representation aspect of mathematical problem solving especially difficult (Hutchinson, 1993; Montague & Applegate, 1993; Zawaiza & Gerber, 1993).

Recall that problem representation involves both translating and integrating the problem statements into a representation that has a quantitative meaning (see the Problem-Solving Processes section of chapter 3). For instance, consider the following word problem from Exhibit 3.1: "Amy had two candies. Mary gave her three more candies. How many candies does Amy have now?" The translation of the first sentence in this problem might be achieved by representing Amy's candies with two blocks. A separate set of three blocks could be used to represent the number of Mary's candies. Next, the representations for Amy's candies and Mary's candies need to be integrated; that is, the child needs to understand their relationship as related to the goal of the problem. In this example, the child needs to understand that to solve the problem correctly, she must add Mary's group of blocks to Amy's group of blocks.

Mathematically disabled adolescents have difficulties with the translation and integration features of problem solving (Hutchinson, 1993). Remediation techniques for MD students who have difficulties with problem representation involve a combination of instruction on using both problem-solving scripts and diagrams, as well as improving basic computational skills (Hutchinson, 1993; Schloss et al., 1990; Zawaiza & Gerber, 1993). The usefulness of diagramming in mathematical problem solving was discussed in chapter 3 (see Figure 3.3). Basically, graphically representing the relationship between important features of the problem greatly reduces the memory demands of the problem and greatly reduces the frequency of problem-solving errors (A. B. Lewis, 1989). Many MD students apparently do not spontaneously use diagrams to aid in their problem solving and therefore need to be taught how and when to use diagrams.

Zawaiza and Gerber (1993) sought to improve the problem-solving skills of MD junior college students by using problem-solving scripts and by teaching them how to use diagrams. Scripts explicitly state the sequence of problem solving for the student; that is, they provide the metacognitive organization to problem solving that appears to occur automatically with skilled problem solvers. In other words, the use of scripts appears to provide the structure to problem solving that many MD children do not achieve automatically (Montague & Applegate, 1993). In this study, instruction was provided on the different types of word problems that the student might encounter, such as compare problems, and the basic goal associated with each problem type. Next, the instructor demonstrated the problem-solving sequence and highlighted important features of each problem-solving step. Finally, corrective feedback was provided while the students practiced specific skills, such as identifying different types of word problems. Zawaiza and Gerber (1993) found that the use of such scripts, combined with specific instruction on how to diagram the relational features of word problems, significantly reduced the frequency of errors associated with solving algebraic word problems for these MD junior college students. Similar results were reported by Hutchinson (1993) for a group of MD high school students.

In all, it appears that the mathematical problem-solving skills of MD children and adolescents can be improved, if the student is instructed on the sequence of steps in problem solving. This instruction should include a discussion of why the step is important and how it is related to the overall goal of the problem, combined with practicing the skills associated with each step. The use of diagrams facilitates the problem-solving performance of nondisabled students (A. B. Lewis, 1989) and also appears to be useful for improving the performance of MD students. However, MD students might need more explicit instruction and practice on how to diagram important features of word problems than do nondisabled students.

Summary

Psychological research suggests that the mathematical performance of children and adolescents with mathematics anxiety or specific mathematical disabilities can be improved, sometimes substantially. Mathematics anxiety involves heightened emotionality and worry, which are centered on many different types of mathematical activities. In combination, emotionality and worry directly interfere with mathematical problem solving,

Exhibit 8.2

Suggestions for Treating Mathematical Disabilities

Semantic Memory

Do not expect the child to be able to memorize basic arithmetic facts.

Teach alternative ways of performing basic calculations, such as the "rule of nines" procedure.

Have the child practice these procedures so that they are executed effortlessly and quickly.

Procedural

Initially, focus on the child's conceptual understanding of related areas (see Exhibit 8.1); for instance, the child should understand counting concepts before being taught the counting-on procedure.

Once the child understands the associated concepts, then teach the child to use the most efficient procedure(s).

Have the child practice using these procedures (see Exhibit 8.1).

Visuospatial

For spatially presented information, such as columnar arithmetic, provide prompts to guide the sequence of problem solving.

Provide labels for information whose meaning is determined by spatial position, as illustrated in Figure 8.1.

Mathematical Problem Solving

Assess and remediate any basic skill deficits.

Have the child or adolescent practice identifying different classes of word problems (see chapter 3).

For different classes of word problems:

State the goal for the problem type.

Provide explicit instruction on each problem-solving step.

Have the child or adolescent practice each step, providing corrective feedback as necessary.

Teach the child or adolescent how to diagram relational information.

---

reduce feelings of mathematical competence, and result in a general avoidance of mathematical endeavors. As noted in chapter 6, the avoidance of mathematical classes and activities can, in the long run, limit the individual's career options in adulthood (Wise, 1985). Fortunately, mathematics anxiety can be effectively treated with systematic desensitization in combination with techniques that are designed to improve basic mathematical skills.

Mathematics anxiety and mathematical disabilities are often confused. As noted in the introductory chapter, in one recent undergraduate textbook it was stated that "there is little evidence that children's problems in learning mathematics should be attributed to a specific form of cognitive dysfunction" (Krantz, 1994, p. 430). Rather, it was stated that mathematics anxiety was an important factor contributing to mathematical disabilities. In contrast, I have argued in this section and in chapter 5 that there are different forms of MD, each of which is distinct from mathematics anxiety. The understanding that mathematics anxiety and MD are different phenomena has important practical implications. In particular, the different forms of mathematical disability appear to require remediation techniques that differ from one another and differ from treatment approaches to mathematics anxiety. Remedial suggestions for the different forms of MD are summarized in Exhibit 8.2. Given the current state of knowledge in the MD area, these suggestions should be taken as only a starting point and not the final word on the most effective remedial approaches to these learning disorders.

Conclusion

Mathematics education in the United States is largely driven by fads and political ideology rather than by a coherent understanding of how children acquire mathematical skills, and there is no clear set of national standards for children's mathematical achievement (Stevenson & Stigler, 1992). Current educational approaches to mathematics emphasize problem solving and conceptual knowledge and de-emphasize procedural skills. This philosophy appears to stem, in part, from the belief that the acquisition of basic procedural skills involves rote learning, which, in turn, robs children of their interest in mathematics (Cobb et al., 1992). Moreover, it is believed that mathematical development should occur in a social context and that children should construct their own understanding of mathematical concepts. As noted in the Educational Philosophy section, it is assumed that with the development of appropriate social mathematical environments, "it is possible for students to construct for themselves the mathematical practices that, historically, took several thousand years to evolve" (Cobb et al., 1992, p. 28).

These beliefs are based, for the most part, on a naive view of children's cognitive development and a failure to recognize the difference between biologically primary and biologically secondary skills (Geary, in press). Many, though certainly not all, researchers in mathematics education are currently advocating instructional techniques that are based on the implicit assumption that complex mathematical skills are biologically primary. However, the conditions under which biologically primary skills, such as number and counting knowledge, are acquired are likely to be very different from the conditions under which more complex, biologically secondary skills, such as algebra, are acquired. Contrary to current educational philosophy, mathematics education must focus on both procedural skills and conceptual knowledge, not simply on problem-solving skills. Procedural skills appear to be best learned with drill and practice, whereas conceptual knowledge appears to be most readily acquired under conditions that have the child think of the many different ways in which the problem or class of problems can be solved (Sweller et al., 1983).

In addition to current educational approaches to mathematics, the two most common impediments to children's mathematical development are mathematics anxiety and mathematical disabilities. Mathematics anxiety represents a real fear or apprehension of mathematical activities and, though related, appears to be distinct from test anxiety (Hembree, 1990). Mathematics anxiety includes both heightened emotionality and arousal, as well as the cognitive component of worry during mathematical endeavors. High levels of mathematics anxiety directly influence mathematical performance by making the individual more easily distracted and rushed and by increasing the frequency with which negative thoughts intrude during mathematical activities (Ellis et al., 1993). In combination, these factors result in less attentional and working-memory resources that can be used for the task at hand. Moreover, high levels of mathematics anxiety increase the likelihood that the individual will avoid mathematical activities. At this point, it appears that treatment should include improving basic skills, as well as reducing the emotionality and worry components of mathematics anxiety. The latter appears to be achievable through the use of systematic desensitization and cognitive approaches designed to reduce the frequency of negative attributions (e.g., "I can't do this") during mathematical activities (Hembree, 1990).

Our understanding of mathematical disabilities is rather primitive at this point. Thus, it is not surprising that there are not many well-developed remediation techniques for these disorders (Mayer, 1993). The remediation suggestions offered earlier are based on our current understanding of the cognitive deficits, or delays, that appear to underlie the different forms of MD. The suggestions should not necessarily be taken as prescriptive, but rather as a personal best guess as to how the remediation of MD might be approached from a cognitive perspective. The basic assumption is that the remediation of what appear to be long-term deficits, such as the semantic-memory form of MD, might focus on alternative cognitive systems that can support the same functional skill. The suggestion that the abacus might be useful for facilitating arithmetic-fact retrieval in MD children is based directly on this assumption.

For other types of deficits that appear to be simply developmental delays or very poorly developed skills remediation techniques should focus directly on improving these skills rather than on bypassing a dysfunctional cognitive system. To achieve this goal, a very detailed analysis of the cognitive processes that are necessary to perform at normal levels is first needed (Mayer, 1993; Schloss et al., 1990). Next, fine-grained studies of the performance deficits of MD children are necessary. Once the basic processes that disrupt the performance of MD children have been identified, then these skills and the associated conceptual knowledge should be directly taught to the student. It appears that many of the discoveries that academically normal children make about mathematical concepts and procedures, without the aid of direct instruction, are not discovered as readily by MD students. Thus, explicit and detailed step-by-step instruction, with practice of component skills, seems to be necessary for most MD students. This approach was illustrated by the techniques developed by Zawaiza and Gerber (1993) for the remediation of difficulties with mathematical problem solving.

Psychological research has much to offer mathematics education (Penner, Batsche, Knoff, & Nelson, 1993). Cognitive and developmental researchers should continue to work to better understand the fundamentals of human numerical skills and the biological and environmental influences on their development (Fuson, 1988; Gallistel & Gelman, 1992), as well as to better understand more complex biologically secondary skills, such as mathematical problem solving (Briars & Larkin, 1984). Social, clinical, and educational psychologists can contribute by developing better ways to treat mathematics anxiety, for example, and by empirically testing the efficacy of different instructional techniques for improving children's mathematical learning. Each of these endeavors should be pursued with vigor, but at the same time, we should be more mindful of the great need in the United States for improving mathematics education. Psychology's greatest contribution can come from considering the ways in which our basic and applied research can be more directly related to the goal of improving the mathematical development of American children.

References

Adams, L. T., Kasserman, J. E., Yearwood, A. A., Perfetto, G. A., Bransford, J. D., & Franks, J. J. (1988). Memory access: The effects of fact-oriented versus problem-oriented acquisition. *Memory & Cognition, 16*, 167–175.

Aiken, L. R., Jr. (1976). Update on attitudes and other affective variables in learning mathematics. *Review of Educational Research, 46*, 293–311.

Albert, R. S., & Runco, M. A. (1986). The achievement of eminence: A model based on a longitudinal study of exceptionally gifted boys and their families. In R. J. Sternberg & J. E. Davidson (Eds.), *Conceptions of giftedness* (pp. 332–357). Cambridge, England: Cambridge University Press.

Al-Uqlidisi, A. (1978). *The arithmetic of Al-Uqlidisi* (S. A. Saidan, Trans.). Boston: Reidel. (Original work published 952–953)

Amato, P. R., & Keith, B. (1991). Parental divorce and the well-being of children: A meta-analysis. *Psychological Bulletin, 110*, 26–46.

Anderson, R. C. (1984). Some reflections on the acquisition of knowledge. *Educational Researcher, 13*, 5–10.

Antell, S. E., & Keating, D. P. (1983). Perception of numerical invariance in neonates. *Child Development, 54*, 695–701.

Arkin, R. M., Detchon, C. S., & Maruyama, G. M. (1982). Roles of attribution, affect, and cognitive interference in test anxiety. *Journal of Personality and Social Psychology, 43*, 1111–1124.

Armstrong, J. M. (1981). Achievement and participation of women in mathematics: Results from two national surveys. *Journal for Research in Mathematics Education, 12*, 356–372.

Arnold, P. A., & Gorski, R. A. (1984). Gonadal steroid induction of structural sex differences in the central nervous system. *Annual Review of Neuroscience, 7*, 413–442.

Ashcraft, M. H. (1982). The development of mental arithmetic: A chronometric approach. *Developmental Review, 2*, 213–236.

Ashcraft, M. H. (1987). Children's knowledge of simple arithmetic: A developmental model and simulation. In J. Bisanz, C. J. Brainerd, & R. Kail (Eds.), *Formal methods in developmental psychology: Progress in cognitive development research* (pp. 302–338). New York: Springer-Verlag.

Ashcraft, M. H. (1992). Cognitive arithmetic: A review of data and theory. *Cognition, 44*, 75–106.

Ashcraft, M. H., & Battaglia, J. (1978). Cognitive arithmetic: Evidence for retrieval and decision processes in mental addition. *Journal of Experimental Psychology: Human Learning and Memory, 4,* 527–538.

Ashcraft, M. H., & Christy, K. S. (in press). The frequency of arithmetic facts in elementary texts: Addition and multiplication in Grades 1–6. *Journal for Research in Mathematics Education.*

Ashcraft, M. H., & Faust, M. W. (1994). Mathematics anxiety and mental arithmetic: An exploratory investigation. *Cognition and Emotion, 8,* 97–125.

Ashcraft, M. H., & Fierman, B. A. (1982). Mental addition in third, fourth, and sixth graders. *Journal of Experimental Child Psychology, 33,* 216–234.

Ashcraft, M. H., Fierman, B. A., & Bartolotta, R. (1984). The production and verification tasks in mental addition: An empirical comparison. *Developmental Review, 4,* 157–170.

Ashcraft, M. H., & Stazyk, E. H. (1981). Mental addition: A test of three verification models. *Memory & Cognition, 9,* 185–196.

Ashcraft, M. H., Yamashita, T. S., & Aram, D. M. (1992). Mathematics performance in left and right brain-lesioned children. *Brain and Cognition, 19,* 208–252.

Baddeley, A. D. (1986). *Working memory.* London: Oxford University Press.

Badian, N. A. (1983). Dyscalculia and nonverbal disorders of learning. In H. R. Myklebust (Ed.), *Progress in learning disabilities* (Vol. 5, pp. 235–264). New York: Stratton.

Baenninger, M., & Newcombe, N. (1989). The role of experience in spatial test performance: A meta-analysis. *Sex Roles, 20,* 327–344.

Bahrick, H. P. (1993). Extending the life span of knowledge. In L. A. Penner, G. M. Batsche, H. M. Knoff, & D. L. Nelson (Eds.), *The challenge in mathematics and science education: Psychology's response* (pp. 61–82). Washington, DC: American Psychological Association.

Bahrick, H. P., & Hall, L. K. (1991). Lifetime maintenance of high school mathematics content. *Journal of Experimental Psychology: General, 120,* 22–33.

Baroody, A. J. (1984a). The case of Felicia: A young child's strategies for reducing memory demands during mental addition. *Cognition and Instruction, 1,* 109–116.

Baroody, A. J. (1984b). Children's difficulties in subtraction: Some causes and questions. *Journal for Research in Mathematics Education, 15,* 203–213.

Baroody, A. J. (1984c). A reexamination of mental arithmetic models and data: A reply to Ashcraft. *Developmental Review, 4,* 148–156.

Baroody, A. J. (1987a). The development of counting strategies for single-digit addition. *Journal for Research in Mathematics Education, 18,* 141–157.

Baroody, A. J. (1987b). Problem size and mentally retarded children's judgment of commutativity. *American Journal of Mental Deficiency, 91,* 439–442.

Baroody, A. J. (1989). Kindergartners' mental addition with single-digit combinations. *Journal for Research in Mathematics Education, 20,* 159–172.

Baroody, A. J. (1992). The development of kindergartners' mental-addition strategies. *Learning and Individual Differences, 4,* 215–235.

Baroody, A. J., & Gannon, K. E. (1984). The development of the commutativity principle and economical addition strategies. *Cognition and Instruction, 1,* 321–339.

Baroody, A. J., & Ginsburg, H. P. (1986). The relationship between initial meaningful and mechanical knowledge of arithmetic. In J. Hiebert (Ed.), *Conceptual and procedural knowledge: The case of mathematics* (pp. 75–112). Hillsdale, NJ: Erlbaum.

Baroody, A. J., Ginsburg, H. P., & Waxman, B. (1983). Children's use of mathematical structure. *Journal for Research in Mathematics Education, 14,* 156–168.

Becker, J. (1993). Young children's numerical use of number words: Counting in many-to-one situations. *Developmental Psychology, 29,* 458–465.

Beilin, H. (1968). Cognitive capacities of young children: A replication. *Science, 162*, 920–921.

Belmont, J. M. (1989). Cognitive strategies and strategic learning: The socio-instructional approach. *American Psychologist, 44*, 142–148.

Benbow, C. P. (1986). Physiological correlates of extreme intellectual precocity. *Neuropsychologia, 24*, 719–725.

Benbow, C. P. (1988). Sex differences in mathematical reasoning ability in intellectually talented preadolescents: Their nature, effects, and possible causes. *Behavioral and Brain Sciences, 11*, 169–232.

Benbow, C. P. (1992). Academic achievement in mathematics and science of students between ages 13 and 23: Are there differences among students in the top one percent of mathematical ability? *Journal of Educational Psychology, 84*, 51–61.

Benbow, C. P., & Stanley, J. C. (1980). Sex differences in mathematical ability: Fact or artifact? *Science, 210*, 1262–1264.

Benbow, C. P., & Stanley, J. C. (1982). Consequences in high school and college of sex differences in mathematical reasoning ability: A longitudinal perspective. *American Educational Research Journal, 19*, 598–622.

Benbow, C. P., & Stanley, J. C. (1983). Sex differences in mathematical reasoning ability: More facts. *Science, 222*, 1029–1031.

Benbow, C. P., Stanley, J. C., Kirk, M. K., & Zonderman, A. B. (1983). Structure of intelligence in intellectually precocious children and their parents. *Intelligence, 7*, 129–152.

Ben-Chaim, D., Lappan, G., & Houang, R. T. (1988). The effect of instruction on spatial visualization skills of middle school boys and girls. *American Educational Research Journal, 25*, 51–71.

Benson, D. F., & Weir, W. F. (1972). Acalculia: Acquired anarithmetria. *Cortex, 8*, 465–472.

Berenbaum, S. A., & Hines, M. (1992). Early androgens are related to childhood sex-typed toy preferences. *Psychological Science, 3*, 203- 206.

Bisanz, J., & LeFevre, J. (1990). Strategic and nonstrategic processing in the development of mathematical cognition. In D. F. Bjorklund (Ed.), *Children's strategies: Contemporary views of cognitive development* (pp. 213–244). Hillsdale, NJ: Erlbaum.

Bishop, J. H. (1989). Is the test score decline responsible for the productivity growth decline? *American Economic Review, 79*, 178–197.

Boissiere, M., Knight, J. B., & Sabot, R. H. (1985). Earnings, schooling, ability, and cognitive skills. *American Economic Review, 75*, 1016–1030.

Boller, F., & Grafman, J. (1983). Acalculia: Historical development and current significance. *Brain and Cognition, 2*, 205–223.

Borkowski, J. G. (1992). Metacognitive theory: A framework of teaching literacy, writing, and math skills. *Journal of Learning Disabilities, 25*, 253–257.

Bouchard, T. J., Jr., Lykken, D. T., McGue, M., Segal, N. L., & Tellegen, A. (1990). Sources of human psychological differences: The Minnesota study of twins reared apart. *Science, 250*, 223–228.

Boysen, S. T. (1993). Counting in chimpanzees: Nonhuman principles and emergent properties of number. In S. T. Boysen & E. J. Capaldi (Eds.), *The development of numerical competence: Animal and human models* (pp. 39–59). Hillsdale, NJ: Erlbaum.

Boysen, S. T., & Berntson, G. G. (1989). Numerical competence in a chimpanzee (*Pan troglodytes*). *Journal of Comparative Psychology, 103*, 23–31.

Boysen, S. T., & Capaldi, E. J. (Eds.). (1993). *The development of numerical competence: Animal and human models*. Hillsdale, NJ: Erlbaum.

Bracey, G. W. (1992, October). The second Bracey report on the condition of public education. *Phi Delta Kappan*, pp. 104–117.

Brainerd, C. J. (1979). *The origins of the number concept*. New York: Praeger.

Bransford, J., & the Cognition and Technology Group of Vanderbilt University. (1993). The Jasper series: Theoretical foundations and data on problem solving and transfer. In L. A. Penner, G. M. Batsche, H. M. Knoff, & D. L. Nelson (Eds.), *The challenge in mathematics and science education: Psychology's response* (pp. 113–152). Washington, DC: American Psychological Association.

Briars, D. J., & Larkin, J. H. (1984). An integrated model of skill in solving elementary word problems. *Cognition and Instruction*, *1*, 245–296.

Briars, D., & Siegler, R. S. (1984). A featural analysis of preschoolers' counting knowledge. *Developmental Psychology*, *20*, 607–618.

Bronfenbrenner, U. (1986). Ecology of the family as a context for human development: Research perspectives. *Developmental Psychology*, *22*, 723–742.

Brophy, J. (1986). Teaching and learning mathematics: Where research should be going. *Journal for Research in Mathematics Education*, *17*, 323–346.

Bruck, M. (1992). Persistence of dyslexics' phonological awareness deficits. *Developmental Psychology*, *28*, 874–886.

Bullock, M., & Gelman, R. (1977). Numerical reasoning in young children: The ordering principle. *Child Development*, *48*, 427–434.

Burnett, S. A., & Lane, D. M. (1980). Effects of academic instruction on spatial visualization. *Intelligence*, *4*, 233–242.

Burnett, S. A., Lane, D. M., & Dratt, L. M. (1979). Spatial visualization and sex differences in quantitative ability. *Intelligence*, *3*, 345–354.

Buss, D. M. (1989). Sex differences in human mate preferences: Evolutionary hypotheses tested in 37 cultures. *Behavioral and Brain Sciences*, *12*, 1–49.

Buss, D. M., & Schmitt, D. P. (1993). Sexual strategies theory: An evolutionary perspective on human mating. *Psychological Review*, *100*, 204–212.

Byrnes, J. P., & Takahira, S. (1993). Explaining gender differences on SAT–Math items. *Developmental Psychology*, *29*, 805–810.

Campbell, J. I. D. (1987). Network interference and mental multiplication. *Journal of Experimental Psychology: Learning, Memory, and Cognition*, *13*, 109–123.

Campbell, J. I. D., & Graham, D. J. (1985). Mental multiplication skill: Structure, process, and acquisition. *Canadian Journal of Psychology*, *39*, 338–366.

Canisia, M. (1962). Mathematical ability as related to reasoning and use of symbols. *Educational and Psychological Measurement*, *22*, 105–127.

Caplan, N., Choy, M. H., & Whitmore, J. K. (1992, February). Indochinese refugee families and academic achievement. *Scientific American*, pp. 36–42.

Carpenter, T. P., Fennema, E., Peterson, P. L., Chiang, C. P., & Loef, M. (1989). Using knowledge of children's mathematics thinking in classroom teaching: An experimental study. *American Educational Research Journal*, *26*, 499–531.

Carpenter, T. P., Hiebert, J., & Moser, J. M. (1981). Problem structure and first-grade children's initial solution processes for simple addition and subtraction problems. *Journal for Research in Mathematics Education*, *12*, 27–39.

Carpenter, T. P., Kepner, H., Corbitt, M. K., Lindquist, M. M., & Reys, R. E. (1980). Solving verbal problems: Results and implications from national assessment. *Arithmetic Teacher*, *28*, 10–12, 44–47.

Carpenter, T. P., & Moser, J. M. (1982). The development of addition and subtraction problem-solving skills. In T. P. Carpenter, J. M. Moser, & T. A. Romberg (Eds.), *Addition and subtraction: A cognitive perspective* (pp. 9–24). Hillsdale, NJ: Erlbaum.

Carpenter, T. P., & Moser, J. M. (1983). The acquisition of addition and subtraction concepts. In R. Lesh & M. Landau (Eds.), *Acquisition of mathematical concepts and processes* (pp. 7–44). New York: Academic Press.

Carpenter, T. P., & Moser, J. M. (1984). The acquisition of addition and subtraction concepts in grades one through three. *Journal for Research in Mathematics Education, 15,* 179–202.

Case, R. (1985). *Intellectual development: Birth to adulthood.* San Diego, CA: Academic Press.

Case, R. (1992). *The mind's staircase: Exploring the conceptual underpinnings of children's thought and knowledge.* Hillsdale, NJ: Erlbaum.

Case, R., & Sowder, J. T. (1990). The development of computational estimation: A neo-Piagetian analysis. *Cognition and Instruction, 7,* 79–104.

Cawley, J. F., & Miller, J. H. (1989). Cross-sectional comparisons of the mathematical performance of children with learning disabilities: Are we on the right track toward comprehensive programming? *Journal of Learning Disabilities, 22,* 250–254.

Ceci, S. J. (1991). How much does schooling influence general intelligence and its cognitive components? A reassessment of the evidence. *Developmental Psychology, 27,* 703–722.

Chein, I. (1939). Factors in mental organization. *Psychological Record, 3,* 71–94.

Chen, C., & Stevenson, H. W. (1993). *Motivation and mathematics achievement of Asian-American high school students.* Manuscript submitted for publication.

Chi, M. T. H., & Koeske, R. D. (1983). Network representation of a child's dinosaur knowledge. *Developmental Psychology, 19,* 29–39.

Chipman, S. F., Krantz, D. H., & Silver, R. (1992). Mathematics anxiety and science careers among able college women. *Psychological Science, 3,* 292–295.

Chipman, S. F., & Thomas, V. G. (1985). Women's participation in mathematics: Outlining the problem. In S. F. Chipman, L. R. Brush, & D. M. Wilson (Eds.), *Women and mathematics: Balancing the equation* (pp. 1–24). Hillsdale, NJ: Erlbaum.

Clement, J. (1982). Algebra word problem solutions: Thought processes underlying a common misconception. *Journal for Research in Mathematics Education, 13,* 16–30.

Clements, M. A., & Del Campo, G. (1990). How natural is fraction knowledge? In L. P. Steffe & T. Wood (Eds.), *Transforming children's mathematics education: International perspectives* (pp. 181–188). Hillsdale, NJ: Erlbaum.

Cobb, P., Yackel, E., & Wood, T. (1992). A constructivist alternative to the representational view of mind in mathematics education. *Journal for Research in Mathematics Education, 23,* 2–33.

Coombs, C. H. (1941). A factorial study of number ability. *Psychometrika, 6,* 161–189.

Coon, H., Carey, G., & Fulker, D. W. (1992). Community influences on cognitive ability. *Intelligence, 16,* 169–188.

Cooney, J. B., & Ladd, S. F. (1992). The influence of verbal protocol methods on children's mental computation. *Learning and Individual Differences, 4,* 237–257.

Cooney, J. B., Swanson, H. L., & Ladd, S. F. (1988). Acquisition of mental multiplication skill: Evidence for the transition between counting and retrieval strategies. *Cognition and Instruction, 5,* 323–345.

Cooper, G., & Sweller, J. (1987). Effects of schema acquisition and rule automation on mathematical problem-solving transfer. *Journal of Educational Psychology, 79,* 347–362.

Cooper, R. G., Jr. (1984). Early number development: Discovering number space with addition and subtraction. In C. Sophian (Ed.), *Origins of cognitive skills: The eighteenth annual Carnegie symposium on cognition* (pp. 157–192). Hillsdale, NJ: Erlbaum.

Crosswhite, F. J., Dossey, J. A., Swafford, J. O., McKnight, C. C., & Cooney, T. J. (1985).

Second International Mathematics Study summary report for the United States. Champaign, IL: Stipes.

Crump, T. (1990). *The anthropology of numbers*. New York: Cambridge University Press.

Crystal, D. S., & Stevenson, H. W. (1991). Mothers' perceptions of children's problems with mathematics: A cross-national comparison. *Journal of Educational Psychology, 83*, 372–376.

Dahmen, W., Hartje, W., Büssing, A., & Sturm, W. (1982). Disorders of calculation in aphasic patients—Spatial and verbal components. *Neuropsychologia, 20*, 145–153.

Dark, V. J., & Benbow, C. P. (1990). Enhanced problem translation and short-term memory: Components of mathematical talent. *Journal of Educational Psychology, 82*, 420–429.

Dark, V. J., & Benbow, C. P. (1991). Differential enhancement of working memory with mathematical versus verbal precocity. *Journal of Educational Psychology, 83*, 48–60.

Davis, H., & Memmott, J. (1982). Counting behavior in animals: A critical examination. *Psychological Bulletin, 92*, 547–571.

Dean, A. L., & Malik, M. M. (1986). Representing and solving arithmetic word problems: A study of developmental interaction. *Cognition and Instruction, 3*, 211–227.

De Corte, E., & Verschaffel, L. (1987). The effect of semantic structure on first graders' strategies for solving addition and subtraction word problems. *Journal for Research in Mathematics Education, 18*, 363–381.

De Corte, E., Verschaffel, L., & De Win, L. (1985). Influence of rewording verbal problems on children's problem representations and solutions. *Journal of Educational Psychology, 77*, 460–470.

DeFries, J. C., Ashton, G. C., Johnson, R. C., Kuse, A. R., McClearn, G. E., Mi, M. P., Rashad, M. N., Vandenberg, S. G., & Wilson, J. R. (1976). Parent-offspring resemblance for specific cognitive abilities in two ethnic groups. *Nature, 261*, 131–133.

DeFries, J. C., Johnson, R. C., Kuse, A. R., McClearn, G. E., Polovina, J., Vandenberg, S. G., & Wilson, J. R. (1979). Familial resemblance for specific cognitive abilities. *Behavior Genetics, 9*, 23–43.

Dehaene, S. (1992). Varieties of numerical abilities. *Cognition, 44*, 1–42.

Dehaene, S., & Cohen, L. (1991). Two mental calculation systems: A case study of severe acalculia with preserved approximation. *Neuropsychologia, 29*, 1045–1074.

Dehaene, S., & Mehler, J. (1992). Cross-linguistic regularities in the frequency of number words. *Cognition, 43*, 1–29.

Detterman, D. K. (1993). The case for the prosecution: Transfer as an epiphenomenon. In D. K. Detterman & R. J. Sternberg (Eds.), *Transfer on trial: Intelligence, cognition, and instruction* (pp. 1–24). Norwood, NJ: Ablex.

Detterman, D. K., Thompson, L. A., & Plomin, R. (1990). Differences in heritability across groups differing in ability. *Behavior Genetics, 20*, 369–384.

Diamond, M. C., Johnson, R. E., & Ehlert, J. (1979). A comparison of cortical thickness in male and female rats—normal and gonadectomized, young and adult. *Behavioral and Neural Biology, 26*, 485–491.

Dockrell, J., & McShane, J. (1993). *Children's learning difficulties: A cognitive approach*. Cambridge, MA: Basil Blackwell.

Dorans, N. J., & Livingston, S. A. (1987). Male-female differences in SAT-Verbal ability among students of high SAT-Mathematical ability. *Journal of Educational Measurement, 24*, 65–71.

Dunbar, R. I. M. (1993). Co-evolution of neocortex size, group size and language in humans. *Behavioral and Brain Sciences, 16*, 681–735.

Dweck, C. S. (1975). The role of expectations and attributions in the alleviation of learned helplessness. *Journal of Personality and Social Psychology, 31*, 674–685.

Dye, N. W., & Very, P. S. (1968). Growth changes in factorial structure by age and sex. *Genetic Psychology Monographs, 78*, 55–88.

Eccles (Parsons), J., Adler, T., & Meece, J. L. (1984). Sex differences in achievement: A test of alternative theories. *Journal of Personality and Social Psychology, 46*, 26–43.

Eccles, J., Wigfield, A., Harold, R. D., & Blumenfeld, P. (1993). Age and gender differences in children's self- and task perceptions during elementary school. *Child Development, 64*, 830–847.

Ehrhardt, A. A., & Meyer-Bahlburg, H. F. L. (1981). Effects of prenatal sex hormones on gender-related behavior. *Science, 211*, 1312–1318.

Eibl-Eibesfeldt, I. (1989). *Human ethology.* New York: Aldine de Gruyter.

Ekstrom, R. B., French, J. W., & Harman, H. H. (1976). *Manual for kit of factor-referenced cognitive tests 1976.* Princeton, NJ: Educational Testing Service.

Ekstrom, R. B., French, J. W., & Harman, H. H. (1979). Cognitive factors: Their identification and replication. *Multivariate Behavioral Research Monographs, 79*(2).

Ellis, H. C., Varner, L. J., & Becker, A. S. (1993). Cognition and emotion: Theories, implications, and educational applications. In L. A. Penner, G. M. Batsche, H. M. Knoff, & D. L. Nelson (Eds.), *The challenge of mathematics and science education: Psychology's response* (pp. 83–111). Washington, DC: American Psychological Association.

Feingold, A. (1988). Cognitive gender differences are disappearing. *American Psychologist, 43*, 95–103.

Fennema, E., Franke, M. L., Carpenter, T. P., & Carey, D. A. (1993). Using children's mathematical knowledge in instruction. *American Educational Research Journal, 30*, 555–583.

Fennema, E., & Sherman, J. (1977). Sex-related differences in mathematics achievement, spatial visualization and affective factors. *American Educational Research Journal, 14*, 51–71.

Fennema, E., & Tartre, L. A. (1985). The use of spatial visualization in mathematics by girls and boys. *Journal of Research in Mathematics Education, 16*, 184–206.

Fennema, E., Wolleat, P. L., Pedro, J. D., & Becker, A. D. (1981). Increasing women's participation in mathematics: An intervention study. *Journal for Research in Mathematics Education, 12*, 3–14.

Ferrini-Mundy, J. (1987). Spatial training for calculus students: Sex differences in achievement and in visualization ability. *Journal for Research in Mathematics Education, 18*, 126–140.

Fleischner, J. E., Garnett, K., & Shepherd, M. J. (1982). Proficiency in arithmetic basic fact computation of learning disabled and nondisabled children. *Focus on Learning Problems in Mathematics, 4*, 47–56.

Fletcher, J. M. (1992). The validity of distinguishing children with language and learning disabilities according to discrepancies with IQ: Introduction to the special series. *Journal of Learning Disabilities, 25*, 546–548.

Flynn, J. R. (1984). Japanese IQ. *Nature, 308*, 222.

Flynn, J. R. (1987). Massive IQ gains in 14 nations: What IQ tests really measure. *Psychological Bulletin, 101*, 171–191.

Frayer, D. W., & Wolpoff, M. H. (1985). Sexual dimorphism. *Annual Review of Anthropology, 14*, 429–473.

Freedman, D. G., & Freedman, N. A. (1969). Differences in behavior between Chinese-American and European-American newborns. *Nature, 224*, 1227.

French, J. W. (1951). The description of aptitude and achievement tests in terms of rotated factors. *Psychometric Monographs* (No. 5).

Fuson, K. C. (1982). An analysis of the counting-on solution procedure in addition. In T. P. Carpenter, J. M. Moser, & T. A. Romberg (Eds.), *Addition and subtraction: A cognitive perspective* (pp. 67–81). Hillsdale, NJ: Erlbaum.

Fuson, K. C. (1988). *Children's counting and concepts of number.* New York: Springer-Verlag.

Fuson, K. C. (1990). Conceptual structures for multiunit numbers: Implications for learning and teaching multidigit addition, subtraction, and place value. *Cognition and Instruction, 7,* 343–403.

Fuson, K. C. (1991). Children's early counting: Saying the number-word sequence, counting objects, and understanding cardinality. In K. Durkin & B. Shire (Eds.), *Language in mathematical education: Research and practice* (pp. 27–39). Milton Keynes, PA: Open University Press.

Fuson, K. C., & Briars, D. J. (1990). Using a base-ten blocks learning/teaching approach for first- and second-grade place-value and multidigit addition and subtraction. *Journal for Research in Mathematics Education, 21,* 180–206.

Fuson, K. C., & Kwon, Y. (1991). Chinese-based regular and European irregular systems of number words: The disadvantages for English-speaking children. In K. Durkin & B. Shire (Eds.), *Language in mathematical education: Research and practice* (pp. 211–226). Milton Keynes, PA: Open University Press.

Fuson, K. C., & Kwon, Y. (1992a). Korean children's single-digit addition and subtraction: Numbers structured by ten. *Journal for Research in Mathematics Education, 23,* 148–165.

Fuson, K. C., & Kwon, Y. (1992b). Korean children's understanding of multidigit addition and subtraction. *Child Development, 63,* 491–506.

Fuson, K. C., Pergament, G. G., Lyons, B. G., & Hall, J. W. (1985). Children's conformity to the cardinality rule as a function of set size and counting accuracy. *Child Development, 56,* 1429–1436.

Fuson, K. C., & Perry, T. (1993, March). *Hispanic children's addition methods: Cultural diversity in children's informal solution procedures.* Paper presented at the Biennial Meeting of the Society for Research in Child Development, New Orleans, LA.

Fuson, K. C., Richards, J., & Briars, D. J. (1982). The acquisition and elaboration of the number word sequence. In C. J. Brainerd (Ed.), *Children's logical and mathematical cognition: Progress in cognitive development research* (pp. 33–92). New York: Springer-Verlag.

Fuson, K. C., Stigler, J. W., & Bartsch, K. (1988). Grade placement of addition and subtraction topics in Japan, Mainland China, the Soviet Union, Taiwan, and the United States. *Journal for Research in Mathematics Education, 19,* 449–456.

Gallagher, J. J. (1993). An intersection of public policy and social science: Gifted students and education in mathematics and science. In L. A. Penner, G. M. Batsche, H. M. Knoff, & D. L. Nelson (Eds.), *The challenge in mathematics and science education: Psychology's response* (pp. 15–47). Washington, DC: American Psychological Association.

Gallistel, C. R., & Gelman, R. (1992). Preverbal and verbal counting and computation. *Cognition, 44,* 43–74.

Garnett, K., & Fleischner, J. E. (1983). Automatization and basic fact performance of normal and learning disabled children. *Learning Disabilities Quarterly, 6,* 223–230.

Geary, D. C. (1989). A model for representing gender differences in the pattern of cognitive abilities. *American Psychologist, 44,* 1155–1156.

Geary, D. C. (1990). A componential analysis of an early learning deficit in mathematics. *Journal of Experimental Child Psychology, 49,* 363–383.

Geary, D. C. (1992). Evolution of human cognition: Potential relationship to the ontogenetic development of behavior and cognition. *Evolution and Cognition, 1,* 93–100.

Geary, D. C. (1993). Mathematical disabilities: Cognitive, neuropsychological, and genetic components. *Psychological Bulletin, 114,* 345–362.

Geary, D. C. (in press). Reflections of evolution and culture in children's cognition: Implications for mathematical development and instruction. *American Psychologist.*

Geary, D. C., Bow-Thomas, C. C., Fan, L., Mueller, J., Turk, A., & Siegler, R. S. (1992, November). *Acquisition of arithmetic skills in Chinese and American children.* Paper presented at the 33rd Annual Meeting of the Psychonomic Society, St. Louis, MO.

Geary, D. C., Bow-Thomas, C. C., Fan, L., & Siegler, R. S. (1993). Even before formal instruction, Chinese children outperform American children in mental addition. *Cognitive Development, 8,* 517–529.

Geary, D. C., Bow-Thomas, C. C., & Yao, Y. (1992). Counting knowledge and skill in cognitive addition: A comparison of normal and mathematically disabled children. *Journal of Experimental Child Psychology, 54,* 372–391.

Geary, D. C., & Brown, S. C (1991). Cognitive addition: Strategy choice and speed-of-processing differences in gifted, normal, and mathematically disabled children. *Developmental Psychology, 27,* 398–406.

Geary, D. C., Brown, S. C, & Samaranayake, V. A. (1991). Cognitive addition: A short longitudinal study of strategy choice and speed-of-processing differences in normal and mathematically disabled children. *Developmental Psychology, 27,* 787–797.

Geary, D. C., & Burlingham-Dubree, M. (1989). External validation of the strategy choice model for addition. *Journal of Experimental Child Psychology, 47,* 175–192.

Geary, D. C., Fan, L., & Bow-Thomas, C. C. (1992). Numerical cognition: Loci of ability differences comparing children from China and the United States. *Psychological Science, 3,* 180–185.

Geary, D. C., Frensch, P. A., & Wiley, J. G. (1993). Simple and complex mental subtraction: Strategy choice and speed-of-processing differences in young and elderly adults. *Psychology and Aging, 8,* 242–256.

Geary, D. C., & Widaman, K. F. (1987). Individual differences in cognitive arithmetic. *Journal of Experimental Psychology: General, 116,* 154–171.

Geary, D. C., & Widaman, K. F. (1992). Numerical cognition: On the convergence of componential and psychometric models. *Intelligence, 16,* 47–80.

Geary, D. C., Widaman, K. F., & Little, T. D. (1986). Cognitive addition and multiplication: Evidence for a single memory network. *Memory & Cognition, 14,* 478–487.

Geary, D. C., Widaman, K. F., Little, T. D., & Cormier, P. (1987). Cognitive addition: Comparison of learning disabled and academically normal elementary school children. *Cognitive Development, 2,* 249–269.

Geary, D. C., & Wiley, J. G. (1991). Cognitive addition: Strategy choice and speed-of-processing differences in young and elderly adults. *Psychology and Aging, 6,* 474–483.

Gelman, R. (1978). Counting in the preschooler: What does and does not develop? In R. S. Siegler (Ed.), *Children's thinking: What develops?* (pp. 213–241). Hillsdale, NJ: Erlbaum.

Gelman, R. (1990). First principles organize attention to and learning about relevant data: Number and the animate-inanimate distinction as examples. *Cognitive Science, 14,* 79–106.

Gelman, R. (1992, November). *"A-half" and "one-have": Not the same meaning*. Paper presented at the 33rd Annual Meeting of the Psychonomic Society, St. Louis, MO.

Gelman, R., & Gallistel, C. R. (1978). *The child's understanding of number*. Cambridge, MA: Harvard University Press.

Gelman, R., & Meck, E. (1983). Preschooler's counting: Principles before skill. *Cognition*, *13*, 343–359.

Ghiglieri, M. P. (1987). Sociobiology of the great apes and the hominid ancestor. *Journal of Human Evolution*, *16*, 319–357.

Gilger, J. W. (1992). Genetics in disorders of language. *Clinics in Communication Disorders*, *2*, 35–47.

Gilger, J. W., Borecki, I. B., Smith, S. D., DeFries, J. C., & Pennington, B. F. (1993). *The etiology of extreme scores for complex phenotypes: An illustration using reading performance*. Manuscript submitted for publication.

Gilger, J. W., & Ho, H. Z. (1989). Gender differences in adult spatial information processing: Their relationship to pubertal timing, adolescent activities, and sex-typing of personality. *Cognitive Development*, *4*, 197–214.

Gilger, J. W., Pennington, B. F., & DeFries, J. C. (1992). A twin study of the etiology of comorbidity: Attention-deficit hyperactivity disorder and dyslexia. *Journal of the American Academy of Child & Adolescent Psychiatry*, *31*, 343–348.

Gillis, J. J., & DeFries, J. C. (1991). Confirmatory factor analysis of reading and mathematics performance measures in the Colorado Reading Project. *Behavior Genetics*, *21*, 572–573.

Ginsburg, H. P. (1982). The development of addition in the contexts of culture, social class, and race. In T. P. Carpenter, J. M. Moser, & T. A. Romberg (Eds.), *Addition and subtraction: A cognitive perspective* (pp. 191–210). Hillsdale, NJ: Erlbaum.

Ginsburg, H. P. (1989). *Children's arithmetic: How they learn it and how you teach it* (2nd ed.). Austin, TX: Pro Ed.

Ginsburg, H. P., Posner, J. K., & Russell, R. L. (1981a). The development of knowledge concerning written arithmetic: A cross-cultural study. *International Journal of Psychology*, *16*, 13–34.

Ginsburg, H. P., Posner, J. K., & Russell, R. L. (1981b). The development of mental addition as a function of schooling and culture. *Journal of Cross-Cultural Psychology*, *12*, 163–178.

Ginsburg, H. P., & Russell, R. L. (1981). Social class and racial influences on early mathematical thinking. *Monographs of the Society for Research in Child Development*, *46*(6, Serial No. 193).

Goldberg, L. R. (1993). The structure of phenotypic personality traits. *American Psychologist*, *48*, 26–34.

Goldman, S. R., Pellegrino, J. W., & Mertz, D. L. (1988). Extended practice of basic addition facts: Strategy changes in learning disabled students. *Cognition and Instruction*, *5*, 223–265.

Goodman, C. H. (1943). A factorial analysis of Thurstone's sixteen Primary Mental Ability tests. *Psychometrika*, *8*, 141–151.

Gould, S. J., & Vrba, E. S. (1982). Exaptation—A missing term in the science of form. *Paleobiology*, *8*, 4–15.

Grafman, J., Passafiume, D., Faglioni, P., & Boller, F. (1982). Calculation disturbance in adults with focal hemispheric damage. *Cortex*, *18*, 37–50.

Gray, E. M. (1991). An analysis of diverging approaches to simple arithmetic: Preferences and its consequences. *Educational Studies in Mathematics*, *22*, 551–574.

Greeno, J. G. (1989). A perspective on thinking. *American Psychologist*, *44*, 134–141.

Greeno, J. G. (1993). For research to reform education and cognitive science. In L. A. Penner, G. M. Batsche, H. M. Knoff, & D. L. Nelson (Eds.), *The challenge in mathematics and science education: Psychology's response* (pp. 153–192). Washington, DC: American Psychological Association.

Greeno, J. G., Riley, M. S., & Gelman, R. (1984). Conceptual competence and children's counting. *Cognitive Psychology, 16,* 94–143.

Greenough, W. T., Black, J. E., & Wallace, C. S. (1987). Experience and brain development. *Child Development, 58,* 539–559.

Groen, G. J., & Parkman, J. M. (1972). A chronometric analysis of simple addition. *Psychological Review, 79,* 329–343.

Groen, G., & Resnick, L. B. (1977). Can preschool children invent addition algorithms? *Journal of Educational Psychology, 69,* 645–652.

Grouws, D. A. (1972). Open sentences: Some instructional considerations from research. *Arithmetic Teacher, 19,* 595–599.

Grouws, D. A., & Cooney, T. J. (1988). *Effective mathematics teaching.* Hillsdale, NJ: Erlbaum and the National Council of Teachers of Mathematics.

Guthrie, G. M. (1963). Structure of abilities in a non-Western culture. *Journal of Educational Psychology, 54,* 94–103.

Halpern, D. F. (1992). *Sex differences in cognitive abilities.* (2nd ed.). Hillsdale, NJ: Erlbaum.

Hamann, M. S., & Ashcraft, M. H. (1985). Simple and complex mental addition across development. *Journal of Experimental Child Psychology, 40,* 49–72.

Hamann, M. S., & Ashcraft, M. H. (1986). Textbook presentations of the basic addition facts. *Cognition and Instruction, 3,* 173–192.

Hammill, D. D., Leigh, J. E., McNutt, G., & Larsen, S. C. (1987). A new definition of learning disabilities. *Journal of Learning Disabilities, 20,* 109–113.

Hardiman, P. T., & Mestre, J. P. (1989). Understanding multiplicative contexts involving fractions. *Journal of Educational Psychology, 81,* 547–557.

Harnisch, D. L., Steinkamp, M. W., Tsai, S. L., & Walberg, H. J. (1986). Cross-national differences in mathematics attitude and achievement among seventeen-year-olds. *International Journal of Educational Development, 6,* 233–244.

Hartje, W. (1987). The effect of spatial disorders on arithmetical skills. In G. Deloche & X. Seron (Eds.), *Mathematical disabilities: A cognitive neuropsychological perspective* (pp. 121–135). Hillsdale, NJ: Erlbaum.

Hatano, G. (1982). Learning to add and subtract: A Japanese perspective. In T. P. Carpenter, J. M. Moser, & T. A. Romberg (Eds.), *Addition and subtraction: A cognitive perspective* (pp. 211–223). Hillsdale, NJ: Erlbaum.

Hatano, G. (1990). Toward the cultural psychology of mathematical cognition. Commentary on Stevenson, H. W., Lee, S. Y., Chen, C., Stigler, J. W., Hsu, C. C., & Kitamura, S. Contexts of achievement: A study of American, Chinese, and Japanese children. *Monographs of the Society for Research in Child Development, 55*(1–2, Serial No. 221).

Hatano, G., Miyake, Y., & Binks, M. G. (1977). Performance of expert abacus operators. *Cognition, 5,* 47–55.

Hécaen, H. (1962). Clinical symptomatology in right and left hemispheric lesions. In V. B. Mountecastle (Ed.), *Interhemispheric relations and cerebral dominance* (pp. 215–243). Baltimore: Johns Hopkins Press.

Hegarty, M., Mayer, R. E., & Green, C. E. (1992). Comprehension of arithmetic word problems: Evidence from students' eye fixations. *Journal of Educational Psychology, 84,* 76–84.

Hembree, R. (1990). The nature, effect, and relief of mathematics anxiety. *Journal for Research in Mathematics Education*, *21*, 33–46.

Hinsley, D. A., Hayes, J. R., & Simon, H. A. (1977). From words to equations: Meaning and representation in algebra word problems. In M. A. Just & P. A. Carpenter (Eds.), *Cognitive processes in comprehension* (pp. 89–106). New York: Wiley.

Hitch, G. J. (1978). The role of short-term working memory in mental arithmetic. *Cognitive Psychology*, *10*, 302–323.

Hitch, G. J., & McAuley, E. (1991). Working memory in children with specific arithmetical learning disabilities. *British Journal of Psychology*, *82*, 375–386.

Howell, R., Sidorenko, E., & Jurica, J. (1987). The effects of computer use on the acquisition of multiplication facts by a student with learning disabilities. *Journal of Learning Disabilities*, *20*, 336–341.

Hudson, T. (1980). *Young children's difficulty with "How many more than are there?" questions.* Unpublished doctoral dissertation, Indiana University, Bloomington.

Hughes, M. (1986). *Children and number: Difficulties in learning mathematics.* New York: Basil Blackwell.

Hunsley, J. (1987). Cognitive processes in mathematics anxiety and test anxiety: The role of appraisals, internal dialogue, and attributions. *Journal of Educational Psychology*, *79*, 388–392.

Hunt, E. (1978). Mechanisms of verbal ability. *Psychological Review*, *85*, 109–130.

Hunt, E., Lunneborg, C., & Lewis, J. (1975). What does it mean to be high verbal? *Cognitive Psychology*, *7*, 194–227.

Husén, T. (1959). *Psychological twin research: A methodological study.* Uppsala, Sweden: Almquist & Wiksells.

Husén, T. (1967). *International study of achievement in mathematics: A comparison of twelve countries* (Vols. 1 & 2). New York: Wiley.

Hutchinson, N. L. (1993). Effects of cognitive strategy instruction on algebraic problem solving of adolescents with learning disabilities. *Learning Disability Quarterly*, *16*, 34–63.

Huttenlocher, J., & Strauss, S. (1968). Comprehension and a statement's relation to the situation it describes. *Journal of Verbal Learning and Verbal Behavior*, *7*, 300–304.

Hutton, L. A., & Levitt, E. (1987). An academic approach to the remediation of mathematics anxiety. In R. Schwarzer, H. M. van der Ploeg, & C. D. Spielberger (Eds.), *Advances in test anxiety research* (Vol. 5, pp. 207–211). Berwyn, PA: Swets North America.

Hyde, J. S., Fennema, E., & Lamon, S. J. (1990). Gender differences in mathematics performance: A meta-analysis. *Psychological Bulletin*, *107*, 139–155.

Hyde, J. S., Fennema, E., Ryan, M., Frost, L. A., & Hopp, C. (1990). Gender comparisons of mathematics attitudes and affect: A meta-analysis. *Psychology of Women Quarterly*, *14*, 299–324.

Hynd, G. W., & Semrud-Clikeman, M. (1989). Dyslexia and brain morphology. *Psychological Bulletin*, *106*, 447–482.

Ilg, F., & Ames, L. B. (1951). Developmental trends in arithmetic. *Journal of Genetic Psychology*, *79*, 3–28.

Irvine, J. J. (1986). Teacher-student interactions: Effects of student race, sex, and grade level. *Journal of Educational Psychology*, *78*, 14–21.

Jackson, M., & Warrington, E. K. (1986). Arithmetic skills in patients with unilateral cerebral lesions. *Cortex*, *22*, 611–620.

James, W. (1950). *The principles of psychology* (Vol. 1). New York: Dover Publications. (Original work published 1890)

Jastak, J. F., & Jastak, S. R. (1978). *Wide Range Achievement Test (WRAT)*. Wilmington, DE: Jastak Associates.

Jerman, M. E., & Mirman, S. (1974). Linguistic and computational variables in problem solving in elementary mathematics. *Educational Studies in Mathematics, 5*, 317–362.

Jerman, M., & Rees, R. (1972). Predicting the relative difficulty of verbal arithmetic problems. *Educational Studies in Mathematics, 4*, 306–323.

Johnson, D. J. (1988). Review of research on specific reading, writing, and mathematical disorders. In J. F. Kavanagh & T. J. Truss, Jr. (Eds.), *Learning disabilities: Proceedings of the national conference* (pp. 79–163). Parkton, MD: York Press.

Johnson, E. S. (1984). Sex differences in problem solving. *Journal of Educational Psychology, 76*, 1359–1371.

Johnson, E. S., & Meade, A. C. (1987). Developmental patterns of spatial ability: An early sex difference. *Child Development, 58*, 725–740.

Johnson, K. R., & Layng, T. V. J. (1992). Breaking the structuralist barrier: Literacy and numeracy with fluency. *American Psychologist, 47*, 1475–1490.

Jones, L. V. (1987). The influence on mathematics test scores, by ethnicity and sex, of prior achievement and high school mathematics courses. *Journal for Research in Mathematics Education, 18*, 180–186.

Jordan, N. C., Huttenlocher, J., & Levine, S. C. (1992). Differential calculation abilities in young children from middle- and low-income families. *Developmental Psychology, 28*, 644–653.

Jordan, N. C., Levine, S. C., & Huttenlocher, J. (in press). Calculation abilities in young children with different patterns of cognitive functioning. *Journal of Learning Disabilities*.

Kail, R. (1992). Processing speed, speech rate, and memory. *Developmental Psychology, 28*, 899–904.

Kaye, D. B. (1986). The development of mathematical cognition. *Cognitive Development, 1*, 157–170.

Keating, D. P., & Bobbitt, B. L. (1978). Individual and developmental differences in cognitive-processing components of mental ability. *Child Development, 49*, 155–167.

Keating, D. P., List, J. A., & Merriman, W. E. (1985). Cognitive processing and cognitive ability: A multivariate validity investigation. *Intelligence, 9*, 149–170.

Kimball, M. M. (1989). A new perspective on women's math achievement. *Psychological Bulletin, 105*, 198–214.

Kintsch, W., & Greeno, J. G. (1985). Understanding and solving arithmetic word problems. *Psychological Review, 92*, 109–129.

Kirby, J. R., & Becker, L. D. (1988). Cognitive components of learning problems in arithmetic. *Remedial and Special Education, 9*, 7–16.

Klahr, D., & Wallace, J. G. (1973). The role of quantification operators in the development of conservation of quantity. *Cognitive Psychology, 4*, 301–327.

Kloosterman, P. (1990). Attributions, performance following failure, and motivation in mathematics. In E. Fennema & G. C. Leder (Eds.), *Mathematics and gender* (pp. 96–127). New York: Teachers College Press.

Kosc, L. (1974). Developmental dyscalculia. *Journal of Learning Disabilities, 7*, 164–177.

Krantz, M. (1994). *Child development: Risk and opportunity*. Belmont, CA: Wadsworth.

Kreutzer, M. A., Leonard, S. C., & Flavell, J. H. (1975). An interview study of children's knowledge about memory. *Monographs of the Society for Research in Child Development, 40*(1, Serial No. 159).

Lampert, M. (1990). When the problem is not the question and the solution is not the

answer: Mathematical knowing and teaching. *American Educational Research Journal,*
27, 29–63.

Lapointe, A. E., Mead, N. A., & Askew, J. M. (1992). *Learning mathematics.* Princeton, NJ:
Educational Testing Service.

Larkin, J. H., McDermott, J., Simon, D. P., & Simon, H. A. (1980). Expert and novice
performance in solving physics problems. *Science, 208,* 1335–1342.

Law, D. J., Pellegrino, J. W., & Hunt, E. B. (1993). Comparing the tortoise and the hare:
Gender differences and experience in dynamic spatial reasoning tasks. *Psychological*
Science, 4, 35–40.

LeFevre, J. A., Bisanz, J., & Mrkonjic, L. (1988). Cognitive arithmetic: Evidence for
obligatory activation of arithmetic facts. *Memory & Cognition, 16,* 45–53.

Leinhardt, G., Seewald, A. M., & Engel, M. (1979). Learning what's taught: Sex differences
in instruction. *Journal of Educational Psychology, 71,* 432–439.

Lester, F. K., Jr. (1983). Trends and issues in mathematical problem-solving research. In
R. Lesh & M. Landau (Eds.), *Acquisition of mathematical concepts and processes* (pp.
229–261). New York: Academic Press.

Levine, M. D. (1987). *Developmental variation and learning disorders.* Cambridge, MA: Educators Publishing Service.

Levine, S. C., Jordan, N. C., & Huttenlocher, J. (1992). Development of calculation abilities
in young children. *Journal of Experimental Child Psychology, 53,* 72–103.

Lewis, A. B. (1989). Training students to represent arithmetic word problems. *Journal of*
Educational Psychology, 81, 521–531.

Lewis, A. B., & Mayer, R. E. (1987). Students' miscomprehension of relational statements
in arithmetic word problems. *Journal of Educational Psychology, 79,* 363–371.

Lewis, C. (1981). Skill in algebra. In J. R. Anderson (Ed.), *Cognitive skills and their acquisition*
(pp. 85–110). Hillsdale, NJ: Erlbaum.

Lindgren, S. D., Richman, L. C., & Eliason, M. J. (1986). Memory processes in reading
disability subtypes. *Developmental Neuropsychology, 2,* 173–181.

Linn, M. C., & Hyde, J. S. (1989). Gender, mathematics, and science. *Educational Researcher,*
18, 17–19, 22–27.

Linn, M. C., & Petersen, A. C. (1985). Emergence and characterization of sex differences
in spatial ability: A meta-analysis. *Child Development, 56,* 1479–1498.

Little, T. D., & Widaman, K. F. (in press). A production task evaluation of individual
differences in mental addition skill development: Internal and external validation
of chronometric models. *Journal of Experimental Child Psychology.*

Loehlin, J. C., & Nichols, R. C. (1976). *Heredity, environment, and personality: A study of 850*
sets of twins. Austin: University of Texas Press.

Lubinski, D., & Humphreys, L. G. (1990). A broadly based analysis of mathematical
giftedness. *Intelligence, 14,* 327–355.

Lubinski, D., & Humphreys, L. G. (1992). Some bodily and medical correlates of mathematical giftedness and commensurate levels of socioeconomic status. *Intelligence,*
16, 99–115.

Lummis, M., & Stevenson, H. W. (1990). Gender differences in beliefs and achievement:
A cross-cultural study. *Developmental Psychology, 26,* 254–263.

Luria, A. R. (1980). *Higher cortical functions in man* (2nd ed.). New York: Basic Books.

Lynn, R. (1982). IQ in Japan and the United States shows a growing disparity. *Nature,*
297, 222–223.

Lynn, R. (1983). Reply to Stevenson and Azuma. *Nature, 306,* 292.

Lytton, H., & Romney, D. M. (1991). Parents' differential socialization of boys and girls:
A meta-analysis. *Psychological Bulletin, 109,* 267–296.

Maccoby, E. E. (1988). Gender as a social category. *Developmental Psychology, 24*, 755–765.

Maccoby, E. E. (1990). Gender and relationships: A developmental account. *American Psychologist, 45*, 513–520.

Maccoby, E. E., & Jacklin, C. N. (1974). *The psychology of sex differences*. Stanford, CA: Stanford University Press.

MacCorquodale, P. (1988). Mexican-American women and mathematics: Participation, aspirations, and achievement. In R. R. Cocking & J. P. Mestre (Eds.), *Linguistic and cultural influences on learning mathematics* (pp. 137–160). Hillsdale, NJ: Erlbaum.

MacDonald, K. B. (1988). *Social and personality development: An evolutionary synthesis*. New York: Plenum.

MacDonald, K. (1992). Warmth as a developmental construct: An evolutionary analysis. *Child Development, 63*, 753–773.

Mandler, G., & Shebo, B. J. (1982). Subitizing: An analysis of its component processes. *Journal of Experimental Psychology: General, 111*, 1–22.

Markman, E. M. (1979). Classes and collections: Conceptual organization and numerical abilities. *Cognitive Psychology, 11*, 395–411.

Marsh, H. W., Smith, I. D., & Barnes, J. (1985). Multidimensional self-concepts: Relations with sex and academic achievement. *Journal of Educational Psychology, 77*, 581–596.

Marshall, S. P., & Smith, J. D. (1987). Sex differences in learning mathematics: A longitudinal study with item and error analyses. *Journal of Educational Psychology, 79*, 372–383.

Masters, M. S., & Sanders, B. (1993). Is the gender difference in mental rotation disappearing? *Behavior Genetics, 23*, 337–341.

Matsuzawa, T. (1985). Use of numbers by a chimpanzee. *Nature, 315*, 57–59.

Mayer, R. E. (1981). Frequency norms and structural analysis of algebra story problems into families, categories, and templates. *Instructional Science, 10*, 135–175.

Mayer, R. E. (1982). Memory for algebra story problems. *Journal of Educational Psychology, 74*, 199–216.

Mayer, R. E. (1985). Mathematical ability. In R. J. Sternberg (Ed.), *Human abilities: An information processing approach* (pp. 127–150). San Francisco: W. H. Freeman.

Mayer, R. E. (1993). Understanding individual differences in mathematical problem solving: Towards a research agenda. *Learning Disabilities Quarterly, 16*, 2–5.

Mayer, R. E., Lewis, A. B., & Hegarty, M. (1992). Mathematical misunderstandings: Qualitative reasoning about quantitative problems. In J. I. D. Campbell (Ed.), *The nature and origins of mathematical skills* (pp. 137–153). Amsterdam: North-Holland.

Mayeske, G. W., Okada, T., Beaton, A. E., Jr., Cohen, W. M., & Wisler, C. E. (1973). *A study of the achievement of our nation's students*. Washington, DC: U.S. Government Printing Office.

McCloskey, M. (1992). Cognitive mechanisms in numerical processing: Evidence from acquired dyscalculia. *Cognition, 44*, 107–157.

McCloskey, M., Aliminosa, D., & Sokol, S. M. (1991). Facts, rules, and procedures in normal calculation: Evidence from multiple single-patient studies of impaired arithmetic fact retrieval. *Brain and Cognition, 17*, 154–203.

McCloskey, M., Caramazza, A., & Basili, A. (1985). Cognitive mechanisms in number processing and calculation: Evidence from dyscalculia. *Brain and Cognition, 4*, 171–196.

McCloskey, M., Harley, W., & Sokol, S. M. (1991). Models of arithmetic fact retrieval: An evaluation in light of findings from normal and brain-damaged subjects. *Journal of Experimental Psychology: Learning, Memory, and Cognition, 17*, 377–397.

McGee, M. G. (1979). Human spatial abilities: Psychometric studies and environmental, genetic, hormonal, and neurological influences. *Psychological Bulletin, 86,* 889–918.

McGuinness, D. (1993). Sex differences in cognitive style: Implications for mathematics performance and achievement. In L. A. Penner, G. M. Batsche, H. M. Knoff, & D. L. Nelson (Eds.), *The challenge of mathematics and science education: Psychology's response* (pp. 251–274). Washington, DC: American Psychological Association.

Meck, W. H., & Church, R. M. (1983). A mode control model of counting and timing processes. *Journal of Experimental Psychology: Animal Behavior Processes, 9,* 320–334.

Meece, J. L., Parsons, J. E., Kaczala, C. M., Goff, S. B., & Futterman, R. (1982). Sex differences in math achievement: Toward a model of academic choice. *Psychological Bulletin, 91,* 324–348.

Meece, J. L., Wigfield, A., & Eccles, J. S. (1990). Predictors of math anxiety and its influence on young adolescents' course enrollment intentions and performance in mathematics. *Journal of Educational Psychology, 82,* 60–70.

Mehler, J., & Bever, T. G. (1967). Cognitive capacity of very young children. *Science, 158,* 141–142.

Meyers, C. E., & Dingman, H. F. (1960). The structure of abilities at the preschool ages: Hypothesized domains. *Psychological Bulletin, 57,* 514–532.

Miller, K. (1984). Child as the measurer of all things: Measurement procedures and the development of quantitative concepts. In C. Sophian (Ed.), *Origins of cognitive skills: The eighteenth annual Carnegie symposium on cognition* (pp. 193–228). Hillsdale, NJ: Erlbaum.

Miller, K. F. (1992, November). *Language and the origin of mathematical abilities: U.S./China comparisons.* Paper presented at the 33rd Annual Meeting of the Psychonomic Society, St. Louis, MO.

Miller, K., & Gelman, R. (1983). The child's representation of number: A multidimensional scaling analysis. *Child Development, 54,* 1470–1479.

Miller, K. F., & Paredes, D. R. (1990). Starting to add worse: Effects of learning to multiply on children's addition. *Cognition, 37,* 213–242.

Miller, K., Perlmutter, M., & Keating, D. (1984). Cognitive arithmetic: Comparison of operations. *Journal of Experimental Psychology: Learning, Memory, and Cognition, 10,* 46–60.

Miller, K. F., & Stigler, J. W. (1987). Counting in Chinese: Cultural variation in a basic cognitive skill. *Cognitive Development, 2,* 279–305.

Miller, M. D., & Linn, R. L. (1989). Cross-national achievement with differential retention rates. *Journal for Research in Mathematics Education, 20,* 28–40.

Miura, I. T. (1987). Mathematics achievement as a function of language. *Journal of Educational Psychology, 79,* 79–82.

Miura, I. T., Kim, C. C., Chang, C. M., & Okamoto, Y. (1988). Effects of language characteristics on children's cognitive representation of number: Cross-national comparisons. *Child Development, 59,* 1445–1450.

Miura, I. T., Okamoto, Y., Kim, C. C., Steere, M., & Fayol, M. (1993). First graders' cognitive representation of number and understanding of place value: Cross-national comparisons—France, Japan, Korea, Sweden, and the United States. *Journal of Educational Psychology, 85,* 24–30.

Montague, M., & Applegate, B. (1993). Middle school students' mathematical problem solving: An analysis of think-aloud protocols. *Learning Disability Quarterly, 16,* 19–32.

Moore, D., Benenson, J., Reznick, J. S., Peterson, M., & Kagan, J. (1987). Effect of auditory numerical information on infants' looking behavior: Contradictory evidence. *Developmental Psychology*, *23*, 665–670.

Moore, E. G., & Smith, A. W. (1987). Sex and ethnic group differences in mathematics achievement: Results from the national longitudinal study. *Journal for Research in Mathematics Education*, *18*, 25–36.

Morales, R. V., Shute, V. J., & Pellegrino, J. W. (1985). Developmental differences in understanding and solving simple mathematics word problems. *Cognition and Instruction*, *2*, 41–57.

Moyer, J. C., Sowder, L., Threadgill-Sowder, J., & Moyer, M. B. (1984). Story problem formats: Drawn versus verbal versus telegraphic. *Journal for Research in Mathematics Education*, *15*, 342–351.

Mullis, I. V. S., Dossey, J. A., Owen, E. H., & Phillips, G. W. (1991). *The state of mathematics achievement: NAEP's 1990 assessment of the nation and the trial assessment of the states*. Washington, DC: U.S. Department of Education.

Murray, J. E. (1949). An analysis of geometric ability. *Journal of Educational Psychology*, *40*, 118–124.

Muth, K. D. (1984). Solving arithmetic word problems: Role of reading and computational skills. *Journal of Educational Psychology*, *76*, 205–210.

Nevin, M. (1973). Sex differences in participation rates in mathematics and science at Irish schools and universities. *International Review of Education*, *19*, 88–91.

Nichols, R. C. (1978). Twin studies of ability, personality, and interests. *Homo*, *29*, 158–173.

Novick, L. R. (1992). The role of expertise in solving arithmetic and algebra word problems by analogy. In J. I. D. Campbell (Ed.), *The nature and origins of mathematical skills* (pp. 155–188). Amsterdam: North-Holland.

Novick, L. R., & Holyoak, K. J. (1991). Mathematical problem solving by analogy. *Journal of Experimental Psychology: Learning, Memory, and Cognition*, *17*, 398–415.

O'Boyle, M. W., Alexander, J. E., & Benbow, C. P. (1991). Enhanced right hemisphere activation in the mathematically precocious: A preliminary EEG investigation. *Brain and Cognition*, *17*, 138–153.

O'Boyle, M. W., & Benbow, C. P. (1990). Enhanced right hemisphere involvement during cognitive processing may relate to intellectual precocity. *Neuropsychologia*, *28*, 211–216.

O'Boyle, M. W., & Hellige, J. B. (1989). Cerebral hemispheric asymmetry and individual differences in cognition. *Learning and Individual Differences*, *1*, 7–35.

Ohlsson, S., & Rees, E. (1991). The function of conceptual understanding in the learning of arithmetic procedures. *Cognition and Instruction*, *8*, 103–179.

Okagaki, L. A., & Frensch, P. A. (1994). Effects of video game playing on measures of spatial performance: Gender effects in late adolescence. *Journal of Applied Developmental Psychology*, *15*, 33–58.

Olson, R. K., Gillis, J. J., Rack, J. P., DeFries, J. C., & Fulker, D. W. (1991). Confirmatory factor analysis of word recognition and process measures in the Colorado Reading Project. *Reading and Writing: An Interdisciplinary Journal*, *4*, 43–56.

Olson, R., Wise, B., Conners, F., Rack, J., & Fulker, D. (1989). Specific deficits in component reading and language skills: Genetic and environmental influences. *Journal of Learning Disabilities*, *22*, 339–348.

Osborne, R. T., & Lindsey, J. M. (1967). A longitudinal investigation of change in the factorial composition of intelligence with age in young school children. *Journal of Genetic Psychology*, *110*, 49–58.

Paredes, D. R., & Miller, K. F. (1993, November). *Notational disparity and numerical com-*

petence: U.S./Chinese comparisons. Paper presented at the 34th Annual Meeting of the Psychonomic Society, Washington, DC.

Pattison, P., & Grieve, N. (1984). Do spatial skills contribute to sex differences in different types of mathematical problems? *Journal of Educational Psychology, 76*, 678–689.

Pawlik, K. (1966). Concepts in human cognition and aptitudes. In R. B. Cattell (Ed.), *Handbook of multivariate experimental psychology* (pp. 535–562). Chicago: Rand McNally.

Pellegrino, J. W., & Goldman, S. R. (1987). Information processing and elementary mathematics. *Journal of Learning Disabilities, 20*, 23–32, 57.

Penner, L. A., Batsche, G. M., Jr., Knoff, H. W., & Nelson, D. L. (Eds.). (1993). *The challenge of mathematics and science education: Psychology's response.* Washington, DC: American Psychological Association.

Pennington, B. F., Gilger, J. W., Olson, R. K., & DeFries, J. C. (1992). External validity of age versus IQ discrepant definitions of reading disability: Lessons from a twin study. *Journal of Learning Disabilities, 25*, 562–573.

Pennington, B. F., Gilger, J. W., Pauls, D., Smith, S. A., Smith, S. D., & DeFries, J. C. (1991). Evidence for major gene transmission of developmental dyslexia. *Journal of the American Medical Association, 266*, 1527–1534.

Pepperberg, I. M. (1987). Evidence for conceptual quantitative abilities in the African grey parrot: Labeling of cardinal sets. *Ethology, 75*, 37–61.

Pepperberg, I. M. (1993, November). *Numerical competence in an African grey parrot.* Paper presented at the 34th Annual Meeting of the Psychonomic Society, Washington, DC.

Perry, M. (1991). Learning and transfer: Instructional conditions and conceptual change. *Cognitive Development, 6*, 449–468.

Perry, M., VanderStoep, S. W., & Yu, S. L. (1993). Asking questions in first-grade mathematics classes: Potential influences on mathematical thought. *Journal of Educational Psychology, 85*, 31–40.

Peterson, P. L., & Fennema, E. (1985). Effective teaching, student engagement in classroom activities, and sex-related differences in learning mathematics. *American Educational Research Journal, 22*, 309–335.

Petitto, A. L. (1990). Development of numberline and measurement concepts. *Cognition and Instruction, 7*, 55–78.

Piaget, J. (1950). *The psychology of intelligence.* London: Routledge & Kegan Paul.

Piaget, J. (1965). *The child's conception of number.* New York: Norton.

Piaget, J., Inhelder, I., & Szeminska, A. (1960). *The child's conception of geometry.* London: Routledge & Kegan Paul.

Plomin, R., DeFries, J. C., & McClearn, G. (1990). *Behavioral genetics: A primer* (2nd ed.). San Francisco: W. H. Freeman.

Ramirez, O. M., & Dockweiler, C. J. (1987). Mathematics anxiety: A systematic review. In R. Schwarzer, H. M. van der Ploeg, & C. D. Spielberger (Eds.), *Advances in test anxiety research* (Vol. 5, pp. 157–175). Berwyn, PA: Swets North America.

Randhawa, B. S., Beamer, J. E., & Lundberg, I. (1993). Role of mathematics self-efficacy in the structural model of mathematics achievement. *Journal of Educational Psychology, 85*, 41–48.

Raymond, C. L., & Benbow, C. P. (1986). Gender differences in mathematics: A function of parental support and student sex typing? *Developmental Psychology, 22*, 808–819.

Rayner, K. (1993). Reading symposium: Introduction. *Psychological Science, 4*, 280–282.

Redfield, D. L., & Rousseau, E. W. (1981). A meta-analysis of experimental research on teacher questioning behavior. *Review of Educational Research, 51*, 237–245.

Reese, H. W., & Overton, W. F. (1970). Models of development and theories of development. In L. R. Goulet & P. B. Baltes (Eds.), *Life-span developmental psychology: Research and theory* (pp. 115–145). New York: Academic Press.

Resnick, L. B. (1983). A developmental theory of number understanding. In H. P. Ginsburg (Ed.), *The development of mathematical thinking* (pp. 109–151). New York: Academic Press.

Resnick, L. B. (1989). Developing mathematical knowledge. *American Psychologist, 44*, 162–169.

Resnick, L. B., & Ford, W. W. (1981). *The psychology of mathematics for instruction*. Hillsdale, NJ: Erlbaum.

Resnick, S. M., Berenbaum, S. A., Gottesman, I. I., & Bouchard, T. J., Jr. (1986). Early hormonal influences on cognitive functioning in congenital adrenal hyperplasia. *Developmental Psychology, 22*, 191–198.

Richman, L. C. (1983). Language-learning disability: Issues, research, and future directions. In M. Wolraich & D. K. Routh (Eds.), *Advances in developmental and behavioral pediatrics* (Vol. 4, pp. 87–107). Greenwich, CT: JAI Press.

Riley, M. S., & Greeno, J. G. (1988). Developmental analysis of understanding language about quantities and of solving problems. *Cognition and Instruction, 5*, 49–101.

Riley, M. S., Greeno, J. G., & Heller, J. I. (1983). Development of children's problem-solving ability in arithmetic. In H. P. Ginsburg (Ed.), *The development of mathematical thinking* (pp. 153–196). New York: Academic Press.

Rivera-Batiz, F. L. (1992). Quantitative literacy and the likelihood of employment among young adults in the United States. *Journal of Human Resources, 27*, 313–328.

Roe, A. (1953). A psychological study of eminent psychologists and anthropologists, and a comparison with biological and physical scientists. *Psychological Monographs: General and Applied, 67*(2, Whole No. 352).

Rourke, B. P. (1989). *Nonverbal learning disabilities: The syndrome and the model*. New York: Guilford Press.

Rourke, B. P., & Finlayson, M. A. J. (1978). Neuropsychological significance of variations in patterns of academic performance: Verbal and visual-spatial abilities. *Journal of Abnormal Child Psychology, 6*, 121–133.

Rourke, B. P., & Strang, J. D. (1978). Neuropsychological significance of variations in patterns of academic performance: Motor, psychomotor, and tactile-perceptual abilities. *Journal of Pediatric Psychology, 3*, 62–66.

Rozin, P., Poritsky, S., & Sotsky, R. (1971). American children with reading problems can easily learn to read English represented by Chinese characters. *Science, 171*, 1264–1267.

Ruff, C. (1987). Sexual dimorphism in human lower limb bone structure: Relationship to subsistence strategy and sexual division of labor. *Journal of Human Evolution, 16*, 391–416.

Rumbaugh, D. M., & Washburn, D. A. (1993). Counting by chimpanzees and ordinality judgments by macaques in video-formatted tasks. In S. T. Boysen & E. J. Capaldi (Eds.), *The development of numerical competence: Animal and human models* (pp. 87–106). Hillsdale, NJ: Erlbaum.

Rushton, J. P. (1992). Cranial capacity related to sex, rank, and race in a stratified random sample of 6,325 U.S. military personnel. *Intelligence, 16*, 401–413.

Russell, R. L., & Ginsburg, H. P. (1984). Cognitive analysis of children's mathematical difficulties. *Cognition and Instruction, 1*, 217–244.

Rutter, M. (1982). Epidemiological-longitudinal approaches to the study of development. In W. A. Collins (Ed.), *Concepts of development: The Minnesota Symposia on Child Psychology* (Vol. 15, pp. 105–144). Hillsdale, NJ: Erlbaum.

Sarason, I. G. (1984). Stress, anxiety, and cognitive interference: Reactions to tests. *Journal of Personality and Social Psychology, 46,* 929–938.

Sattler, J. M. (1988). *Assessment of children* (3rd ed.). San Diego, CA: Sattler.

Saxe, G. B. (1977). A developmental analysis of notational counting. *Child Development, 48,* 1512–1520.

Saxe, G. B. (1981). Body parts as numerals: A developmental analysis of numeration among the Oksapmin of Papua New Guinea. *Child Development, 52,* 306–316.

Saxe, G. B. (1982a). Culture and the development of numerical cognition: Studies among the Oksapmin of Papua New Guinea. In C. J. Brainerd (Ed.), *Children's logical and mathematical cognition: Progress in cognitive development research* (pp. 157–176). New York: Springer-Verlag.

Saxe, G. B. (1982b). Developing forms of arithmetical thought among the Oksapmin of Papua New Guinea. *Developmental Psychology, 18,* 583–594.

Saxe, G. B. (1985). Effects of schooling on arithmetical understandings: Studies with Oksapmin children in Papua New Guinea. *Journal of Educational Psychology, 77,* 503–513.

Saxe, G. B. (1988). The mathematics of child street vendors. *Child Development, 59,* 1415–1425.

Saxe, G. B. (1991). *Culture and cognitive development: Studies in mathematical understanding.* Hillsdale, NJ: Erlbaum.

Saxe, G. B., Guberman, S. R., & Gearhart, M. (1987). Social processes in early number development. *Monographs of the Society for Research in Child Development, 52*(2, Serial No. 216).

Scarr, S., & McCartney, K. (1983). How people make their own environments: A theory of genotype —> environment effects. *Child Development, 54,* 424–435.

Schaeffer, B., Eggleston, V. H., & Scott, J. L. (1974). Number development in young children. *Cognitive Psychology, 6,* 357–379.

Schaie, K. W. (1983). The Seattle longitudinal study: A 21-year exploration of psychometric intelligence in adulthood. In K. W. Schaie (Ed.), *Longitudinal studies of adult psychological development* (pp. 64–135). New York: Guilford Press.

Schloss, P. J., Smith, M. A., & Schloss, C. N. (1990). *Instructional methods for adolescents with learning and behavior problems.* Needham Heights, MA: Allyn & Bacon.

Schoenfeld, A. H. (1985). *Mathematical problem solving.* San Diego, CA: Academic Press.

Schoenfeld, A. H. (1987). What's all the fuss about metacognition? In A. H. Schoenfeld (Ed.), *Cognitive science and mathematics education* (pp. 189–215). Hillsdale, NJ: Erlbaum.

Schratz, M. M. (1978). A developmental investigation of sex differences in spatial (visual-analytic) and mathematical skills in three ethnic groups. *Developmental Psychology, 14,* 263–267.

Sells, L. W. (1980). The mathematics filter and the education of women and minorities. In L. H. Fox, L. Brody, & D. Tobin (Eds.), *Women and the mathematical mystique* (pp. 66–75). Baltimore: Johns Hopkins University Press.

Senk, S., & Usiskin, Z. (1983). Geometry proof writing: A new view of sex differences in mathematics ability. *American Journal of Education, 91,* 187–201.

Serbin, L. A., & Connor, J. M. (1979). Sex typing of children's play preferences and patterns of cognitive performance. *Journal of Genetic Psychology, 134,* 315–316.

Sewell, T. E., Farley, F. H., & Sewell, F. B. (1983). Anxiety, cognitive style, and mathematics achievement. *Journal of Genetic Psychology, 109,* 59–66.

Share, D. L., Moffitt, T. E., & Silva, P. A. (1988). Factors associated with arithmetic-and-reading disability and specific arithmetic disability. *Journal of Learning Disabilities*, *21*, 313–320.

Shaywitz, S. E., Escobar, M. D., Shaywitz, B. A., Fletcher, J. M., & Makuch, R. (1992). Evidence that dyslexia may represent the lower tail of a normal distribution of reading ability. *New England Journal of Medicine*, *326*, 145–150.

Sherman, J. (1980). Mathematics, spatial visualization, and related factors: Changes in girls and boys, Grades 8–11. *Journal of Educational Psychology*, *72*, 476–482.

Sherman, J. (1981). Girls' and boys' enrollments in theoretical math courses: A longitudinal study. *Psychology of Women Quarterly*, *5*, 681–689.

Sherman, J. A. (1982). Mathematics the critical filter: A look at some residues. *Psychology of Women Quarterly*, *6*, 428–444.

Siegel, L. S. (1971a). The development of the understanding of certain number concepts. *Developmental Psychology*, *5*, 362–363.

Siegel, L. S. (1971b). The sequence of development of certain number concepts in preschool children. *Developmental Psychology*, *5*, 357–361.

Siegel, L. (1974). Development of number concepts: Ordering and correspondence operations and the role of length cues. *Developmental Psychology*, *10*, 907–912.

Siegel, L. S., & Ryan, E. B. (1989). The development of working memory in normally achieving and subtypes of learning disabled children. *Child Development*, *60*, 973–980.

Siegler, R. S. (1983). Five generalizations about cognitive development. *American Psychologist*, *38*, 263–277.

Siegler, R. S. (1986). Unities across domains in children's strategy choices. In M. Perlmutter (Ed.), *Perspectives for intellectual development: The Minnesota Symposia on Child Psychology* (Vol. 19, pp. 1–48). Hillsdale, NJ: Erlbaum.

Siegler, R. S. (1987). The perils of averaging data over strategies: An example from children's addition. *Journal of Experimental Psychology: General*, *116*, 250–264.

Siegler, R. S. (1988a). Individual differences in strategy choices: Good students, not-so-good students, and perfectionists. *Child Development*, *59*, 833–851.

Siegler, R. S. (1988b). Strategy choice procedures and the development of multiplication skill. *Journal of Experimental Psychology: General*, *117*, 258–275.

Siegler, R. S. (1989a). Hazards of mental chronometry: An example from children's subtraction. *Journal of Educational Psychology*, *81*, 497–506.

Siegler, R. S. (1989b). Mechanisms of cognitive development. *Annual Review of Psychology*, *40*, 353–379.

Siegler, R. S. (1991). In young children's counting, procedures precede principles. *Educational Psychology Review*, *3*, 127–135.

Siegler, R. S. (1993). Adaptive and nonadaptive characteristics of low-income children's mathematical strategy use. In L. A. Penner, G. M. Batsche, H. M. Knoff, & D. L. Nelson (Eds.), *The challenge in mathematics and science education: Psychology's response* (pp. 341–366). Washington, DC: American Psychological Association.

Siegler, R. S., & Crowley, K. (1991). The microgenetic method: A direct means for studying cognitive development. *American Psychologist*, *46*, 606–620.

Siegler, R. S., & Crowley, K. (in press). Constraints on learning in non-privileged domains. *Cognitive Psychology*.

Siegler, R. S., & Jenkins, E. (1989). *How children discover new strategies*. Hillsdale, NJ: Erlbaum.

Siegler, R. S., & Kotovsky, K. (1986). Two levels of giftedness: Shall ever the twain meet? In R. J. Sternberg & J. E. Davidson (Eds.), *Conceptions of giftedness* (pp. 417–435). Cambridge, England: Cambridge University Press.

Siegler, R. S., & Robinson, M. (1982). The development of numerical understandings. In H. Reese & L. P. Lipsitt (Eds.), *Advances in child development and behavior* (Vol. 16, pp. 241–312). New York: Academic Press.

Siegler, R. S., & Shrager, J. (1984). Strategy choice in addition and subtraction: How do children know what to do? In C. Sophian (Ed.), *Origins of cognitive skills* (pp. 229–293). Hillsdale, NJ: Erlbaum.

Siegler, R. S., & Taraban, R. (1986). Conditions of applicability of a strategy choice model. *Cognitive Development, 1*, 31–51.

Sierpinska, A., Kilpatrick, J., Balacheff, N., Howson, A. G., Sfard, A., & Steinbring, H. (1993). What is research in mathematics education, and what are its results? *Journal for Research in Mathematics Education, 24*, 274–278.

Silver, E. A. (1987). Foundations of cognitive theory and research for mathematics problem-solving instruction. In A. H. Schoenfeld (Ed.), *Cognitive science and mathematics education* (pp. 33–60). Hillsdale, NJ: Erlbaum.

Silverman, I. W., & Rose, A. P. (1980). Subitizing and counting skills in 3-year-olds. *Developmental Psychology, 16*, 539–540.

Sokol, S. M., McCloskey, M., Cohen, N. J., & Aliminosa, D. (1991). Cognitive representations and processes in arithmetic: Inferences from the performance of brain-damaged subjects. *Journal of Experimental Psychology: Learning, Memory, and Cognition, 17*, 355–376.

Song, M. J., & Ginsburg, H. P. (1987). The development of informal and formal mathematical thinking in Korean and U.S. children. *Child Development, 58*, 1286–1296.

Sophian, C., & Adams, N. (1987). Infants' understanding of numerical transformations. *British Journal of Developmental Psychology, 5*, 257–264.

Spearman, C. (1927). *The abilities of man*. London: Macmillan.

Spiers, P. A. (1987). Acalculia revisited: Current issues. In G. Deloche & X. Seron (Eds.), *Mathematical disabilities: A cognitive neuropsychological perspective* (pp. 1–25). Hillsdale, NJ: Erlbaum.

Stanley, J. C., Huang, J., & Zu, X. (1986). SAT-M scores of highly selected students in Shanghai tested when less than 13 years old. *College Board Review, 140*, 10–13, 28–29.

Starkey, P. (1992). The early development of numerical reasoning. *Cognition, 43*, 93–126.

Starkey, P., & Cooper, R. G., Jr. (1980). Perception of numbers by human infants. *Science, 210*, 1033–1035.

Starkey, P., & Gelman, R. (1982). The development of addition and subtraction abilities prior to formal schooling in arithmetic. In T. P. Carpenter, J. M. Moser, & T. A. Romberg (Eds.), *Addition and subtraction: A cognitive perspective* (pp. 99–116). Hillsdale, NJ: Erlbaum.

Starkey, P., Spelke, E. S., & Gelman, R. (1983). Detection of intermodal numerical correspondences by human infants. *Science, 222*, 179–181.

Starkey, P., Spelke, E. S., & Gelman, R. (1990). Numerical abstraction by human infants. *Cognition, 36*, 97–127.

Starkey, P., Spelke, E. S., & Gelman, R. (1991). Toward a comparative psychology of number. *Cognition, 39*, 171–172.

Stazyk, E. H., Ashcraft, M. H., & Hamann, M. S. (1982). A network approach to mental multiplication. *Journal of Experimental Psychology: Learning, Memory, and Cognition, 8*, 320–335.

Steele, J. (1989). Hominid evolution and primate social cognition. *Journal of Human Evolution, 18*, 421–432.

Steffe, L. P. (1990). Inconsistencies and cognitive conflict: A constructivist's view. *Focus on Learning Problems in Mathematics*, *12*, 99–109.

Steffe, L. P. (1992). Schemes of action and operation involving composite units. *Learning and Individual Differences*, *4*, 259–309.

Steffe, L. P., Thompson, P. W., & Richards, J. (1982). Children's counting in arithmetical problem solving. In T. P. Carpenter, J. M. Moser, & T. A. Romberg (Eds.), *Addition and subtraction: A cognitive perspective* (pp. 83–97). Hillsdale, NJ: Erlbaum.

Steffe, L. P., von Glasersfeld, E., Richards, J., & Cobb, P. (1983). *Children's counting types: Philosophy, theory, and application*. New York: Praeger.

Steffe, L. P., & Wood, T. (Eds.). (1990). *Transforming children's mathematics education: International perspectives*. Hillsdale, NJ: Erlbaum.

Steinberg, R. M. (1985). Instruction on derived facts strategies in addition and subtraction. *Journal for Research in Mathematics Education*, *16*, 337–355.

Steinkamp, M. W., Harnisch, D. L., Walberg, H. J., & Tsai, S. L. (1985). Cross-national gender differences in mathematics attitude and achievement among 13-year-olds. *Journal of Mathematical Behavior*, *4*, 259–277.

Stern, E. (1993). What makes certain arithmetic word problems involving comparison of sets so difficult for children? *Journal of Educational Psychology*, *85*, 7–23.

Sternberg, R. J. (1977). *Intelligence, information processing, and analogical reasoning: The componential analysis of human abilities*. Hillsdale, NJ: Erlbaum.

Sternberg, R. J., & Davidson, J. E. (1986). Conceptions of giftedness: A map of the terrain. In R. J. Sternberg & J. E. Davidson (Eds.), *Conceptions of giftedness* (pp. 3–18). Cambridge, England: Cambridge University Press.

Sternberg, R. J., & Gardner, M. K. (1983). Unities in inductive reasoning. *Journal of Experimental Psychology: General*, *112*, 80–116.

Steuer, F. B. (1994). *The psychological development of children*. Pacific Grove, CA: Brooks/Cole.

Stevenson, H. W. (1992a). Don't deceive children through a feel-good approach. *School Administrator*, *49*, 23–30.

Stevenson, H. W. (1992b, December). Learning from Asian schools. *Scientific American*, *267*, 70–76.

Stevenson, H. W., & Azuma, H. (1983). IQ in Japan and the United States. *Nature*, *306*, 291–292.

Stevenson, H. W., & Bartsch, K. (1992). An analysis of Japanese and American textbooks in mathematics. In R. Leetsma & H. Walberg (Eds.), *Japanese educational productivity* (pp. 103–133). Ann Arbor, MI: Center for Japanese Studies.

Stevenson, H. W., Chen, C., & Lee, S. Y. (1993). Mathematics achievement of Chinese, Japanese, and American children: Ten years later. *Science*, *259*, 53–58.

Stevenson, H. W., Chen, C., Lee, S. Y., & Fuligni, A. J. (1991). Schooling, culture, and cognitive development. In R. J. Sternberg & L. Okagaki (Eds.), *Influences on the development of children's thinking* (pp. 243–268). Hillsdale, NJ: Erlbaum.

Stevenson, H. W., Lee, S. Y., Chen, C., Lummis, M., Stigler, J., Fan, L., & Ge, F. (1990). Mathematics achievement of children in China and the United States. *Child Development*, *61*, 1053–1066.

Stevenson, H. W., Lee, S. Y., Chen, C., Stigler, J. W., Hsu, C. C., & Kitamura, S. (1990). Contexts of achievement: A study of American, Chinese, and Japanese children. *Monographs of the Society for Research in Child Development*, *55*(1–2, Serial No. 221).

Stevenson, H. W., Lee, S. Y., & Stigler, J. W. (1986). Mathematics achievement of Chinese, Japanese, and American children. *Science*, *231*, 693–699.

Stevenson, H. W., Parker, T., Wilkinson, A., Hegion, A., & Fish, E. (1976). Longitudinal study of individual differences in cognitive development and scholastic achievement. *Journal of Educational Psychology, 68*, 377–400.

Stevenson, H. W., & Stigler, J. W. (1992). *The learning gap: Why our schools are failing and what we can learn from Japanese and Chinese education.* New York: Summit Books.

Stevenson, H. W., Stigler, J. W., Lee, S. Y., Lucker, G. W., Kitamura, S., & Hsu, C. C. (1985). Cognitive performance and academic achievement of Japanese, Chinese, and American children. *Child Development, 56*, 718–734.

Stigler, J. W. (1984). "Mental abacus": The effect of abacus training on Chinese children's mental calculation. *Cognitive Psychology, 16*, 145–176.

Stigler, J. W., Fuson, K. C., Ham, M., & Kim S. M. (1986). An analysis of addition and subtraction word problems in American and Soviet elementary mathematics textbooks. *Cognition and Instruction, 3*, 153–171.

Stigler, J. W., Lee, S. Y., Lucker, G. W., & Stevenson, H. W. (1982). Curriculum and achievement in mathematics: A study of elementary school children in Japan, Taiwan, and the United States. *Journal of Educational Psychology, 74*, 315–322.

Stigler, J. W., Lee, S. Y., & Stevenson, H. W. (1987). Mathematics classrooms in Japan, Taiwan, and the United States. *Child Development, 58*, 1272–1285.

Stigler, J. W., & Perry, M. (1988). Mathematics learning in Japanese, Chinese, and American classrooms. In G. B. Saxe & M. Gearhart (Eds.), *Children's mathematics* (pp. 27–54). (New Directions for Child Development, Series No. 41.) San Francisco: Jossey-Bass.

Strang, J. D., & Rourke, B. P. (1985). Arithmetic disability subtypes: The neuropsychological significance of specific arithmetical impairment in childhood. In B. P. Rourke (Ed.), *Neuropsychology of learning disabilities: Essentials of subtype analysis* (pp. 167–183). New York: Guilford Press.

Strauss, M. S., & Curtis, L. E. (1981). Infant perception of numerosity. *Child Development, 52*, 1146–1152.

Strauss, M. S., & Curtis, L. E. (1984). Development of numerical concepts in infancy. In C. Sophian (Ed.), *Origins of cognitive skills: The Eighteenth Annual Carnegie Symposium on Cognition* (pp. 131–155). Hillsdale, NJ: Erlbaum.

Sutaria, S. D. (1985). *Specific learning disabilities: Nature and needs.* Springfield, IL: Charles C Thomas.

Svenson, O., & Broquist, S. (1975). Strategies for solving simple addition problems: A comparison of normal and subnormal children. *Scandinavian Journal of Psychology, 16*, 143–151.

Svenson, O., & Hedenborg, M. L. (1979). Strategies used by children when solving simple subtractions. *Acta Psychologica, 43*, 477–489.

Svenson, O., Hedenborg, M. L., & Lingman, L. (1976). On children's heuristics for solving simple additions. *Scandinavian Journal of Educational Research, 20*, 161–173.

Svenson, O., & Sjöberg, K. (1983). Evolution of cognitive processes for solving simple additions during the first three school years. *Scandinavian Journal of Psychology, 24*, 117–124.

Sweller, J., Mawer, R. F., & Ward, M. R. (1983). Development of expertise in mathematical problem solving. *Journal of Experimental Psychology: General, 112*, 639–661.

Symonds, P. M., & Chase, D. H. (1929). Practice vs. motivation. *Journal of Educational Psychology, 20*, 19–35.

Tartre, L. A. (1990). Spatial skills, gender, and mathematics. In E. Fennema & G. C. Leder (Eds.), *Mathematics and gender* (pp. 27–59). New York: Teachers College Press.

Temple, C. M. (1989). Digit dyslexia: A category-specific disorder in developmental dyscalculia. *Cognitive Neuropsychology, 6*, 93–116.

Temple, C. M. (1991). Procedural dyscalculia and number fact dyscalculia: Double dissociation in developmental dyscalculia. *Cognitive Neuropsychology, 8,* 155–176.

Terman, L. M. (1954). The discovery and encouragement of exceptional talent. *American Psychologist, 9,* 221–230.

Thompson, L. A., Detterman, D. K., & Plomin, R. (1991). Associations between cognitive abilities and scholastic achievement: Genetic overlap and environmental differences. *Psychological Science, 2,* 158–165.

Thorndike, E. L. (1911). *Individuality.* Cambridge, MA: Riverside Press.

Thurstone, L. L. (1938). Primary mental abilities. *Psychometric Monographs* (No. 1).

Thurstone, L. L., & Thurstone, T. G. (1941). Factorial studies of intelligence. *Psychometric Monographs* (No. 2).

Tobias, S. (1978). *Overcoming math anxiety.* New York: Norton.

Travers, K. J., & Westbury, I. (1989). *The IEA study of mathematics: I. Analysis of mathematics curricula.* Oxford, England: Pergamon Press.

Trivers, R. L. (1972). Parental investment and sexual selection. In B. Campbell (Ed.), *Sexual selection and the descent of man 1871–1971* (pp. 136–179). Chicago: Aldine-Atherton.

Tsang, S. L. (1984). The mathematics education of Asian Americans. *Journal for Research in Mathematics Education, 15,* 114–122.

Tsang, S. L. (1988). The mathematics achievement characteristics of Asian-American students. In R. R. Cocking & J. P. Mestre (Eds.), *Linguistic and cultural influences on learning mathematics* (pp. 123–136). Hillsdale, NJ: Erlbaum.

U.S. Department of Education. (1991). *America 2000: An education strategy.* Washington, DC: Author.

Vandenberg, S. G. (1959). The primary mental abilities of Chinese students: A comparative study of the stability of a factor structure. *Annals of the New York Academy of Sciences, 79,* 257–304.

Vandenberg, S. G. (1962). The Hereditary Abilities Study: Hereditary components in a psychological test battery. *American Journal of Human Genetics, 14,* 220–237.

Vandenberg, S. G. (1966). Contributions of twin research to psychology. *Psychological Bulletin, 66,* 327–352.

VanLehn, K. (1990). *Mind bugs: The origins of procedural misconceptions.* Cambridge, MA: MIT Press.

van Loosbroek, E., & Smitsman, A. W. (1990). Visual perception of numerosity in infancy. *Developmental Psychology, 26,* 916–922.

Vergnaud, G. (1983). Multiplicative structures. In R. Lesh & M. Landau (Eds.), *Acquisition of mathematics concepts and processes* (pp. 127–174). New York: Academic Press.

Verschaffel, L., De Corte, E., & Pauwels, A. (1992). Solving compare problems: An eye movement test of Lewis and Mayer's consistency hypothesis. *Journal of Educational Psychology, 84,* 85–94.

Very, P. S. (1967). Differential factor structures in mathematical ability. *Genetic Psychology Monographs, 75,* 169–207.

von Glaserfeld, E., & Steffe, L. P. (1991). Conceptual models in educational research and practice. *Journal of Educational Thought, 25,* 91–103.

Wagner, R. K., & Torgesen, J. K. (1987). The nature of phonological processing and its causal role in the acquisition of reading skills. *Psychological Bulletin, 101,* 192–212.

Walker, D. F., & Schaffarzick, J. (1974). Comparing curricula. *Review of Educational Research, 44,* 83–111.

Warrington, E. K. (1982). The fractionation of arithmetic skills: A single case study. *Quarterly Journal of Experimental Psychology, 34A,* 31–51.

Washburn, D. A., & Rumbaugh, D. M. (1991). Ordinal judgments of numerical symbols by macaques (*Macaca mulatta*). *Psychological Science, 2*, 190–193.

Washburne, C., & Vogel, M. (1928). Are any number combinations inherently difficult? *Journal of Educational Research, 17*, 235–255.

Wasserman, E. A. (1993). Comparative cognition: Beginning the second century of the study of animal intelligence. *Psychological Bulletin, 113*, 211–228.

Wechsler, D. (1949). *Wechsler Intelligence Scale for Children*. New York: Psychological Corporation.

Wechsler, D. (1967). *Manual for the Wechsler Preschool and Primary Scale of Intelligence*. New York: Psychological Corporation.

Wechsler, D. (1974). *Manual for the Wechsler Intelligence Scale for Children—Revised*. New York: Psychological Corporation.

Welsh, M. C., Pennington, B. F., & Groisser, D. B. (1991). A normative-developmental study of executive function: A window on prefrontal function in children. *Developmental Neuropsychology, 7*, 131–149.

Wenger, R. H. (1987). Cognitive science and algebra learning. In A. H. Schoenfeld (Ed.), *Cognitive science and mathematics education* (pp. 217–251). Hillsdale, NJ: Erlbaum.

Wheeler, L. R. (1939). A comparative study of the difficulty of the 100 addition combinations. *Journal of Genetic Psychology, 54*, 295–312.

Widaman, K. F., Geary, D. C., Cormier, P., & Little, T. D. (1989). A componential model for mental addition. *Journal of Experimental Psychology: Learning, Memory, and Cognition, 15*, 898–919.

Widaman, K. F., & Little, T. D. (1992). The development of skill in mental arithmetic: An individual differences perspective. In J. I. D. Campbell (Ed.), *The nature and origins of mathematical skills* (pp. 189–253). Amsterdam: North-Holland.

Widaman, K. F., Little, T. D., Geary, D. C., & Cormier, P. (1992). Individual differences in the development of skill in mental addition: Internal and external validation of chronometric models. *Learning and Individual Differences, 4*, 167–213.

Wilson, M., & Daly, M. (1985). Competitiveness, risk taking, and violence: The young male syndrome. *Ethology and Sociobiology, 6*, 59–73.

Wilson, R. S. (1978). Synchronies in mental development: An epigenetic perspective. *Science, 202*, 939–948.

Wise, L. L. (1985). Project TALENT: Mathematics course participation in the 1960s and its career consequences. In S. F. Chipman, L. R. Brush, & D. M. Wilson (Eds.), *Women and mathematics: Balancing the equation* (pp. 25–58). Hillsdale, NJ: Erlbaum.

Witelson, S. E. (1987). Neurobiological aspects of language in children. *Child Development, 58*, 653–688.

Wolpe, J. (1958). *Psychotherapy by reciprocal inhibition*. Stanford, CA: Stanford University Press.

Wolters, M. D. (1983). The part-whole schema and arithmetical problems. *Educational Studies in Mathematics, 14*, 127–138.

Woltz, D. J. (1988). An investigation of the role of working memory in procedural skill acquisition. *Journal of Experimental Psychology: General, 117*, 319–331.

Wood, R. (1976). Sex differences in mathematics attainment at GCE ordinary level. *Educational Studies, 2*, 141–160.

Woods, S. S., Resnick, L. B., & Groen, G. J. (1975). Experimental test of five process models for subtraction. *Journal of Educational Psychology, 67*, 17–21.

Wynn, K. (1990). Children's understanding of counting. *Cognition, 36*, 155–193.

Wynn, K. (1992a). Addition and subtraction by human infants. *Nature, 358*, 749–750.

Wynn, K. (1992b). Children's acquisition of the number words and the counting system. *Cognitive Psychology, 24*, 220–251.

Wynn, K. (1992c). Evidence against empiricist accounts of the origins of numerical knowledge. *Mind & Language, 7*, 315–332.

Young, R. M., & O'Shea, T. (1981). Errors in children's subtraction. *Cognitive Science, 5*, 153–177.

Zaslavsky, C. (1973). *Africa counts: Number and pattern in African culture*. Boston: Prindle, Weber, & Schmidt.

Zawaiza, T. R. W., & Gerber, M. (1993). Effects of explicit instruction on math word-problem solving by community college students with learning disabilities. *Learning Disability Quarterly, 16*, 64–79.

Zentall, S. S., & Ferkis, M. A. (1993). Mathematical problem solving for youth with ADHD, with and without learning disabilities. *Learning Disability Quarterly, 16*, 6–18.

Zentall, S. S., & Meyer, M. (1987). Self-regulation of stimulation for ADD-H children during reading and vigilance task performance. *Journal of Abnormal Child Psychology, 15*, 519–536.

Index

A

Abacus, 280
Ability factor, defined, 132
Abstraction, 3, 4, 24, 269, 286. *See also specific concepts*
Accumulator, 8, 9
Acquired dyscalculia, 156
Action schema, defined, 109
Adding-on procedure, 63, 70
Addition, xi, 49–61, 184
   addition reference, 61, 65, 66
   chimpanzees and, 32–33
   complex, 57–58, 63
   confidence criterion, 83, 84, 85
   counting-all procedure, 89, 109, 184
   counting-on procedure, 53, 89, 165
   fractions and, 47
   image-based strategy, 41
   implicit sense, 91
   infants and, x, 6, 39
   language and, 41
   of large numbers, 13
   manipulatives and, 48–49, 103
   mathematically disabled children, 165
   memorization and, 54
   multiplication and, 73
   numerosity and, 39, 41–42
   ordinality and, 10
   procedures for, 87, 88, 89
   simple, 49–57, 62
   spatial abilities and, 143
   strategies for, 62, 88, 102
   subtraction and, 61, 65–66
   verbal counting, 42, 50
   word problems, 95–106

   See also Arithmetic; *specific procedures*
Adjacency, 27
Adolescence, xiii, 197, 232
Advanced mathematics courses, 210
African-American children, 242–243
Age problems, 118, 126
Agraphia, 171–172
Agricultural societies, 39
Alexia, 171–172
Algebra, 116–130
   arithmetic and, 116–117
   automation and, 125
   bugs and, 117, 122
   developmental patterns, 124–127
   expressions in, 127
   problem-solving errors, 117, 125
   problem translation, 118
   relational representations, 121
   research on, 124–127
   rule automation, 125
   schemas in, 117–118
   solution options, 120–121
   translation errors, 119, 127
   word problems, xi, 116–130, 197, 228
   working-memory demands, 120
Alphabet, 16
America 2000: An Education Strategy (USDE), xiv, 231–232, 237, 238, 257
American children
   African-American children, 242–243
   American culture and, 256, 258, 259
   Asian children and, 237, 241, 242, 255, 268
   curriculum and, 248, 267

317

About the Author

David C. Geary is currently an associate professor and former director of the Training Program in Experimental Psychology at the University of Missouri–Columbia. His research primarily focuses on individual and developmental differences in numerical and arithmetical skills. His research with children includes cognitive studies of mathematically disabled children, as well as studies of the cognitive factors that influence cross-national differences in early arithmetical skills. He has also conducted research in the area of cognitive aging, comparing younger and older adults on strategic and speed-of-processing skills associated with basic arithmetic. His interests also include gender differences and evolutionary psychology. He received an award for excellence in intelligence research from the Mensa Education and Research Foundation in 1992 for his cross-cultural research.

DATE DUE